Emerging Trends in Microbial Electrochemical Technologies for Sustainable Mitigation of Water Resources Contamination

Rangabhashiyam Selvasembian · Joyabrata Mal ·
Sovik Das · Dakeshwar Kumar Verma ·
Ioannis Anastopoulos
Editors

Emerging Trends in Microbial Electrochemical Technologies for Sustainable Mitigation of Water Resources Contamination

Microbial Electrochemical Technologies in Wastewater Treatment

Springer

Editors
Rangabhashiyam Selvasembian
Department of Environmental Science
and Engineering, School of Engineering
and Sciences
SRM University
Amaravati, Andhra Pradesh, India

Sovik Das
Department of Civil Engineering
Indian Institute of Technology Delhi
Hauz Khas, India

Ioannis Anastopoulos
Department of Agriculture
University of Ioannina
Arta, Greece

Joyabrata Mal
Department of Biotechnology
Motilal Nehru National Institute
of Technology
Prayagraj, Uttar Pradesh, India

Dakeshwar Kumar Verma
Department of Chemistry
Govt. Digvijay Autonomous Postgraduate
College
Rajnandgaon, Chhattisgarh, India

ISBN 978-3-031-74635-2 ISBN 978-3-031-74636-9 (eBook)
https://doi.org/10.1007/978-3-031-74636-9

This Springer imprint is published by the registered company Springer Nature Switzerland AG
The registered company address is: Gewerbestrasse 11, 6330 Cham, Switzerland

Contents

Water Pollution: Transportation, Fate and Its Hazardous Impacts on the Biosphere

A. Bharani, P. Papitha, Raveena Ravi, and Nikil Shanmugam

Abstract Our planet's water resources are under siege. Transportation of pollutants from various sources contaminates rivers, lakes, and oceans, disrupting the delicate balance of the biosphere. This abstract explores the pathways by which pollutants travel (transport), their persistence and transformation in water (fate), and the devastating consequences they have on all living things (hazardous impacts) According to a UN report, by 2050, almost half of the world's population will be living in areas experiencing water scarcity. This scarcity is further exacerbated by pollution, rendering freshwater sources unusable for drinking, irrigation, and aquatic life. This work will delve into the various modes of transportation, such as runoff from agricultural lands laden with fertilizers and pesticides, or industrial discharges containing heavy metals and toxic chemicals. It will then explore the fate of these pollutants, including how they degrade, accumulate in sediments, or travel through the food chain. Finally, the chapter will highlight the hazardous impacts on the biosphere, including disruption of ecosystems, endangerment of aquatic species, and potential human health risks. By understanding the transport, fate, and hazardous impacts of water pollution, we can develop strategies to mitigate this growing threat. This chapter aims to raise awareness of this critical issue and pave the way for a cleaner, healthier future for our water resources and the biosphere they sustain.

Keywords Environmental contaminants · Biosphere impacts · Heavy metals · Fate of pollutants · Biosphere impacts

A. Bharani (✉) · R. Ravi
Department of Environmental Sciences, Tamil Nadu Agricultural University, Coimbatore, Tamil Nadu 641 003, India
e-mail: bharani.a@tnau.ac.in

P. Papitha
Guest Faculty, School of Agriculture and Animal Sciences, Gandhigram Rural Institute, Dindigul, Tamil Nadu, India

N. Shanmugam
Environmental Protection and Management, The University of Edinburgh, Edinburgh, UK

R. Selvasembian et al. (eds.), *Emerging Trends in Microbial Electrochemical Technologies for Sustainable Mitigation of Water Resources Contamination*,
https://doi.org/10.1007/978-3-031-74636-9_1

1 Introduction

Water is a fundamental component, playing a critical role in the origin and sustenance of life. Historically, all major civilizations flourished near water sources. Similarly, most modern cities are situated near rivers or other water bodies, which initially fueled population growth and industrial development. However, industries such as textiles, dyeing, tanning, sugar production, distilleries, paper, and pulp are now significant contributors to water pollution through the disposal of hazardous waste. The type, volume, and frequency of industrial waste vary depending on the specific industry and the population using its products (Adekunle and Eniola 2008). Globally, it's estimated that industries discharge 300–400 million tons of heavy metals, solvents, toxic sludge, and other pollutants into water bodies annually (UNEP 2010). Additionally, many industries contribute to thermal pollution by releasing hot water from cooling systems into waterways. Mining activities further exacerbate water quality issues.

A staggering 80% of the vast amount of wastewater generated by these industries remains untreated, as per data given by UNESCO (2021) furnished in Table 1. These industrial effluents, particularly from textiles, tanning, and dyeing processes, harbor a multitude of highly toxic chemicals, leading to widespread contamination of global water bodies (Awaleh and Soubaneh 2014). Water pollutants can be categorized into dissolved and undissolved forms.

Dissolved pollutants comprise substances that dissociate into cations and anions when in water. These can be further classified into organic and inorganic categories. Organic pollutants are carbon-based compounds that may also contain elements like hydrogen, oxygen, nitrogen, and sulfur. They can be further subdivided into three main types: (i) oxygen-demanding contaminants, (ii) synthetic organic contaminants, and (iii) crude oil and petroleum products. Inorganic pollutants, on the other hand, are non-biodegradable and persist in the environment (Breida et al. 2019).

Table 1 Quantum of water generated from industries

S. no.	Industry	Quantum of water generated standards (m′/ton)
1.	Textile	120 m^3/ton fibre
2.	Pulp and paper	175 m^3/ton paper
	Large	!50 m^3/ton paper
	Small	
3.	Tannery	28 m^3/ton raw hide
4.	Kraft mill	40 m^3/ton
5.	Sugar	0.4 m^3/ton cane

Anjaneyulu et al. (2005)

2 Transportation of Water Pollutants Impacts on Biosphere

Transportation of water pollutants has a devastating impact on the biosphere, the interconnected web of ecosystems and their resident lifeforms (Windsor et al. 2019). These contaminants, encompassing chemicals, fertilizers, pathogens, heavy metals, and more, infiltrate various water bodies like rivers, lakes, seas, and groundwater. The consequences extend beyond direct harm to individual aquatic organisms, such as fish, amphibians, and invertebrates, which can occur through exposure to contaminants in the water. Pollutants and heavy metals can disrupt their reproductive function, feeding behaviour, and respiratory systems, potentially leading to population declines or extinctions (Authman et al. 2015). Furthermore, a concerning process known as bioaccumulation causes many contaminants to concentrate within the tissues of aquatic creatures over time (Thanigaivel et al. 2022).

This bioaccumulation has a ripple effect throughout the food web, leading to a phenomenon called biomagnification. As larger organisms consume smaller ones, they ingest the accumulated contaminants. Predatory organisms at higher trophic levels end up with even higher concentrations of these pollutants in their bodies. This poses a significant threat to top predators, including humans, who consume fish and other seafood. The biomagnification of water pollutants can result in a number of health problems, including developmental issues, hormonal disruption, and even cancer.

2.1 Effects of Organic Water Pollutants on Environment

Organic water contaminants pose a significant threat to aquatic ecosystems. These pollutants encompass a diverse range of organic substances, including synthetic pesticides, industrial chemicals, pharmaceuticals, personal care products, and more. Human activities are the primary source of these contaminants (Pal et al. 2014). Contamination by organic pollutants degrades water quality, rendering it less desirable for both human consumption and the health of aquatic ecosystems. Changes in taste, odor, and appearance can significantly reduce the palatability of water. Volatile Organic Compounds (VOCs), such as benzene, toluene, and xylene, can contaminate groundwater and surface water. These pollutants originate from various sources, including industrial discharges and oil spills (Ite et al. 2018). As mentioned previously, these pollutants directly threaten the health and survival of marine organisms and have a demonstrably negative impact on human well-being.

Urban runoff is a significant source of Polycyclic Aromatic Hydrocarbons (PAHs), which are hazardous byproducts of incomplete combustion, processes (Anyahara 2021). PAHs can cause persistent contamination of water bodies, posing a long-term threat to aquatic life. A study by Chen et al. (2020) evaluated the health and ecological risks of 16 PAHs using approaches based on incremental lifetime cancer risk (ILCR), hazard quotients (HQ), and potentially impacted fraction (PAF). Their findings suggest that anthracene (Ant) and benzo (a) pyrene (BaP) pose the greatest ecological risks based on HQ and PAF methods. Additionally, the study revealed health risks associated with both non-carcinogenic PAHs (ranging from 1.01×10^{-10} to 1.60×10^{-9}) and carcinogenic PAHs (ranging from 5.03×10^{-7} to 4.74×10^{-5}). Natural organic matter, derived from the decomposition of plants and animals, plays a vital role in aquatic ecosystems. However, excessive organic matter and nutrients from sources like sewage and agricultural runoff can lead to eutrophication. This process, characterized by algal blooms and oxygen depletion, has a detrimental effect on water quality (Khan and Mohammad 2014). Chlorinated solvents used in industrial processes pose a contamination threat to groundwater (Kaown et al. 2014).

2.2 *Effects of Non Metallic Cum Inorganic Water Pollutants on Environment*

Diverse origins characterize non-metallic inorganic water pollutants. These contaminants stem from industrial discharges, agricultural runoff, and improper waste disposal practices. Notably, emissions of sulfur dioxide (SO_2) and nitrogen oxides (NOx) from industrial facilities can significantly contribute to the formation of acid precipitation (Hung et al. 2020). Acid precipitation disrupts the delicate pH equilibrium of water bodies, hindering plant growth and impacting aquatic life. Research by Du et al. (2017) demonstrated a direct correlation between acid precipitation and reduced chlorophyll content in leaves. Their experiment revealed an average decrease of 6.7% chlorophyll per unit decrease in pH. Furthermore, these pollutants can infiltrate the subsurface and contaminate groundwater resources. This contamination not only jeopardizes drinking water supplies but also disrupts the health of groundwater-dependent ecosystems. For Example: In, Nitrates (NO_3^-) excessive use of fertilizers in agriculture can lead to nitrate leaching into groundwater. High nitrate concentrations in drinking water pose a health risk, particularly for infants (methemoglobinemia), Fluoride (F^-) is beneficial in low concentrations (dental health), and high levels naturally occurring in some geological formations or from industrial discharges can cause dental fluorosis and in Ammonia (NH_4) improper disposal of animal waste and runoff from agricultural fields can elevate ammonia levels in water. Ammonia can be toxic to aquatic organisms, especially fish.

Inorganic pollutants disrupt the aquatic ecosystems by altering pH, oxygen content and nutrient balance in water bodies (Bhateria and Jain 2016) and these pollutants when bioaccumulated leads to death of fish, loss of aquatic biodiversity, and algal bloom proliferation. In Northwest China, groundwater samples were collected from private wells in rural areas where the groundwater is the main source of drinking water. Statistical analysis indicates that the mean concentration of NO_3–N was 12.15 mg/L where 40.7% of ground water samples are unfit for drinking purpose because of high nitrate concentration. It may be attributed to the increased fertilizer rate, intensive irrigation, shallow groundwater depth, high soil permeability, poor sanitation facilities and bare infrastructures (Chen et al. 2016).

For aquatic organisms, several manufacturing pesticides and chemical compounds are potentially hazardous. These poisons may accumulate along the food chain, posing health risks to higher trophic level animals including humans. Plankton was found to have the greatest concentrations of Al, Cr, Mn, Fe, Ni, Se, Cd, and Pb during the dry season in Argentina compared to the cumulative amounts detected in shrimp and fish muscle (Griboff et al. 2018). Non-metallic pollutants such as methane, can contribute to greenhouse gas emissions and climate change when released into water bodies (Bano and Arshad 2018). As toxicity threatens more than 200 million people globally, geogenic arsenic pollution is a growing problem. Both organic and inorganic forms of arsenic occur, although the inorganic forms are more dangerous and predominate in nature. In the aquatic environment, the biogeochemical cycling of As is influenced by factors such as pH, redox potential, microorganisms, and particularly algae, as well as its interactions with other elements (such as iron and sulphur) (Hussain et al. 2019). In conclusion, aquatic ecosystems face a multitude of threats from non-metallic pollutants. These pollutants can directly harm organisms, accumulate in the food chain, and contribute to broader environmental issues like climate change. Urgent action is needed to address these threats and protect the health of aquatic ecosystems and humans who depend on them.

2.3 *Effects of Metal Complexes Based Water Pollutants on Environment*

Depending on the metal, its quantity, the complex's structure, and the particular environmental circumstances, metal complexes in water pollution may have a variety of consequences on the environment (Schweitzer and Noblet 2018). Marine organisms are sometimes extremely hazardous to certain metal complexes, such as those that include heavy metals like mercury, lead, cadmium, and copper. These complexes have the ability to interfere with fish, invertebrates, and microbial metabolism even at low concentrations, which might result in fish deaths and reduced biodiversity in the impacted water bodies (Jaishankar et al. 2014). ROS are highly reactive molecules containing oxygen. Examples include superoxide radicals ()$_2^{-}$-), hydrogen peroxide (H2O2), and hydroxyl radicals (OH$^-$). ROS are naturally produced in organisms during cellular respiration. However, under stress conditions, their levels can become excessive. Metal complexes are formed when metal ions bind to other molecules, often organic compounds. Certain metal complexes, particularly those involving transition metals like copper, iron, and chromium, can react with water and dissolved oxygen to generate ROS through processes like Fenton chemistry and Haber–Weiss reactions. This ROS generation can have detrimental effects on aquatic organisms: Damaging cellular components like proteins, lipids, and DNA, disrupting normal cellular functions and metabolic processes and contributing to oxidative stress, a condition linked to various cellular malfunctions and diseases.

ROS are produced by toxic metals, according to reports, and these ROS cause aquatic organisms exposed to them to lose their protein, lipid, and DNA content (Okereafor et al. 2020). Redox cycling occurs with hazardous metals that are redox active, such as Fe, Cu, and Cr, while covalent electron sharing occurs with poisonous metals that are redox inactive such as Pb, Cd, and Hg, with the primary antioxidant enzymes in cells (Thiols). ROS such as the hydroxyl radical (OH), superoxide radical (O_2^-), or hydrogen peroxide (H_2O_2), are produced by both classes of hazardous metals and lower cells' antioxidants in nature defense (Jaishankar et al. 2014). Metallic pollutants, such as copper, zinc, and nickel, can exert detrimental effects on aquatic plants. These metals can interfere with photosynthesis, thereby hindering plant growth and altering nutrient uptake mechanisms (Zheng et al. 2020). Given that aquatic plants are the foundation of the food chain (primary producers), such impacts can have cascading effects throughout the entire ecosystem. The persistence of long-lived metal complexes in the environment leads to chronic contamination of water bodies. This persistence significantly complicates and increases the expense of remediation efforts. Bioaccumulation, the process by which metal complexes accumulate in tissues, can occur in aquatic organisms (Moiseenko and Gashkina 2020). As predators at higher trophic levels consume contaminated prey, this bioaccumulation can be further magnified through biomagnification. This biomagnification ultimately poses health risks to humans and animals that consume these contaminated fish or other aquatic life. Studies have demonstrated reduced ATPase activity in the brain of the freshwater fish *Oreochromis niloticus* following both chronic and acute nickel exposure (Atli 2018). By altering the equilibrium of numerous species, metal complex pollution may cause aquatic ecosystems to collapse. Sensitive species may decline or become extinct, while metal-tolerant species can become more prevalent, changing the ecosystem's structure and behaviour. Metal complexes may attach to sediment particles, possibly contaminating the sediment over time and endangering benthic species that dwell in or on the sediment. In marine sediments at various distances from the Indigirka River Delta, zinc is the most prevalent trace metal, with amounts ranging from 85.47 to 128.88 mg kg^{-1} (Sevastyanov et al. 2020).

3 Fate of Water Pollutants on the Biosphere

The fate of water contaminants in the biosphere, the collective term encompassing all living organisms and their surroundings significantly impacts ecosystems, human health, and the environment. The specific trajectory of these contaminants within the biosphere is highly dependent on several factors, including the pollutant type, source, chemical composition, and the receiving environment (Arias-Estévez et al. 2008). Eutrophication, a process characterized by excessive nutrient (nitrogen and phosphorous) loading, disrupts aquatic ecosystems (Khan and Mohammad 2014). Common sources of these nutrients include agricultural runoff and inadequately treated sewage. Eutrophication triggers algal blooms and depletes dissolved oxygen levels, ultimately leading to mortality events among fish and other aquatic organisms. Persistent contaminants can further exacerbate this issue. These contaminants may facilitate the establishment of invasive species. Invasive species can outcompete native biota, disrupt ecological stability, and cause significant environmental damage by altering habitat conditions (Singh and Gupta 2016).

3.1 Fate of Nanoplastics in Water Bodies and Its Impacts on Biosphere

Environmental science and research are becoming more concerned with the fate of nanoplastics in aquatic environments and their effects on the biosphere (Al-Thawadi 2020). Nanoplastics are extremely tiny plastic particles, usually fewer than 100 nm in size, having diameters in the nanometer range. They may come through the decomposition of larger plastic items, from industrial processes, or from the intentional production of nanoplastic item. The major reasons for intentional nanoplastic production include: enhanced drug delivery in which nanoparticles can be engineered to deliver medications directly to targeted cells within the body, potentially improving treatment efficacy and reducing side effects, improved material properties in which nanoplastics can be incorporated into materials to enhance their strength, durability, or conductivity. For example, they might be used in lightweight composites or conductive films and environmental remediation where certain nanoplastics are being explored for their potential to capture and remove pollutants from contaminated environments. They are buoyant and tiny, making them suitable for transportation across water (Kooi et al. 2018). They may be dispersed across aquatic environments by currents, winds, and water movement. Many aquatic organisms, from little plankton to larger fish and mammals, frequently consume nanoplastics. This may result in bioaccumulation, a process through which plastics build up in the tissues of species as they ascend the food chain.

Nanoplastics may disrupt normal biological processes, result in inflammation and oxidative stress, and even harm aquatic animals reproductive and developmental systems and these impacts on aquatic life are toxicological (Coisne and Engelhardt 2011). It can alter the physical and chemical properties of marine ecosystem and they may alter the nutrient cycle, the habitat of benthic organisms, which are important to the general health of aquatic ecosystems, and the sediment structure (Al-Thawadi 2020). Nanoplastics inhibit algal growth, reduce chlorophyll and photosynthesis, induce ultrastructural changes, and affect gene expression in algae and the effects of energy flow can alter the productivity of aquatic organisms (Zhao et al. 2023). Microbes in water bodies may interact with nanoplastics, changing the composition of microbial populations (da Costa et al. 2016).

Some bacteria have the ability to break down plastics, which might alter the microbial community of water bodies. Some bacteria, such as *Ideonella sakaiensis* and

Pseudomonas aeruginosa, possess the enzymatic machinery to degrade certain plastics, potentially leading to alterations in the microbial community composition of water bodies. For Example: *Ideonella sakaiensis* bacterium, discovered in 2016, was the first reported organism capable of efficiently degrading PET (polyethylene terephthalate), a common plastic used in water bottles and clothing and in *Pseudomonas aeruginosa*, versatile bacterium has been shown to degrade various types of plastics, including polyurethanes and polyesters. However, its plastic degradation abilities are generally considered weaker compared to *Ideonella sakaiensis*. Nanoplastics have the ability to persist in the atmosphere for long periods of time, possibly increasing their influence over time, in contrast to larger plastic products that may degrade over time into smaller fragments (Bouwmeester et al. 2015). *Artemia franciscana* larvae's motility was reduced and their moulting behaviour was affected in the presence of 40 nm-sized Polystyrene particles (Tiwari et al. 2020). Recent studies have revealed the presence of polystyrene nanoparticles in the brains of various species, raising concerns about potential neurotoxicity such as Rainbow trout (*Oncorhynchus mykiss*) and fathead minnow (*Pimephales promelas*) were found to have polystyrene nanoparticles in their brains after exposure through contaminated food and water. *Daphnia Magna*, freshwater crustacean also accumulated polystyrene nanoparticles in its body, including the brain, when exposed to them in its environment. These findings imply that nano-sized particles can cross the blood–brain barrier, a highly selective permeability barrier necessary for the protection of the cerebral cortex from systematic toxins and for maintaining a healthy equilibrium of neuronal activity (Almutairi et al. 2016). Reduced predacious performance and efficiency of common goby (*Pomato schistusmicrops*) juveniles are caused by fish accidentally ingesting polyethylene particles that are micron-sized and shaped like spheres as food particles (de Sá et al. 2015).

3.2 Fate of Microplastics in Water Bodies and Its Impacts on Biosphere

The biosphere, encompassing both aquatic and terrestrial ecosystems, is significantly impacted by the fate of microplastics in lakes and rivers. Microplastics, generally smaller than 5 mm in size, may come from a variety of sources, including as the decomposition of larger plastic objects, microbeads in personal care products, and industrial operations (Eerkes-Medrano et al. 2015). One of the primary causes of the rising concentration of microplastic in different water bodies, including rivers, ponds, lakes, and the ocean, is the increase in human population and the availability of microplastic (Zhou et al. 2018). More than 100 fish species were tested for microplastic levels, and fibers and hard fragments were found in the stomach cells. Additionally, fish samples taken from areas near urban populations revealed higher levels of microplastic ingestion (Chan et al. 2019). The microplastic that is present in the water bodies is exposed to and consumed by aquatic animals. Since no enzyme is available to break down this synthetic polymer into more basic, non-toxic chemical components after ingestion, it does not undergo any enzymatic breakdown through a metabolic route. The temperature of the turtle's habitat may be impacted by plastic garbage, and this change in habitat conditions has a detrimental effect on the ability of sea turtles to reproduce (Nelms et al. 2016). The population of algae, which many aquatic species feed on and use as a source of energy, may be impacted by the presence of microplastics in aquatic bodies. It has been well shown that numerous species of algae, including *Chlorella sp., Scenedesmus*, and *Dunaliella sp.*, absorb and accumulate microplastics. Some harmful consequences of the microplastics on algae include oxidative stress and a decrease in photosynthetic activity (Gola et al. 2021).

4 Hazardous Impact of Industrial Pollution on Water Bodies

Industrial activity plays a crucial role in national economic development. However, these benefits often come at a cost to the environment. Industrial processes rely heavily on water and various chemicals, leading to the generation of significant waste streams. These effluents and solid wastes, often containing toxic materials, can pollute water bodies if not managed responsibly. Table 2 provides a breakdown of chemicals used and released by specific industries. The leather industry exemplifies the environmental challenges associated with industrial production. While it contributes significantly to economic growth (Moktadir et al. 2018; Alemu et al. 2019; Kumar and Singh 2021), its tanning processes are highly polluting. These processes involve large quantities of water, salts, and tides. Consequently, the leather industry generates vast amounts of saline effluent containing organic and inorganic substances, along with suspended putrescible matter.

Chromium, a heavy metal pollutant used in tanning, poses a particular threat due to its toxicity. Chromium exists in various oxidation states, ranging from -2 to $+6$. The hexavalent form [Cr(VI)] is especially concerning due to its carcinogenic and mutagenic properties. Balancing economic development with environmental protection is a critical challenge. Stringent regulations and the adoption of cleaner production technologies are essential for minimizing industrial waste and its associated environmental impact. Industries, like the leather industry, need to implement sustainable practices to reduce their reliance on harmful chemicals and manage effluent disposal responsibly. By prioritizing both economic growth and environmental sustainability, it can ensure a healthier future for our planet.

Industrial activity is a cornerstone of economic development. However, these benefits often come at an environmental cost. Industries rely heavily on water and various chemicals, generating significant waste streams. These effluents and solid wastes, frequently containing toxic materials, can pollute water bodies if not managed responsibly. But only the $+6$ and $+3$ oxidation states are commonly found in the

Table 2 Chemicals used in various industries

Type of industry	Chemicals used	Contaminants
Tanning	NaOH, NaCl, pentachlorophenol, sodium sulphide, enzymes, milk of lime, chlorides, H_2SO_4, chromium, formic acid, ammonium chloride, ammonium sulphate, non-identical metallic salts, and organic chemicals (Mannucci et al. 2010; UNIDO 2000)	Metallic and non-metallic contaminants like Cr(III), Cr(VI), chlorides, sulphates, sulphides$^-$, ammonium nitrogen, iron, and calcium (Kassim et al. 2022)
Textile and dyeing	HCl, H_2SO_4, Phosphoric Acid, NaCl, Sodium Carbonate, Sodium Sulphate, Sodium Bicarbonate, NaOH, KOH, Sulphur Dioxide, Trisodium phosphate, Benzene, Acetic Acid, Oxalic Acid, Cellulose, Starch, Sodium hydrosulphite, dyes, detergents, hydrogen peroxide (Kabir and Koh 2018)	Chlorine, chlorides, nitrogen, nitrate nitrogen, Free ammonia, sodium carbonate, sodium hydroxide, sodium chloride, phosphate, sulphates, sulphides, sulphur trioxide, oil and grease, dyes, heavy metals (Yaseen and Scholz 2019)
Paper and pulp	NaOH, sodium carbonate, sodium sulphide, chlorine dioxide or elemental chlorine, H_2O_2, H_2SO_4, dyes, resins, fillers such as clay, titanium dioxide and sizing agents like rosin and starch (Kamali and Khodaparast 2015; Avsar and Demirer 2008)	Chlorinated phenols, resin acids, chlorinated lignins, unsaturated fatty acids and other organics, which contains chlorine/sulphur compounds, phosphorus and nitrogen compounds (Kumar et al. 2022)
Sugar and distillery	CaOH phosphoric acid, CO_2, polyelectrolytes, SO_2, NaOH, HCl sodium carbonate, lead sub acetate (Jadhav et al. 2013; Kushwaha 2015)	Chloride, sulphate, nitrate, calcium and magnesium, melanoidins, caramels, phenolics, and furfurals (Fito et al. 2019)
Mining	Cyanide, H_2SO_4, nitric acid, ammonium nitrate and fuel oil, heavy metals such as Hg, Pb and acetylene (Conant and Fadem 2008)	Sulphide, pyrite, cyanide, sulphuric acid, heavy metals (da Silva-Rego et al. 2022)

environment. Cr(III) is highly insoluble, relatively immobile, and most thermodynamically stable while Cr(VI) is highly soluble, mobile and can be easily reduced. Hence, Cr(VI) is found to be mutagenic, carcinogenic and 100 times more toxic than Cr(III), and carcinogenic in nature (Rojas et al. 2005). Cr(VI) compounds at high concentration are found to be phytotoxic decrease the growth of plants by interfering in its metabolism (Davies et al. 2002). The textile industry is responsible for 7% of the world's export (Desore and Narula 2018) but is found to use various types of chemicals and enormous quantity of water for various processes (Hossain et al. 2018). As discussed, textile industry is not an exception in discharging 80% of its emissions in the form of untreated effluent into water bodies (Bhatia and Devraj 2017).

The textile industry contributes to release of coloured waste water attributed to the use of dyes to colour the fibres (Senthivelan et al. 2016). These dyes are synthetic in nature and non-biodegradable organic compounds (Orts et al. 2018), imparting colour to a given substrate (Shamey and Zhao 2014) because of the presence of chromophore groups in its molecular structures. These are highly soluble in water and hence removal is difficult (Hassan and Carr 2018). However, azo dyes do not bind completely to the fabric and are released into wastewater. This waste water used for irrigation in agriculture affects the germination and growth of plants (Rehman et al. 2018) and soil microbial communities (Imran et al. 2015). The textile dyes persist in the environment and are toxic, mutagenic and carcinogenic (Khatri et al. 2018).

The growing demand for paper led to the development of paper and pulp industry. The raw material for making paper comes from the trees and deforestation is an important phenomenon linked to making paper. The production of cellulose pulp from non-wood fibres has increased in recent year's which are responsible for generation of large amounts of wastewater when some sulphur-containing reagent is used for cooking of raw materials. These effluents are highly coloured, mainly due to their tannin content. Isopimaric, sand acopimaric, levopimaric, abietic, dehydroabietic, neoabietic and palustric acids are resin acids which can be generally found in pulping wastewaters (Vepsäläinen et al. 2011). Wastewater from bleaching process normally includes wood debris and soluble wood materials. Unsaturated fatty acids (such as oleic acids, linoleic acid, and linolenic acid), diterpene alcohols, juvaniones, chlorinated resin acids (Ali and Sreekrishnan 2001; Pokhrel and Viraraghavan 2004). The lignin fraction remains after bleaching in the generated wastewater in addition to soluble silicates widely known as water glass (Huuha et al. 2010). Black liquors, from the kraft process, generally contain aqueous solution of sodium hydroxide and sodium sulphide, used to break up and remove the lignin, and they are known to adversely affect the treatment facilities and aquatic life (Betancur et al. 2009). Some halogenated polycyclic aromatic hydrocarbons have been detected in kraft pulp mill products and discharges (Koistinen et al. 1994). Chlorophenols, adsorbable organic halogens (AOX), and extractable organic halogens (EOXs) as well as a trace of toxic substances such as DDT, polychlorinated biphenyls, and polychlorinated dibenzodioxines are found in the effluents after the chlorine-based bleaching (Karrasch et al. 2006). In addition, dioxins, furans, chlorinated lignosulfonic acids, and chlorinated resin acids are found to be produced during the bleaching process (Ali and Sreekrishnan 2001).

The sugar industry is always integrated with ethanol distillery and more recently with bioplastic production unit in which the sugar is converted into lactic acid and polymerizes into biopolymer. These units put together create water pollution by discharging a large amount of wastewater as effluent. The waste water from sugar industry has large amount of solids, chloride, sulphate, nitrate, calcium and magnesium (Fito et al. 2019; Samuel and Muthukaruppan 2011). These effluents become foul smelling and render the water bodies black as a result of precipitation of iron and dissolved salts (Elayaraj 2014). Distillery spent wash is the most complex among the waste waters namely fermented sludge, spent lees released from distilleries. The spent wash has enormous organic matter in terms of high COD (80,000–160,000 mg/L), high temperature, high ash content, high acidity (pH 3.7–4.5) with a high content of dissolved inorganic salts (Basu et al. 2015; Fito et al. 2017). The spent wash is deep dark brown in colour due melanoidins, caramels, phenolics, and furfurals (Dos

Reis et al. 2015). Hence the water bodies contaminated with the release of this kind of effluents are unfit for human consumption as well as aquatic life (Hampannavar and Shivayogimath 2010).

Mining activities affects the ground water as well as the river water. The seepage from tailings and waste rock impoundments and mine effluents generated during processing of ore are the main sources of pollution due to mining. The waste water called as acid mine drainage (AMD) is formed due to oxidative dissolution of sulphide minerals, lowering the pH of the water. AMD water also contains heavy metals which affect aquatic life by bioaccumulation (Dutta et al. 2020). AMD due to its acidic nature contaminate drinking water. It interferes in the lifecycle of aquatic plants and animals and corrodes structures such as bridges, railways and ships. Coal mines and metal mines are responsible for sulphate emission into groundwater, due to the presence of large quantities of sulphide minerals (Miao et al. 2013). The presence of elevated concentrations of heavy metals in groundwater and surface water near to mining sites is evident from many studies. The high metal concentrations and acidity observed in waters from mining areas depend on several parameters and processes, such as the buffering capabilities of the host rock, ore deposits depth and permeability, and the type and abundance of the metal-bearing sulphides (Mohapatra and Kirpalani 2017).

Heavy metals such as manganese, iron, arsenic, aluminium, chromium, lead and zinc from the metal mines contaminate the surface and groundwater samples. Some of the potentially harmful elements with significant high concentrations mostly exceeded the WHO water guidelines for human consumption in discharge water (Salvarredy-Aranguren et al. 2008; Bhattacharya et al. 2012). Mobility studies on the movement of heavy metals from tailings to groundwater in a mining activity contaminated site indicated that the contamination of groundwater became worse and strongly visible due to the red coloring of bed stream sediments as a consequence of abundant Fe(III)-hydroxides precipitation (Concas et al. 2006). However, in certain cases, the impact of mining has been limited by precipitation and adsorption of the metals by iron oxides, carbonates, clays and organic materials (Smedley and Kinniburgh 2002; Sracek et al. 2012).

4.1 Agricultural Production and Allied Activities

Agriculture involving crop production, livestock and aquaculture accounts for 70% of the water consumption globally and also have become a source of water pollution in order to feed a growing global population projected to reach 9.8 billion by 2050 (UNDESA 2017). The sources are categorized into point and non-point sources (Chen et al. 2017). Agriculture is responsible for the discharge of large quantities of agrochemicals such as pesticides, herbicides and fertilizers, nutrients, organic matter, drug residues, sediments, saline drainage inducing salinization and alkalinization, microplastics and pathogens (Evans et al. 2019). When applied, these fertilizers are either taken up by crops, remain in the soil, or enter the aquatic environments. Nitrate has a high mobility and low affinity for adsorbing onto clay particles, which facilities its passage into groundwater. Phosphates which are not as soluble as nitrates may get adsorbed onto soil particles and pollute the water through soil erosion. Animal wastes are also rich source of nutrients and can be used as fertilizer. In aquaculture industry, excess nutrients and animal wastes may pollute the water. Table 3 summarizes most common water contaminants in agriculture.

Generally, pests become resistant to pesticide when used for a long period of time, and therefore new pesticides need to be developed continuously, requiring new treatment methods and/or modification of existing ones. Many of the herbicides are not easily degraded and remain in the environment for extended periods of time (Youssef and Younes 2019). Drugs, dyes, herbicides, pesticides, and other products of daily use have been detected in ground/underground water. Irrigation and rain also facilitate transportation of pesticides into ground/underground water especially those which are soluble in water (Sharma et al. 2019). Health issues such as hematologic and hormonal abnormalities, infertility, foetal malformation, neurological diseases, and cancer are commonly observed among those exposed to these pesticides. The underlying mechanisms of these effects are genotoxic, neurotoxic, and endocrine-disrupting actions (Monneret 2017; Gundogan et al. 2018; Jokanovi 2018).

Agriculture pesticides are generally complex molecules and undergo several different reactions to become completely degraded. Pesticide transformation/conversion to other molecules starts when introduced at a site as shown in Fig. 1. Degradation is significantly more complex for insoluble pesticides compared to soluble pesticides. For Example: Soluble pesticide, Malathion is an organophosphate insecticide readily dissolves in water and is easily degraded by soil microbes. Studies have shown a half-life (the time it takes for half of the pesticide to degrade) of malathion in soil to be around 7–14 days and Insoluble pesticide such as DDT is persistent organic pollutant (POP) is highly insoluble in water. Due to its limited interaction with microbes, DDT degradation is exceptionally slow. Research suggests a half-life for DDT in soil can range from several years to decades. The degradation reactions may be fast or slow and the intermediate molecules formed during the degradation may have different properties and follow different pathways to be mineralized, which depends on the type, physical, and chemical properties of the pesticide. Heat and photodegradation imposed by solar radiation play a major role in the breakdown of pesticide molecules, primarily through oxidation, reduction, and hydrolysis reactions. Microorganisms play a key role in the biodegradation of pesticides where complex biochemical reactions occur (Rad et al. 2022).

Vaccines, antibiotics, growth hormones, and drug consumption have increased in recent decades by humans or use in livestock, which can reach water resources in different ways. Agriculture emits hormones and antibiotics, when manure is inappropriately stored, handled and applied as fertilizer, and from aquaculture; and through the application of certain pesticides. Use of antimicrobials in agriculture, not just to treat infections but also to promote growth in livestock and fish, consumption is greater than all human consumption and is increasing (Van Boeckel et al. 2015). Heavy metals which exist in pesticides and animal feed are considered as emerging pollutants and more than 700 of them and their metabolites have been detected in

Table 3 Categories of major water pollutants in agriculture and the relative contributions of the three main agricultural production systems (Mateo-Sagasta et al. 2017)

Pollutant category	Indicators/examples	Relative contribution		
		Crops	Livestock	Aquaculture
Nutrients	Primarily nitrogen and phosphorus present in chemical and organic fertilizers as well as animal excreta and normally found in water as nitrate, ammonia, or phosphate	***	*	***
Pesticides	Herbicides, insecticides, fungicides, and bactericides, including organophosphates, carbamates, pyrethroids, organochlorine pesticides, and others (many, such as DDT, are banned in most countries but are still being used illegally and persistently)	***	–	–
Salts	E.g., ions of sodium, chloride, potassium, magnesium, sulphate, calcium, and bicarbonate. Measured in water, either directly as total dissolved solids or indirectly as electric conductivity	***	*	*
Sediment	Measured in water as total suspended solids or nephelometric turbidity units—especially from pond drainage during harvesting	***	***	*
Organic matter	Chemical or biochemical oxygen demanding substances (e.g., organic materials such as plant matter and livestock excreta), which use up dissolved oxygen in water when they degrade	*	***	**
Pathogens	Bacteria and pathogen indicators., e.g., Escherichia coli, total coliforms, faecal coliforms, and enterococci	*	***	*
Metals	E.g., selenium, lead, copper, mercury, arsenic, and manganese	*	*	*
Emerging pollutants	E.g., drug residues, hormones, and feed additives	–	***	**

*Represents low contribution; **represents moderate contribution; ***represents high contribution

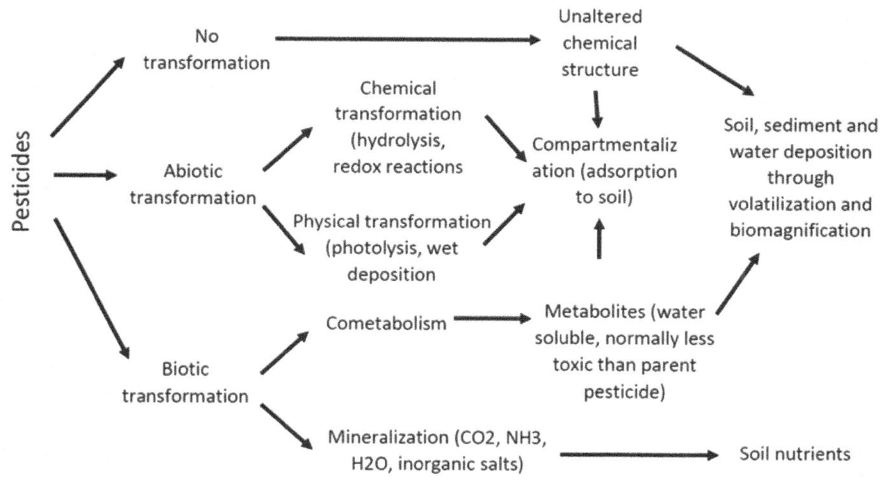

Fig. 1 Transformation of pesticides in environment (Adapted from Ikehat and El-Din 2005)

European aquatic environments (Navarro-Ortega et al. 2015). Agricultural activities may reintroduce these pollutants into aquatic environments through the generated wastewater.

Agriculture is responsible for the spread of CEC (Cation Exchange Capacity) from domestic and industrial sources where biosolids and manure are used as soil conditioners and fertilizer (Pepper et al. 2018) and wastewater for irrigation (Lesser et al. 2018). Studies have explored the fate and transport of pharmaceuticals to waterbodies (Koreje et al. 2018). Plastics, particularly microplastics, are another CEC expected to accumulate in agricultural areas where plastics are used in production processes (Horton et al. 2017).

4.2 *Eutrophication*

Eutrophication is described as a condition in which a high level of nutrient concentrations of nitrogen and phosphorus result in proliferous algal blooms, thus degrading the water quality in the aquatic ecosystems (Bali and Gueddari 2019). Eutrophication is a process may be natural or anthropogenic in origin. Natural eutrophication process takes a long geological time to render eutrophic status to a water body, while in cultural eutrophication there is a quick rise in the eutrophic status of water body due to anthropogenic reasons (Sonarghare et al. 2020).

Many reasons have been recognized for the process of eutrophication. They are addition of nutrients into water bodies from land due to rainfall and storm, the organic matter builds up as sediments due to ageing in water bodies that result in higher cyanobacteria blooms and phytoplankton growth Yang 2022). In addition to this, nitrate and phosphate from fertilizers, animal feeds and dung, applied to plants, run off from fertilizer production units, mining and direct discharge of aquaculture effluent contribute to eutrophication (Knight et al. 2022; Qi et al. 2022). Eutrophication results in algal blooms that reduce sunlight penetration into the water system thereby decreasing the photosynthesis of submerged plants. Moreover, due to the death and decay of algae the amount of dissolved oxygen is depleted leading to the death of submerged aquatic species living in the water bodies (Costa et al. 2018; Rathore et al. 2016; Kim et al. 2020; Hwang 2020). The low dissolved oxygen levels lead to reduction in number of fishes thereby leading to diminish in the population of birds and marine animals that depend on fish for survival (Rathore et al. 2016; Park et al. 2020; Hwang 2020).

The presence of nitrates in drinking water result in the formation of compounds like nitrosamides and nitrosamine categorized to be potentially carcinogenic. Moreover, the nitrates and nitrites induce methemoglobinemia with symptoms that include suffocation, lethargy and blue baby syndrome and in some cases leading to anoxia and death at 5 to 10% concentration of methemoglobin (Matei and Racoviteanu 2021). The cyanobacteria dominant in the water bodies due to its nitrogen fixation ability in low nitrogen to phosphorus ratio, produce toxins most commonly hepatotoxin microcystin (MCs) found to bioaccumulate in aquatic species which on consumption affect human health (Rathore et al. 2016; Usman et al. 2022). These toxins are found to be highly toxic to organs like kidney, liver, heart, gastrointestinal tract, the nervous system as well as gonads of invertebrates. The concentration of these microcystins due to cooking processes together with its resistance to oxidation and hydrolysis at neutral pH has increased the negative impact of eutrophication on human health (Usman et al. 2022; Lone et al. 2015) and have led to higher occurrences of colorectal cancer (Koto et al. 2022).

4.3 River Pollution

Rivers are the important source of fresh water used for agriculture, industrial and domestic purposes. However, scientific reports suggest unbearable pollution in rivers. Organic matter, nutrients, heavy metals, and pathogens from domestic and industrial wastewater are classified as traditional pollutants while classes of pharmaceuticals and personal care products (PPCPs), plasticizers, surfactants, fire retardants, nanomaterials, and pesticides are classified as traditional pollutants (Kumar et al. 2022). Organic pollution due to nutrients and organic matter might result from natural and anthropogenic activities like farming as well as waste water release (Malai et al. 2014). Seasonal fluctuations in precipitation resulted in improved water quality during the rainy season due to dilution, while poorer water quality was observed during the summer months. Heavy metals including zinc (Zn), arsenic (As), copper (Cu), lead (Pb), cadmium (Cd), mercury (Hg), nickel (Ni), and chromium (Cr) are released into water bodies by natural phenomenon like leaching and weathering and anthropogenic sources like sewage and untreated waste water from industries as well as farming activities in which case mostly the levels are beyond the limits prescribed by WHO (Paul 2017). These heavy metals accumulate in the sediments and are harmful to aquatic organisms. Metals in high concentration can be toxic which is aggravated when reacting with organic compounds to form organometallic compounds. Moreover, these heavy metals impact human health when leached into ground and surface waters, become chelated with clay and organic matter in the soil and absorbed by plants (Gabrielyan et al. 2018). Pathogens importantly coliforms serve as indicators of faecal pollution which pose a great threat to human health. These pathogens gain entry into rivers through sewers, waste water treatment plants and from agriculture activities (Pandey et al. 2014).

Another class of pollutants called emerging pollutants has been detected in water bodies as a result of advanced detection technologies. The wastewater from industrial and agricultural activities released into rivers remain the main source of these pollutants These pollutants have been classified into pharmaceuticals and personal care products (PPCPs), plasticizers, surfactants, fire retardants, nanomaterials, and pesticides (Kumar et al. 2022). Some emerging and persistent organic pollutants such as polycyclic aromatic hydrocarbons (PAHs), organochlorine pesticides (OCPs), polychlorinated biphenyls (PCBs), and perfluoroalkyl and polyfluoroalkyl substances (PFASs) were found in rivers. Some of these pollutants called as endocrine-disrupting chemicals (EDCs), include PCBs, bisphenol A, pharmaceutical products, pesticides, phthalates, polybrominated compounds, alkylphenol ethoxylates, and alkylphenols stimulate or inhibit hormone production and metabolism (Arman et al. 2021; Anh et al. 2023).

4.4 Wetland Ecosystems

Wetlands are defined as "areas of marsh, fen, peatland or water, whether natural or artificial, permanent or temporary, with water that is static or flowing, fresh, brackish or salt, including areas of marine water the depth of which atlow tide does not exceed six metres" by the Ramsar convention on wetlands. Cowardin et al. (1979) categorized wetlands into marine (coastal wetlands), estuarine (including deltas, tidal marshes, and mangrove swamps), lacustarine (lakes), riverine (along rivers and streams), and palustrine ('marshy'—marshes, swamps and bogs) based on their hydrological, ecological and geological characteristics. Most of the natural water bodies such as rivers, lakes, coastal lagoons, mangroves, peat land, coral reefs and manmade wetlands such as ponds, farm ponds, irrigated fields, sacred groves, salt pans, reservoirs, gravel pits, sewage farms and canals constitute the wetland ecosystem. Wetlands provide water for irrigation, domestic needs, recreation, fisheries, non-timber forest products, carbon sequestration, flood control, groundwater recharge, nutrient removal, toxics retention and biodiversity maintenance (Turner et al. 2000).

Significant areas of wetlands worldwide have shrunk due to urbanization, land-use changes, agriculture, infrastructure development, industrial effluent and agricultural runoff pollution, climate change, and variability (Bassi et al. 2014). This has resulted in a substantial loss of biodiversity, with 21% of bird species, 37% of mammal species, and 20% of freshwater fish species either extinct or globally threatened. Agricultural runoff and untreated wastewater from various sources like sewers and industries degrade wetlands, leading to eutrophication (Liu and Diamond 2005; Prasad et al. 2002). Eutrophication not only disrupts the wetland ecosystem but also increases the cost of purifying water from affected lakes for drinking or industrial applications. Wetland loss due to anthropogenic activities has severe ecological and economic consequences. Protecting and restoring wetlands is crucial for maintaining biodiversity, mitigating the effects of climate change, and ensuring access to clean water resources.

5 Impacts on Zooplankton and Phytoplankton and Fishes

Industrial effluents, agricultural runoff and sewage discharge cause widespread water pollution, including toxic contamination, organic pollution, and eutrophication, leading to severe ecological damage (Miao et al. 2012). Heavy metals from industrial effluents combine with suspended solids in domestic wastewater, forming harmful muck (Mohamed et al. 2013). The diversity of toxic and persistent pollutants makes their removal from water bodies and sediments a costly endeavor (Doust et al. 1994). Wastewater releases can alter the diversity of aquatic organisms and even drive some species to extinction (Niu et al. 2019). Pollutants can induce mutations and affect germ cells, causing genetic changes that may be transmitted to future generations (Obiakor et al. 2012). Agricultural runoff contributes to eutrophication, which can harm the health and reproductive capacity of fish and other aquatic animals. Pesticides and their byproducts disrupt the aquatic food chain from top predators to the base level (Islam and Tanaka 2004), causing harmful mutations and serious diseases in marine organisms, ultimately affecting their tissues, biochemistry, and development.

Animal manure remains the source of pathogens and alters the turbidity, odour, and colour of the water body. Turbidity blocks sunlight which prevents the use of sunlight for photosynthesis by marine plants. Oil also threatens the coral reefs that are home to many marine creatures. It clogs up fish gills, consumes plankton, and also hurts the sea birds. The discharge of hot or cold water from industries and power plants affect the behaviour and reproduction of animals such as fish. Aquatic organisms cannot tolerate sudden temperature changes and excessive temperature can lead to the release of immature eggs or affect the development of certain eggs. Water temperature may not be important to pure water due to the wide range of temperature tolerances in aquatic life. However, the temperature of polluted water may significantly affect the DO concentration (Rahman et al. 2021).

Phytoplankton or microalgae are very sensitive to contaminants like heavy metals and a change in its species diversity is observed in the marine ecosystem (Hosmani

2014). *Wolffiaglobosa* a flowering plant commonly known as Asian water meal or duckweed shows sensitivity to cadmium is an important tool for identifying cadmium contamination. Hence some species like *Euglena clastica, Trachelonanas, Phacus-tortus* which react quickly to environmental changes are viewed as indicators of marine pollution (Parmar et al. 2016). The freshwater ecosystem rich in biodiversity are considered as the world's most threatened ecosystem and vulnerable habitat due to its fast rate of decline in case of interferences than the marine water or land ecosystem (Obiakor et al. 2012). Zooplanktons like *Cyclops, Mesocylops, Aheyella,* etc. are sensitive to contaminants and serve as zone-based pollution indicators (Hosmani 2014). Several invertebrates that live near the benthic region and diatoms can act as bioindicators (Khatri and Tyagi 2015). Coral reefs involving symbiotic association between plant and animal are sensitive to change in the ecosystem conditions and act as bioindicators of water quality (Cooper et al. 2009).

A reduction in the population of fishes is attributed to the heavy metal pollution which accumulates in the active tissues such as gills, liver, and kidneys due to the induction/occurrence of metal-binding proteins called metallothioneins (MTs) in these tissues upon exposure to heavy metals. Waterborne heavy metals are incorporated in fish and enter the human body through the food chain and therefore have impact on human health (Dwivedi et al. 2015). Furthermore, toxic heavy metals also affect fish health and well-being. It has been reported that pollution of river with heavy metals-containing wastewater induced stress in the freshwater fish *Channa punctatus*, making it weak and more vulnerable to diseases (Javed and Usmani 2015). Heavy metal pollution is regarded as one of the possible causes for population declines of freshwater fish and other aquatic species in freshwater ecosystems.

The discharge of pollutants and emerging contaminants from various sources has a cascading effect on aquatic ecosystems, impacting organisms at all levels, including phytoplankton, zooplankton, and fishes. In phytoplanktons, reduced growth and productivity which causes pollutants like heavy metals and some pharmaceuticals can inhibit photosynthesis, hinder nutrient uptake, and ultimately reduce phytoplankton growth and overall primary production at the base of the food web, shifts in species composition for emerging contaminants like endocrine-disrupting chemicals (EDCs) can disrupt hormonal balance, potentially favoring more resistant phytoplankton species and altering the overall community structure and bioaccumulation of toxins for some pollutants, like microplastics, can be ingested by phytoplankton, potentially concentrating toxins as they move up the food chain.

In case of Zooplankton, reduced feeding and reproduction in pollutants can harm zooplankton's ability to filter food particles, impacting their growth and reproduction. Additionally, EDCs can disrupt zooplankton development and hormone signaling, increased susceptibility to predation in pollutants can weaken zooplankton, making them easier prey for fish, thus disrupting the food web balance and biomagnification of toxins: Pollutants like PFAS (per- and polyfluoroalkyl substances) can biomagnify through the food chain, reaching high concentrations in zooplankton and impacting their health. In Fishes, disrupted development and reproduction for EDCs can mimic natural hormones, leading to abnormal development, impaired fertility, and skewed sex ratios in fish populations, Organ damage and disease for pollutants like heavy metals can accumulate in fish tissues, causing organ damage, immune system suppression, and increased susceptibility to diseases and behavioral alterations for pollutants can affect fish behavior, impacting their ability to find food, avoid predators, and navigate their environment. In case of microplastics ingestion, fish can ingest microplastics, leading to physical blockages in their digestive system, reduced nutrient uptake, and potential leaching of harmful chemicals. Overall, the combined effects of pollutants and emerging contaminants can disrupt the delicate balance of aquatic ecosystems. By harming phytoplankton, zooplankton, and fish populations, these contaminants can ultimately lead to reduced biodiversity, impaired ecosystem services, and potential threats to human health through contaminated fish consumption.

6 Conclusions

Water being a very essential component for sustaining life in this world is being polluted by indiscriminate release of waste waters from industries, agricultural fields and domestic sources. There are a variety of contaminants from these sources which are harmful and toxic challenging the health and life of humans as well as other organisms. Though stringent control measures are followed by some countries, many countries lack in appropriate waste water treatment measures and facilities which involves good amount of cost. Hence cost-effective technologies together with implementation of rules and regulations for water treatment will come a long way in saving the water bodies of the world.

7 Future Aspects

- Emerging Contaminants and their Evolving Threats: Expand the focus to include not just traditional pollutants but also emerging threats like pharmaceuticals, microplastics, and nanomaterials. Explore how their unique properties affect transport, fate, and ecological impacts.
- Climate Change and Water Pollution Dynamics: Discuss how climate change, with factors like rising temperatures, altered precipitation patterns, and increased extreme weather events, will influence the transport, fate, and impact of water pollution.
- Biodegradation and Bioremediation Technologies: Shift the focus to explore how advancements in biodegradation and bioremediation technologies can aid in mitigating future water pollution challenges. Discuss potential solutions using microbes, enzymes, or phytoremediation for specific pollutants.
- Predictive Modeling and Risk Assessment: Emphasize the role of advanced modeling and risk assessment tools in predicting the transport and fate of pollutants. This can help identify areas at risk and prioritize preventive measures.

- Sustainable Water Management and Policy Solutions: Explore future directions for sustainable water management practices and policy changes that can minimize pollution at the source, improve wastewater treatment, and promote water reuse

References

Adekunle AS, Eniola ITK (2008) Impact of industrial effluents on quality of segment of Asa river within an industrial estate in Ilorin, Nigeria. N Y Sci J 1(1):17–21

Alemu A, Lemma B, Gabbiye N (2019) Adsorption of chromium(III) from aqueous solution using vesicular basalt rock. Cogent Environ Sci 5(1):1650416

Ali M, Sreekrishnan TR (2001) Aquatic toxicity from pulp and paper mill effluents: a review. Adv Environ Res 5(2):175–196

Almutairi MMA et al (2016) Factors controlling permeability of the blood–brain barrier. Cell Mol Life Sci 73:57–77

Al-Thawadi S (2020) Microplastics and nanoplastics in aquatic environments: challenges and threats to aquatic organisms. Arab J Sci Eng 45(6):4419–4440

Anh NT, Nhan NT, Schmalz B, Luu TL (2023) Influences of key factors on river water quality in urban and rural areas: a review. Case Stud Chem Environ Eng 2023:100424

Anjaneyulu Y, Sreedhara Chary N, Samuel Suman Raj D (2005) Decolourization of industrial effluents–available methods and emerging technologies—a review. Rev Environ Sci Bio/technol 4:245–273

Anyahara JN (2021) Effects of polycyclic aromatic hydrocarbons (PAHs) on the environment: a systematic review. Int J Adv Acad Res. https://doi.org/10.46654/ij.24889849

Arias-Estévez M et al (2008) The mobility and degradation of pesticides in soils and the pollution of groundwater resources. Agric Ecosyst Environ 123(4):247–260

Arman NZ et al (2021) A review on emerging pollutants in the water environment: existences, health effects and treatment processes. Water 13(22):3258

Atli G (2018) The effect of waterborne mercury and nickel on the ATPases and AChE activities in the brain of freshwater fish (oreochromis niloticus) depending on the Ca 2+ concentrations. Turk J Fish Aquat Sci 19(5):363–371

Authman MMN, Zaki MS, Khallaf EA, Abbas HH (2015) Use of fish as bio-indicator of the effects of heavy metals pollution. J Aquacult Res Develop 6(4):1–13

Avşar E, Demirer GN (2008) Cleaner production opportunity assessment study in SEKA balikesir pulp and paper mill. J Clean Prod 16(4):422–431

Awaleh MO, Soubaneh YD (2014) Waste water treatment in chemical industries: the concept and current technologies. Hydrol Current Res 5(1):1–12

Bali M, Gueddari M (2019) Removal of phosphorus from secondary effluents using infiltration–percolation process. Appl Water Sci 9:1–8

Bano I, Arshad M (2018) Climatic changes impact on water availability. In: Perspectives on water usage for biofuels production: aquatic contamination and climate change, pp 39–54

Bassi N, Dinesh Kumar M, Sharma A, Pardha-Saradhi P (2014) Status of wetlands in India: a review of extent, ecosystem benefits, threats and management strategies. J Hydrol Region Stud 2:1–19

Basu S et al (2015) Integrated treatment of molasses distillery wastewater using microfiltration (MF). J Environ Manag 158:55–60

Betancur M, Bonelli PR, Velásquez JA, Cukierman AL (2009) Potentiality of lignin from the Kraft pulping process for removal of trace nickel from wastewater: effect of demineralisation. Bioresour Technol 100(3):1130–1137

Bhateria R, Jain D (2016) Water quality assessment of lake water: a review. Sustain Water Resour Manag 2:161–173

Bhatia SC, Devraj S (2017) Pollution control in textile industry. WPI Publishing, London
Bhattacharya P et al (2012) Hydrogeochemical study on the contamination of water resources in a part of Tarkwa mining area, Western Ghana. J Afr Earth Sci 66:72–84
Boeckel V, Thomas P et al (2015) Global trends in antimicrobial use in food animals. Proc Natl Acad Sci 112(18):5649–5654
Bouwmeester H, Hollman PCH, Peters RJB (2015) Potential health impact of environmentally released micro-and nanoplastics in the human food production chain: experiences from nanotoxicology. Environ Sci Technol 49(15):8932–8947
Breida M et al (2019) Pollution of water sources from agricultural and industrial effluents: special attention to NO3⁻, Cr(VI), and Cu(II). Water Chem 39:1
Chan HS, Hamsun CD, Not C (2019) Evidence for non-selective ingestion of microplastic in demersal fish. Mar Pollut Bull 149:110523
Chen J, Hao W, Qian H (2016) Groundwater nitrate contamination and associated health risk for the rural communities in an agricultural area of Ningxia, northwest China. Expos Health 8:349–359
Chen Y, Wen X, Wang B, Nie P (2017) Agricultural pollution and regulation: How to subsidize agriculture? J Clean Prod 164:258–264
Chen J et al (2020) Development of human health ambient water quality criteria of 12 polycyclic aromatic hydrocarbons (PAH) and risk assessment in China. Chemosphere 252:126590
Coisne C, Engelhardt B (2011) Tight junctions in brain barriers during central nervous system inflammation. Antioxid Redox Signal 15(5):1285–1303
Conant J, Fadem P (2008) A community guide to environmental health. Hesperian Foundation, London
Concas A et al (2006) Mobility of heavy metals from tailings to stream waters in a mining activity contaminated site. Chemosphere 63(2):244–253
Cooper TF, Gilmour JP, Fabricius KE (2009) Bioindicators of changes in water quality on coral reefs: review and recommendations for monitoring programmes. Coral Reefs 28:589–606
Cowardin LM (1979) Classification of wetlands and deepwater habitats of the United States. Fish and wildlife service, US Department of the Interior
da Costa JP, Santos PSM, Duarte AC, Rocha-Santos T (2016) (Nano) plastics in the environment—sources, fates and effects. Sci Total Environ 566:15–26
da Costa JA et al (2018) Eutrophication in aquatic ecosystems: a scientometric study. Acta Limnol Brasil 30:1
da Silva-Rêgo LL, de Almeida LA, Gasparotto J (2022) Toxicological effects of mining hazard elements. Energy Geosci 3(3):255–262
Davies FT Jr et al (2002) Mycorrhizal fungi increase chromium uptake by sunflower plants: influence on tissue mineral concentration, growth, and gas exchange. J Plant Nutr 25(11):2389–2407
de Sá LC, Luís LG, Guilhermino L (2015) Effects of microplastics on juveniles of the common goby (pomatoschistus microps): confusion with prey, reduction of the predatory performance and efficiency, and possible influence of developmental conditions. Environ Pollut 196:359–362
Desore A, Narula SA (2018) An overview on corporate response towards sustainability issues in textile industry. Environ Develop Sustain 20:1439–1459
Doust JL, Schmidt M, Doust LL (1994) Biological assessment of aquatic pollution: a review, with emphasis on plants as biomonitors. Biol Rev 69(2):147–186
Du E et al (2017) Direct effect of acid rain on leaf chlorophyll content of terrestrial plants in China. Sci Total Environ 605:764–769
Dutta M et al (2020) Acid mine drainage in an Indian high-sulfur coal mining area: cytotoxicity assay and remediation study. J Hazardous Mater 389:121851
Dwivedi AC, Tiwari A, Mayank P (2015) Seasonal determination of heavy metals in muscle, gill and liver tissues of Nile tilapia, oreochromis niloticus (Linnaeus, 1758) from the tributary of the Ganga River, India. Zool Ecol 25(2):166–171
Eerkes-Medrano D, Thompson RC, Aldridge DC (2015) Microplastics in freshwater systems: a review of the emerging threats, identification of knowledge gaps and prioritisation of research needs. Water Res 75:63–82

Elayaraj B (2014) Physico-chemical analysis of sugar factory effluent stress on seedling growth of black gram (Vigna mungo (L.) Hepper) varieties. Int Lett Nat Sci 12(1):1

Evans AEV et al (2019) Agricultural water pollution: key knowledge gaps and research needs. Curr Opin Environ Sustain 36:20–27

Fito J, Tefera N, Van Hulle SWH (2017) Adsorption of distillery spent wash on activated bagasse fly ash: kinetics and thermodynamics. J Environ Chem Eng 5(6):5381–5388

Fito J, Tefera N, Kloos H, Van Hulle SWH (2019) Physicochemical properties of the sugar industry and ethanol distillery wastewater and their impact on the environment. Sugar Tech 21:265–277

Gabrielyan AV, Shahnazaryan GA, Minasyan SH (2018) Distribution and identification of sources of heavy metals in the Voghji River basin impacted by mining activities (Armenia). J Chem 2018:1

Gola D et al (2021) The impact of microplastics on marine environment: a review. Environ Nanotechnol Monit Manag 16:100552

Griboff J, Horacek M, Wunderlin DA, Monferran MV (2018) Bioaccumulation and trophic transfer of metals, as and Se through a freshwater food web affected by antrophic pollution in Córdoba, Argentina. Ecotoxicol Environ Saf 148:275–284

Gundogan K et al (2018) Evaluation of chromosomal DNA damage, cytotoxicity, cytostasis, oxidative DNA damage and their relationship with endocrine hormones in patients with acute organophosphate poisoning. Mutat Res/genet Toxicol Environ Mutag 825:1–7

Hampannavar US, Shivayogimath CB (2010) Anaerobic treatment of sugar industry wastewater by upflow anaerobic sludge blanket reactor at ambient temperature. Int J Environ Sci 1(4):631–639

Hassan MM, Carr CM (2018) A critical review on recent advancements of the removal of reactive dyes from dyehouse effluent by ion-exchange adsorbents. Chemosphere 209:201–219

Horton AA et al (2017) Microplastics in freshwater and terrestrial environments: evaluating the current understanding to identify the knowledge gaps and future research priorities. Sci Total Environ 586:127–141

Hosmani S (2014) Freshwater plankton ecology: a review. J Res Manag Technol 3:1–10

Hossain MS et al (2018) Reuse of textile mill ETP sludge in environmental friendly bricks—effect of gamma radiation. Radiat Phys Chem 151:77–83

Hung Y-T, Wang LK, Shammas NK (2020) 3 handbook of environment and waste management-volume 3: acid rain and greenhouse gas pollution control. World Scientific, Singapore

Hussain MM et al (2019) Biogeochemical cycling, speciation and transformation pathways of arsenic in aquatic environments with the emphasis on algae. In: Comprehensive analytical chemistry. Elsevier, London, pp 15–51

Huuha TS, Kurniawan TA, Sillanpää MET (2010) Removal of silicon from pulping whitewater using integrated treatment of chemical precipitation and evaporation. Chem Eng J 158(3):584–592

Hwang S-J (2020) Eutrophication and the ecological health risk. Int J Environ Res Public Health 17(17):6332

Ikehata K, El-Din MG (2005) Aqueous pesticide degradation by ozonation and ozone-based advanced oxidation processes: a review (part I). Ozone Sci Eng 27(2):83–114

Imran M et al (2015) Microbial biotechnology for decolorization of textile wastewaters. Rev Environ Sci Bio/technol 14:73–92

Islam MS, Tanaka M (2004) Impacts of pollution on coastal and marine ecosystems including coastal and marine fisheries and approach for management: a review and synthesis. Mar Pollut Bull 48(7–8):624–649

Ite AE et al (2018) Petroleum hydrocarbons contamination of surface water and groundwater in the Niger delta region of Nigeria. J Environ Pollut Hum Health 6(2):51–61

Jadhav PG, Vaidya NG, Dethe SB (2013) Characterization and comparative study of cane sugar industry wastewater. Int J Chem Phys Sci 2(2):19–25

Jaishankar M et al (2014) Toxicity, mechanism and health effects of some heavy metals. Interdiscip Toxicol 7(2):60

Javed M, Usmani N (2015) Stress response of biomolecules (carbohydrate, protein and lipid profiles) in fish Channa punctatus inhabiting river polluted by thermal power plant effluent. Saudi J Biol Sci 22(2):237–242

Jokanović M (2018) Neurotoxic effects of organophosphorus pesticides and possible association with neurodegenerative diseases in man: a review. Toxicology 410:125–131

K'oreje KO et al (2018) Occurrence, fate and removal of pharmaceuticals, personal care products and pesticides in wastewater stabilization ponds and receiving rivers in the Nzoia Basin, Kenya. Sci Total Environ 637:336–348

Kabir SMM, Koh J (2018) Dyeing chemicals chemistry and technology of natural and synthetic dyes and pigments. BoD–Books on Demand, London

Kamali M, Khodaparast Z (2015) Review on recent developments on pulp and paper mill wastewater treatment. Ecotoxicol Environ Saf 114:326–342

Kaown D et al (2014) Identification of multiple sources of groundwater contamination by dual isotopes. Groundwater 52(6):875–885

Karrasch B et al (2006) Effects of pulp and paper mill effluents on the microplankton and microbial self-purification capabilities of the Biobio River, Chile. Sci Total Environ 359(1–3):194–208

Kassim NS, Abu SAISMG, Bohari FL, Abidin NAZ (2022) Assessment of heavy metals in wastewater plant effluent and lake water by using atomic absorption spectrophotometry. Mater Today Proc 66:3961–3954

Khan MN, Mohammad Fi (2014) Eutrophication: challenges and solutions. Eutrophic Causes Conseq Control 2:1–15

Khatri N, Tyagi S (2015) Influences of natural and anthropogenic factors on surface and groundwater quality in rural and urban areas. Front Life Sci 8(1):23–39

Khatri J, Nidheesh PV, Anantha Singh TS, Kumar MS (2018) Advanced oxidation processes based on zero-valent aluminium for treating textile wastewater. Chem Eng J 348:67–73

Kim K, Yoon Y, Cho H, Hwang S-J (2020) Molecular probes to evaluate the synthesis and production potential of an odorous compound (2-methylisoborneol) in cyanobacteria. Int J Environ Res Public Health 17(6):1933

Koistinen J, Paasivirta J, Nevalainen T, Lahtiperä M (1994) Chlorinated fluorenes and alkylfluorenes in bleached Kraft pulp and pulp mill discharges. Chemosphere 28(12):2139–2150

Kooi M et al (2018) Modeling the fate and transport of plastic debris in freshwaters: review and guidance. Freshw Microplast Emerg Environ Contamin 2018:125–152

Koto Y et al (2022) Microcystin-LR Incorporated into colonic cells through probenecid-sensitive transporters leads to upregulated MCP-1 expression induced by JNK activation. Toxicol Rep 9:937–944

Kumar S, Singh D (2021) Municipal solid waste incineration bottom ash: a competent raw material with new possibilities. Innov Infrastruct Solut 6:1–15

Kumar R et al (2022) A review on emerging water contaminants and the application of sustainable removal technologies. Case Stud Chem Environ Eng 6:100219

Kushwaha JP (2015) A review on sugar industry wastewater: sources, treatment technologies, and reuse. Desalin Water Treat 53(2):309–318

Lesser E et al (2018) Survey of 218 organic contaminants in groundwater derived from the world's largest untreated wastewater irrigation system: Mezquital Valley, Mexico. Chemosphere 198:510–521

Liu J, Diamond J (2005) China's environment in a globalizing world. Nature 435(7046):1179–1186

Lone Y, Koiri RK, Bhide M (2015) An overview of the toxic effect of potential human carcinogen microcystin-LR on testis. Toxicol Rep 2:289–296

Malaj E et al (2014) Organic chemicals jeopardize the health of freshwater ecosystems on the continental scale. Proc Natl Acad Sci 111(26):9549–9554

Mannucci A, Munz G, Mori G, Lubello C (2010) Anaerobic treatment of vegetable tannery wastewaters: a review. Desalination 264(1–2):1–8

Matei A, Racoviteanu G (2021) Review of the technologies for nitrates removal from water intended for human consumption. In: IOP conference series: earth and environmental science. IOP Publishing, New York, p 12024
Mateo-Sagasta J, Zadeh SM, Turral H, Burke J (2017) Water pollution from agriculture: a global review. Exec Summary 2017:1
Miao Y, Fan C, Guo J (2012) China's water environmental problems and improvement measures. Environ Resour Econ 3:43–44
Miao Z, Carroll KC, Brusseau ML (2013) Characterization and quantification of groundwater sulfate sources at a mining site in an arid climate: the monument valley site in Arizona, USA. J Hydrol 504:207–215
Mohamed AG, El Safty AM, Siha MS (2013) Current situation of water pollution and its effect on aquatic life in Egypt. Egypt J Occup Med 37(1):95–115
Mohapatra DP, Kirpalani DM (2017) Process effluents and mine tailings: sources, effects and management and role of nanotechnology. Nanotechnol Environ Eng 2:1–12
Moiseenko TI, Gashkina NA (2020) Distribution and bioaccumulation of heavy metals (Hg, Cd and Pb) in fish: influence of the aquatic environment and climate. Environ Res Lett 15(11):115013
Moktadir MA et al (2018) Drivers to sustainable manufacturing practices and circular economy: a perspective of leather industries in Bangladesh. J Clean Prod 174:1366–1380
Monneret C (2017) What is an endocrine disruptor? C R Biol 340(9–10):403–405
Navarro-Ortega A et al (2015) Managing the effects of multiple stressors on aquatic ecosystems under water scarcity. the GLOBAQUA project. Sci Total Environ 503:3–9
Nelms SE et al (2016) Plastic and marine turtles: a review and call for research. ICES J Mar Sci 73(2):165–181
Niu A et al (2019) Impact of water quality on the microbial diversity in the surface water along the three gorge reservoir (TGR), China. Ecotoxicol Environ Saf 181:412–418
Obiakor MO, Okonkwo JC, Nnabude PC, Ezeonyejiaku CD (2012) Eco-genotoxicology: micronucleus assay in fish erythrocytes as in situ aquatic pollution biomarker: a review. J Anim Sci Adv 2(1):123–133
Okereafor U et al (2020) Toxic metal implications on agricultural soils, plants, animals, aquatic life and human health. Int J Environ Res Public Health 17(7):2204
Orts F et al (2018) Electrochemical treatment of real textile wastewater: trichromy procion HEXL®. J Electroanal Chem 808:387–394
Pal A et al (2014) Emerging contaminants of public health significance as water quality indicator compounds in the urban water cycle. Environ Int 71:46–62
Pandey PK et al (2014) Contamination of water resources by pathogenic bacteria. AMB Express 4:1–16
Park M-H, Park C-H, Sim YB, Hwang S-J (2020) Response of scenedesmus quadricauda (chlorophyceae) to salt stress considering nutrient enrichment and intracellular proline accumulation. Int J Environ Res Public Health 17(10):3624
Parmar TK, Rawtani D, Agrawal YK (2016) Bioindicators: the natural indicator of environmental pollution. Front Life Sci 9(2):110–118
Paul D (2017) Research on heavy metal pollution of river Ganga: a review. Ann Agrar Sci 15(2):278–286
Pepper IL, Brooks JP, Gerba CP (2018) Antibiotic resistant bacteria in municipal wastes: Is there reason for concern? Environ Sci Technol 52(7):3949–3959
Pokhrel D, Viraraghavan T (2004) Treatment of pulp and paper mill wastewater—a review. Sci Total Environ 333(1–3):37–58
Prasad SN et al (2002) Conservation of wetlands of India—a review. Trop Ecol 43(1):173–186
Qi J et al (2022) Nutrient thresholds required to control eutrophication: Does it work for natural alkaline lakes? Water 14(17):2674
Rad SM, Ray AK, Barghi S (2022) Water pollution and agriculture pesticide. Clean Technol 4(4):1088–1102

Rahman A, Jahanara I, Jolly YN (2021) Assessment of physicochemical properties of water and their seasonal variation in an urban river in Bangladesh. Water Sci Eng 14(2):139–148
Rathore SS, Chandravanshi P, Chandravanshi A, Jaiswal K (2016) Eutrophication: impacts of excess nutrient inputs on aquatic ecosystem. IOSR J Agric Vet Sci 9(10):89–96
Rehman K et al (2018) Effect of reactive black 5 azo dye on soil processes related to C and N cycling. PeerJ 6:e4802
Reis D, Marques C et al (2015) Evaluation of hydrogen and methane production from sugarcane vinasse in an anaerobic fluidized bed reactor. Int J Hydrog Energy 40(27):8498–8509
Rojas G et al (2005) Adsorption of chromium onto cross-linked chitosan. Separ Purific Technol 44(1):31–36
Salvarredy-Aranguren MM, Probst A, Roulet M, Isaure M-P (2008) Contamination of surface waters by mining wastes in the milluni valley (cordillera real, bolivia): mineralogical and hydrological influences. Appl Geochem 23(5):1299–1324
Samuel S, Muthukkaruppan SM (2011) Physico-chemical analysis of sugar mill effluent, contaminated soil and its effect on seed germination of paddy (*Oryza sativa* L.). Int J Pharmaceut Biol Arch 2(5):1469–1472
Schweitzer L, Noblet J (2018) Water contamination and pollution. In: Green chemistry. Elsevier, London, pp 261–290
Senthivelan T, Kanagaraj J, Panda RC (2016) Recent trends in fungal laccase for various industrial applications: an eco-friendly approach—a review. Biotechnol Bioprocess Eng 21:19–38
Sevastyanov VS et al (2020) Accumulation of organic matter, heavy metals, and rare-earth elements in marine sediment at different distance from the Indigirka river Delta. Geochem Int 58:1313–1320
Shamey R, Zhao X (2014) Modelling, simulation and control of the dyeing process. Elsevier, London
Sharma A et al (2019) Worldwide pesticide usage and its impacts on ecosystem. SN Appl Sci 1:1–16
Singh MR, Gupta A (2016) Water pollution-sources, effects and control. Department of Botany, Nagaland University, Centre for Biodiversity, London, pp 1–16
Smedley PL, Kinniburgh DG (2002) A review of the source, behaviour and distribution of arsenic in natural waters. Appl Geochem 17(5):517–568
Sracek O et al (2012) Mining-related contamination of surface water and sediments of the Kafue river drainage system in the Copperbelt District, Zambia: an example of a high neutralization capacity system. J Geochem Explor 112:174–188
Sonarghare PC, Masram SC, Sonparote UR, Khaparde KP, Kharkate SK (2020) Causes and effects of eutrophication on aquatic life (a review). Int J Env Rehab Conserv XI (SP2):213–218. https://eoi.citefactor.org/10.11208/essence.20.11.SP2.147
Thanigaivel S et al (2022) Ecological disturbances and abundance of anthropogenic pollutants in the aquatic ecosystem: critical review of impact assessment on the aquatic animals. Chemosphere 2022:137475
Tiwari N, Santhiya D, Sharma JG (2020) Microbial remediation of micro–nano plastics: current knowledge and future trends. Environ Pollut 265:115044
Turner RK et al (2000) Ecological–economic analysis of wetlands: scientific integration for management and policy. Ecol Econ 35(1):7–23
UNDESA (2017) World population prospects: the 2017 revision, key findings and advance tables. Working paper no. ESA/P/WP/248. United Nations, Department of Economic and Social Affairs, Population Division
UNEP (2010) Clearing the water: a focus on water quality solutions. Nirobi, Kenya
UNESCO (United Nations Educational Scientific and Cultural Organization) (2021) The united nations world water development report 2021: valuing water. Water Politics 206
UNIDO (2000) The scope for decreasing pollution load in leather processing (US/RAS/92/120/11-51). United Nations Industrial Development Organization—Regional Programme for Pollution Control in the Tanning Industry in South-East Asia
Usman MO, Aturagaba G, Ntale M, Nyakairu GW (2022) A review of adsorption techniques for removal of phosphates from wastewater. Water Sci Technol 2022:wast2022382

Vepsäläinen M et al (2011) Removal of toxic pollutants from pulp mill effluents by electrocoagulation. Separ Purific Technol 81(2):141–150
Windsor FM et al (2019) A catchment-scale perspective of plastic pollution. Global Change Biol 25(4):1207–1221
Yang C (2022) Review on the causes of eutrophication in water. In: 2022 6th international seminar on education, management and social sciences (ISEMSS 2022). Atlantis Press, London, pp 246–252
Yaseen DA, Scholz M (2019) Textile dye wastewater characteristics and constituents of synthetic effluents: a critical review. Int J Environ Sci Technol 16:1193–1226
Youssef L, Younes G, Al-Oweini R (2019) Photocatalytic degradation of atrazine by heteropolyoxotungstates. J Taibah Univ Sci 13(1):274–279
Zhao Y et al (2023) Research advances on impacts micro/nanoplastics and their carried pollutants on algae in aquatic ecosystems: a review. Aquat Toxicol 2023:106725
Zheng X, Como S, Huang L, Magni P (2020) Temporal changes of a food web structure driven by different primary producers in a subtropical eutrophic lagoon. Mar Environ Res 161:105128
Zhou Q et al (2018) The distribution and morphology of microplastics in coastal soils adjacent to the Bohai sea and the yellow sea. Geoderma 322:201–208

Water Pollutants: Conventional Detection and Characterization Approaches of Water Pollutants

Jéssica Mesquita do Nascimento and Camila Budim Lopes

Abstract This publication presents the main conventional techniques for detecting organic and inorganic pollutants present in wastewater. The authors introduce the fundamentals of some analytical techniques used in monitoring water quality, monitoring and highlight the potential risks to living beings arising from the presence of these pollutants in living beings, soil and plants. In addition to presenting new data on the detection of emerging pollutants in wastewater.

Keywords Organic pollutants · Inorganic pollutants · Analytical techniques

Abbreviations

BOD	Biochemical oxygen demand
TOC	Total organic carbon
pH	Hydrogen ion potential
COD	Chemical oxygen demand
TDS	Total dissolved solids
DO	Dissolved oxygen
SEE	Single entity electrochemistry
SECE	Single entity collision electrochemistry
LC	Liquid chromatography
GC	Gas chromatography
MS	Mass spectrometry
DBPs	Disinfection byproducts

J. M. do Nascimento (✉)
Department of Environmental Engineering, Federal University of Rondônia, Ji-Paraná, Rondônia, Brazil
e-mail: jessica.nascimento@unir.br

C. B. Lopes
Federal Institute of Education, Science and Technology of Rondônia, Jaru, Rondônia, Brazil
e-mail: camila.lopes@ifro.edu.br

R. Selvasembian et al. (eds.), *Emerging Trends in Microbial Electrochemical Technologies for Sustainable Mitigation of Water Resources Contamination*,
https://doi.org/10.1007/978-3-031-74636-9_2

THMs	Trihalomethanes
ECD	Electron capture detector
NPs	Nanoparticles
HAAs	Haloacetic acids
POPs	Persistent organic pollutants
SPE	Solid phase extraction
ESI	Electrospray ionization
PFCs	Perfluorinated compounds
BFRs	Brominated flame retardants
PCBs	Polychlorinated biphenyls
AAS	Atomic absorption spectrophotometry
FAAS	Flame atomic absorption spectrophotometry
ICP-OES	Inductively coupled plasma optical emission spectrophotometry
ETAAS	Electrothermal atomization atomic absorption spectrophotometry
CVG-AAS	Chemical vapor generation atomic absorption spectrophotometry
EPs	Emerging pollutants
EDCs	Endocrine disrupting chemicals
CECs	Contaminants of emerging concern
PCPs	Personal care products
PPs	Pharmaceutical pollutants
ASs	Artificial sweeteners
MPs	Microplastics
WTPs	Wastewater treatment plants
SAC	Saccharin
CYC	Cyclamate
ASP	Aspartame
ACE	Acesulfame
SUC	Sucralose
NEO	Neotame
NHDC	Neohesperidin dihydrochalcone
ALI	Alitame
ADV	Advantame
STPs	Sewage treatment plants
PE	Polyethylene
PS	Polystyrene
PP	Polypropylene
PES	Polyester
PA	Polyamide
PET	Polyethylene terephthalate
SAN	Styrene-acrylonitrile
PU	Polyurethane
PVC	Polyvinyl chloride
ALK	Alkyd resin
PPO	Polyphenylene oxide
EVA	Ethylene vinyl acetate

FIA	Flow injection analysis
SIA	Sequential injection analysis

1 Introduction

Water pollution can lead to serious environmental disasters. Such as the destruction of fauna and flora. To prevent such destructions, regular monitoring of water quality is very important. Some estimates show that by 2050, over six million people will suffer from shortage fresh water. One of the main reasons supporting this data is the increasing water pollution. It is estimated that around 80% of wastewater is discharged back into the environment without any prior treatment. Rivers, for example, are the main receivers of sewage, in addition to suffering damage from agriculture, energy sectors (construction of hydroelectric plants), tourism, and transport (Wieczorek et al. 2024).

Millions of people worldwide suffer from shortage of fresh water, at least once a year. At the same time, hundreds of thousands suffer from climate related issues resulting from excessive rainfall. In addition to these aggravating factors, part of the water is contaminated by chemical pollutants, such as potentially toxic metals, plastic waste, organic pollutants, pharmaceutical products, and pathogenic microorganisms. All these described problems are driven by rapid industrial, population, and economic growth, which in turn, increases the release of contaminated wastewater into water bodies. Global warming and climate changes in rainfall patterns further aggravate the global water crisis (Young et al. 2021).

Monitoring water quality becomes essential since drinking water is essential for maintaining life (Wieczorek et al. 2024). Contaminants in water ecosystems can arise from natural phenomena, or anthropogenic contributions, are divided into organic pollutants, such as dyes and pharmaceuticals, and inorganic pollutants, such as potentially toxic metals (Lu et al. 2024).

Some organic contaminants are not biodegradable and remain in the environment for a long time. The bioaccumulation of these contaminants in living organisms occurs via the food chain and directly affects the balance of the ecosystem (Kaur and Batra 2023).

Given the presented data and the importance of expanding and exploring research on water pollution, it is imperative to focus on conventional approaches to detecting water pollutants and their respective characterizations.

In the literature, several authors explore different techniques for identifying and quantifying organic and inorganic pollutants present in wastewater. In studies by Gefenienė et al. (2024), for example, HPLC, GC/FID, ICP-OES and GC/MS techniques were used to detect polycyclic aromatic hydrocarbons (PAHs), toxic metals such as Cr, Ni, Cu and Zn, volatile organic compounds (VOCs), and Gasoline range organic, analyzing wastewater from firefighting operations involving tires (Gefenienė et al. 2024).

Studies focused on elemental quantification of toxic metals in wastewater usually use the techniques of ICP–MS, ICP–MS/MS, ICP–TOF–MS, HR–ICP–MS, MH–ICP–MS and MC–ICP–MS (Balaram et al. 2023).

In the assessment of pollution reduction in Victoria Harbor, and adjacent aquaculture area of Hong Kong, through the bioaccumulation profile of organic and inorganic pollutants in green-lipped mussel *Perna viridis*, the bioaccumulation of PAHs (LMW and HMW). and heavy metals (Cd, Cr, Cu, Pb and Zn) in the soft tissues of *P. viridis*, was revealed. However organic pollutants such as PCBs and OCPs were not detected using ICP-OES, ICP–MS and Gas chromatography electron ionization-quadrupole mass spectrometer (Leung et al. 2024).

Other widely studied techniques that are also are electrochemical. In studies by conducted (Chen et al. 2024) the construction of a synergistic fuel cell device with double photoelectrode (DPFC), combined with a p-type organic semiconductor, a polythiophene (pTTh) photocathode, and a printed molecular photoanode (MI)-CdS/ZnO NRs, allowed the application of the photoanode in the quantitative detection of the pollutant atrazine (ATZ).

While (Shanlee et al. 2023) used electrochemical sensors based on neodymium molybdate/reduced graphene oxide ($Nd_2Mo_3O_{12}$/RGO) for the detection of the carcinogenic pollutant metol (ML).

The synthesis of layered double hydroxide nanocomposites/conducting polymers for electrochemical detection of wastewater pollutants revealed that the developed electrode presents low cost and good sensitivity, in accordance with the comparisons studied for Pb removal, using the new electrode, with conventional ICP and AAS techniques (Kamel et al. 2024).

2 Conventional Water Pollutant Detection Approaches

Different methodologies are used for pollutant detection (Fig. 1), such as mass spectrometry, electrochemical method, chromatography, inductively coupled plasma emission spectrophotometry, and others (Kaur and Batra 2023).

2.1 Physicochemical Analysis

Information about water quality is essential for assessing its environmental quality, whether in water bodies, groundwater, or aquifers. All of these need continuously monitoring to ensure the maintenance of the aquatic ecosystem. There are simpler and faster methods (water quality monitoring kits) that are based on physical–chemical analyses, such as Biochemical Oxygen Demand (BOD), alkalinity, turbidity, colorimetric method, and others (Fig. 2) (Ramírez et al. 2023).

The physicochemical characterizations of water aim to identify and quantify the elements and chemical species present in this sampling environment. Consequently,

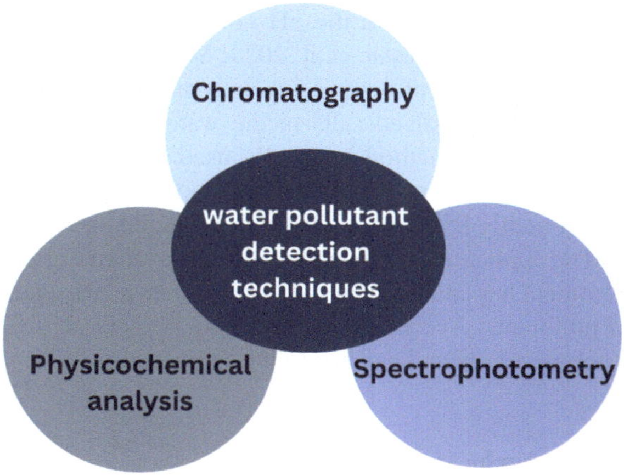

Fig. 1 Main techniques for detecting water pollutants

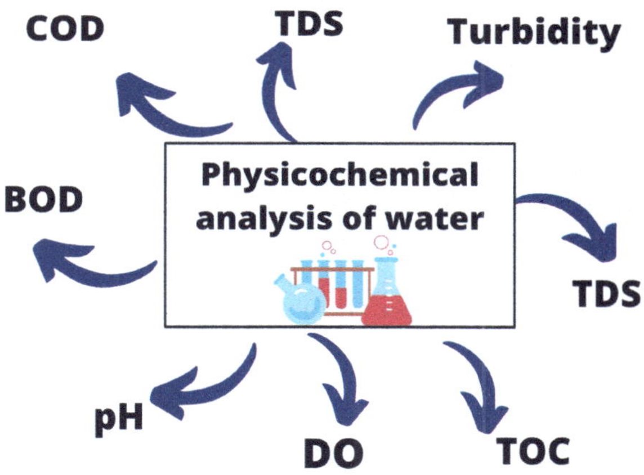

Fig. 2 Main physical–chemical analyses of water

the results will allow the assessment of the level of impact arising from anthropogenic activities (Ramírez et al. 2023).

2.1.1 pH

The pH (Hydrogen ion Potential) of a solution is defined as the negative logarithm base 10 (log10) of the concentration of hydrogen ions (H+). Therefore, pH is a measure of the intensity of acidity ($pH < 7.0$), alkalinity ($pH > 7.0$), or neutrality ($pH = 7.0$) of a water sample (Honikel and Puolanne 2024). pH is an important parameter of water quality. pH ranges above 6.0, for example, directly influence the precipitation of potentially toxic metals through the formation of metal complexes (Parron et al. 2011). In addition, it has a direct influence on the physiology of several aquatic species, with a the pH range of 10, which tends to be lethal for the vast majority of fish (Dessie et al. 2024; Mramba and Kahindi 2023; Rho et al. 2023).

The pH range of natural waters typically falls within the basic range, due to the presence of dissolved salts in the water, such as bicarbonate and carbonate. The potentiometric method is usually used to determine pH. There are benchtops and portable pH meters on the market. Laptops are excellent for taking to the field and consequently for carrying out measurements in situ (Ramírez et al. 2023).

The application of smart functional covalent organic frameworks (COF) for water treatment has enabled reversible colorimetric pH sensing and photocatalytic removal of pollutants. According to studies by Zhou et al. (2023), COFs have a rapid visual response to pH and can be synthesized easily.

2.1.2 Turbidity

Turbidity is a physical property associated with the reduction in water transparency, due to the presence of suspended materials that interfere with the passage of light. In other words, it causes changes in the optical property of water, which is related to the scattering and absorption of light (Parron et al. 2011; Qian et al. 2023).

Water turbidity can be caused by the presence of effluents containing residues of dyes, and organic and inorganic matter. The crystallinity of a natural water body is one of the parameters that indicate its quality and productivity. Highly turbid waters, for example, drastically reduce the aquatic ecosystem (Mramba and Kahindi 2023). This parameter can be evaluated using a device called a turbidimeter, the result is expressed in nephelometric units (NTU) (Qian et al. 2023).

2.1.3 Total Organic Carbon (TOC)

Organic carbon present in industrial and domestic effluents can be extremely toxic and persistent, with the potential for bioaccumulation in living beings, which leads to several negative impacts on the health of animals and humans (Rho et al. 2023; Shetty and Goyal 2022).

The organic matter present in water contains carbon, and this content can be quantified directly by analyzing the Total Organic Carbon Content (TOC), or indirectly by Biochemical Oxygen Demand (BOD), Chemical Oxygen Demand (COD), turbidity, and color. TOC is a well-reported parameter in the literature, which is reliable for estimating the amount of organic carbon present in a water sample (Parron et al. 2011; Rho et al. 2023; Manninen et al. 2023). The universal method used to measure TOC is based on oxidizing the organic content of the water sample to convert it into quantifiable forms (Shetty and Goyal 2022).

2.1.4 Biochemical Oxygen Demand (BOD), and Chemical Oxygen Demand (COD)

Biochemical Oxygen Demand (BOD) and Chemical Oxygen Demand (COD) are used to determine the presence of organic matter in water. Both parameters indicate the consumption, or the demand for oxygen required to stabilize the organic matter present in the water. The difference between the two methodologies is that BOD is based on biochemical oxidation, while COD corresponds to chemical oxidation (Parron et al. 2011).

The (BOD) indicates the degree of water pollution by biodegradable organic matter content, measured by the amount of dissolved oxygen consumed in microbial metabolism. In general, BOD is an analysis used to evaluate the pollution and quality of effluents, or wastewater (Islam et al. 2023). These methodologies have disadvantages such as the presence of interferers that impair the result (COD), and long analysis time (BOD) (Singh et al. 2021; Nguyen et al. 2014).

2.1.5 Total Dissolved Solids (TDS)

Total Dissolved Solids (TDS) are the sum of all dissolved chemical components in water. They are based the measurement of the concentration of ionic substances present in the sample. This analysis is very important for evaluating the quality of drinking water for human consumption, and for determining chemical pollutants, both organic and inorganic nature, present in aquatic ecosystems (Herawati et al. 2023). The organic and inorganic ions determined in this analysis may include carbonates, bicarbonates, chlorides, sulfates, phosphates, nitrates, organic matter, and metals such as calcium, magnesium, and sodium (Parron et al. 2011; Herawati et al. 2023).

One way to measure this parameter is by converting the sample's electrical conductivity into an approximate concentration of total dissolved solids, or by using instruments equipped with a conductivity meter that performs the conversion (Parron et al. 2011).

Water quality can have a significant I impact on the productivity of the livestock sector (Patra et al. 2023). With water scarcity in many regions, some producers are already using brackish water for animals drinking. However, some studies indicate that the consumption of this water, which has a high TDS content, can affect the use of nutrients, blood profiles, and ruminal fermentation in sheep and goats differently (Patra et al. 2024).

2.1.6 Dissolved Oxygen (DO)

The analysis of Dissolved Oxygen (DO) is essential for assessing the level of oxygen available to aerobic microorganisms that live in natural waters. It is also crucial for fish, which are directly affected by change in DO in the water. Therefore, it has become one of the essential environmental indicators for assessing water quality (Abouelsaad et al. 2022).

DO levels can be increased by aquatic plants performing photosynthesis, but a decrease in DO can occur when water temperature rises, or when water eutrophication occurs (Bailey and Ahmadi 2014; Sánchez et al. 2007).

To determine DO in water, volumetric or potentiometric methods are commonly used, but there is equipment available in the market that accurately measure the concentration of oxygen dissolved in water. The choice of the best method depends on factors such as the presence of interferers and desired accuracy (Parron et al. 2011).

2.2 *Electrochemical Techniques*

Electrochemical techniques are widely used in the areas of common environmental pollutants measurements. These methodologies have advantages such as low cost and easy miniaturization (Kaur and Batra 2023). The use of electrochemical methods for detecting pollutants has the advantage of being less prone to interference due to the turbidity or color of the sample, which tends to harm methods based on colorimetry, which is preferable for environmental samples. Electrochemical sensors are economical, fast, sensitive, and simple to use. Despite the ease of scalability of bare electrodes, they are generally not sensitive enough to achieve feasibility (Coronado-Apodaca et al. 2023).

The main disadvantages are low sensitivity and selectivity (Kaur and Batra 2023). A promising technique is Single Entity Electrochemistry (SEE), which can be considered as a process of studying a single thing (one particle or one reaction) at a time using an electrochemical approach (Kaur and Batra 2023).

The most recent research is directed towards the use of Single Entity Collision Electrochemical (SECE) technique also known as nano-impacts. This technique can reveal information that other techniques cannot (Kaur and Batra 2023).

The technique is based on the fact that, microscopically, the nanoparticle initially reaches the electron tunneling distance. Subsequently, a Faradaic reaction occurs, leading to charge transfer, and generating a measurable current signal. By recognizing current transients (generally low, on the Pico ampere scale), information such as size distribution, physicochemical and catalytic properties, redox reactions, and dynamic characteristics involving unique entities at the electrical contact interface can be interpreted. This promising technology has expanded the field of detection from the nanoscale to individual molecules (Kaur and Batra 2023).

Photofuel cells (PFCs) are considered a new generation of environmentally friendly photovoltaic devices. In studies by Chen et al. (2024) the construction of a synergistic fuel cell device with a double photoelectrode (DPFC), based on the synergistic reaction of the water/oxygen cycle, combined with an organic p-type semiconductor, a polythiophene (pTTh), and a molecularly imprinted photoanode (MI)-CdS/ZnO NRs, besidesallowing high stability, and excellent photoactivity in an efficient 2-electron catalytic pathway, enabled the application of the photoanode in the quantitative detection of the environmental pollutant atrazine (ATZ).

The use of high-performance electrochemical sensors based on neodymium molybdate/reduced graphene oxide ($Nd_2Mo_3O_{12}$/RGO) has shown rapid detection of organic pollutants in water, for example, the carcinogenic agent in photographic developer, metol (ML) (Shanlee et al. 2023).

The synthesis of an electrochemical sensor based on strontium-doped lanthanum trisulfide showed improved electrochemical detection performance for the detection of the herbicide mesotrione. The method developed by the authors (Rajaji et al. 2022) showed good sensitivity, selectivity, stability and applicability.

The production of an electrode for electrochemical detection of pollutants in wastewater, based on double hydroxide nanocomposites in layers/conducting polymers, revealed that the electrode produced has a low cost when compared to conventional ICP and AAS techniques, as well as good sensitivity for the metal under study, which in this case was lead (Kamel et al. 2024).

2.3 *Chromatographic Techniques Coupled to Mass Spectrometry*

The evaluation of pollutants, such as those emerging in wastewater samples, is generally carried out using liquid chromatography (LC) or gas chromatography (GC), coupled with mass spectrometry (MS) technique. Tandem mass spectrometry is one of the most used techniques due to its sensitivity and selectivity in complex matrices such as environmental samples (Coronado-Apodaca et al. 2023).

Water analysis can be highly complex, especially when the matrix contains multiple pollutants that share mass/charge ratios, elute similarly, and/or affect the matrix in similar ways. Therefore, the necessity for the development and utilization of new technologies in the contaminants detection is evident (Gefenienė et al. 2024; Leung et al. 2024; Nguyen et al. 2014; Coronado-Apodaca et al. 2023; Elseblani et al. 2023; Omotola et al. 2024).

2.3.1 Gas Chromatography Coupled to Mass Spectrometry (GC/MS)

The safety of drinking water supply is essential for the population because if the quality of the water is compromised, it can serve as a vehicle for pathogenic agents. Therefore, to prevent of waterborne transmission diseases, most drinking water is chemically disinfected before consumption. However, chemical disinfection has the disadvantage of forming Potentially Harmful Disinfection Byproducts (DBPs). These by-products have been detected in treated water samples and identified as potential cytotoxic, genotoxic, and carcinogenic agents (Allard and Lauritsen 2023).

Volatile disinfection by-products, such as Trihalomethanes (THMs), were traditionally quantified using Gas Chromatography (GC) followed by Electron Capture Detection (ECD) or mass spectrometry (MS). However, direct injection of water samples is not recommended in GC systems, as it degrades the column. Consequently, samples would need to undergo prior treatment for solvent extraction. Another necessary step before analysis would be pre-concentration, as these compounds are generally found at levels ranging from ng L^{-1} to a few hundred $\mu g\ L^{-1}$ (Allard and Lauritsen 2023).

Non-volatile DBPs or polar compounds have been quantified by derivatization followed by GC/ECD, coupled with MS, or by liquid chromatography coupled with MS. For example, Haloacetic acids (HAAs), cannot be directly analyzed with GC/ECD due to their polarity (Allard and Lauritsen 2023).

Gas chromatography-mass spectrometry (GC/MS) is an analytical technique that has been available for a few decades. GC/MS can be used to analyze the metabolites of steroid hormones and their precursors. This technique can be easily employed in both comprehensive (full scan) and targeted (ion-selected) modes (Krone et al. 2010).

Among the numerous groups of pollutants released into the environment, halogenated organic contaminants are of great concern, because most of them cause serious damage to living organisms. They act as endocrine disruptors, causing neurotoxicity and immunotoxicity, as well as adverse effects on the. reproductive system, and organs such as the kidneys, liver, and thyroid (Ayala-Cabrera et al. 2021).

Some of these halogenated organic contaminants are also classified as Persistent Organic Pollutants (POPs) because they can accumulate in adipose tissues and are highly resistant to degradation, besides beingprone to long-distance transport in the environment. GC-based methods are generally used to determine the total concentration of the 16 potential Hexabromocyclododecanes (HBCDs) with six enantiomers and diastereomeric pairs (Ayala-Cabrera et al. 2021).

GC is one of the most widelyused analytical techniques for the analysis of chiral pesticides and volatile organic pollutants in environmental matrices. This is due to its simplicity, high efficiency, short run times, good sensitivity, absence of liquid mobile phases, and low consumption of organic solvent (green analytical technique), in addition to its easy hyphenation with MS (Abouelsaad et al. 2022; Arenas et al. 2021).

The evaluation of wastewater from distillery industries showed that the physical–chemical parameters values of the wastewater were 13,268 mg L-1 (BOD), 25,144 mg L-1 (COD), 25,144 mg L-1 (TS) and 6634 mg L-1 (phosphate), while the pH was alkaline. In addition to these parameters, the authors also evaluated the detection of some pollutants, such as quercetin 7,3′,4′-trimethoxy, octadecadienoic acid, propanoic acid, glycocholic acid methyl ester, canthaxanthin by GC–MS (Chowdhary et al. 2022).

The application of non-targeted analysis using dispersive liquid–liquid microextraction, combined with gas chromatography with mass spectrometry (DLLME-GC–MS) allowed the monitoring of 118 environmental pollutants in the Mar Menor lagoon located in southeastern Spain (Peñalver et al. 2022).

The chemical analysis of firefighting wastewater after a real tire fire showed the presence of several organic and inorganic pollutants. Using HPLC, GC/FID, ICP-OES and GC/MS techniques, fifteen polycyclic aromatic hydrocarbons (PAHs) were identified, along with a high presence of the toxic metals Cr, Ni, Cu and Zn, volatile organic compounds (VOCs) and Gasoline range organics (Gefenienė et al. 2024).

When assessing the reduction of pollution in Victoria Harbor, and the adjacent aquaculture area of Hong Kong, through investigating the bioaccumulation of organic and inorganic pollutants in green-lipped mussel *Perna viridis*, by ICP-OES, ICP–MS and Gas chromatography electron ionization-quadrupole mass spectrometer, the results revealed the bioaccumulation of PAHs (LMW and HMW) and heavy metals (Cd, Cr, Cu, Pb and Zn) in the soft tissues of *P. viridis*, while organic pollutants such as PCBs and OCPs were not detectable (Leung et al. 2024).

2.3.2 Liquid Chromatography Coupled to Mass Spectrometry (LC/MS)

Liquid chromatography (LC) methods are ideally suited for the analysis of polar and thermally labile DBPs. The most common method used for DBPs consists of solid phase extraction (SPE)-LC–electrospray ionization (ESI)–MS/MS (Allard and Lauritsen 2023).

The use of Liquid Chromatography has allowed the expansion of environmental analyses in recent decades by being capable of separating a range of halogenated contaminants, including polar, ionic, non-volatile, and thermally labile compounds, which were more difficult to analyze by GC. Combined mass spectrometry, a powerful technique, especially when coupled with chromatography, it has made it possible to overcome the lack of confirmation of classical selective detection systems (such as electron capture detector, fluorescent detection, or even electrochemical detection) in determining halogenated organic contaminants (Ayala-Cabrera et al. 2021).

Some time ago, it was commonly assumed that Liquid Chromatography—Mass Spectrometry (LC–MS) was the preferred technique for determining UV filters in cosmetic products, while Gas Chromatography-Mass Spectrometry (GC–MS) was technique favored for analyzing environmental and biological samples. However, the literature shows that the work focuses on using (GC–MS), (LC–MS) and (LC/MS/MS) techniques forquantify residues of UV filter, an emerging pollutant (Oubahmane et al. 2023).

The choice between GC and LC in UV filter separation is mainly based on the physicochemical properties of the target analytes, and not on their concentration level, or the complexity of the sample. Therefore, LC–MS/MS can be the technique chosen to analyze polar and semipolar UV light-absorbing compounds, in addition to their transformation products with greater molar mass. Likewise, GC is the technique selected for the identification and quantification of volatile or thermostable volatile compounds (Oubahmane et al. 2023).

In general, the analytical methods reported for the determination of chiral pollutants in environmental samples are based on liquid chromatography (LC) and gas chromatography (GC), the vast majority of which are coupled to mass spectrometry (MS), due to high selectivity, and sensitivity provided (Arenas et al. 2021).

LC and GC techniques are the most used for enantiomeric determination. Chiral LC has been applied to the enantiomeric determination of pharmaceuticals and chiral pesticides, while chiral GC has been applied to volatile and thermally stable enantiomers of pesticides, polycyclic musks, Perfluorinated Compounds (PFCs), Brominated Flame Retardants (BFRs), and Polychlorinated Biphenyls (PCBs) (Arenas et al. 2021).

However, LC is the analytical technique commonly used for the enantiomeric analysis of chiral compounds in environmental samples. This is due to the high number of commercially available chiral columns and the high versatility provided by the elution modes (normal, reverse, polar organic, or polar ionic elution modes). However, a reversed-phase mode is preferred as it is better compatible with MS detectors (Arenas et al. 2021).

LC and GC techniques are the most commonly used for enantioselective analysis of pesticides. However, LC is the most used, and is usually the main choice for the enantioselective analysis of pesticides, due to the large number of commercially available columns (Arenas et al. 2021).

The identification of metallic and organic contaminants transported by microplastics in the Lebanese coastal environment was assessed by ICP–MS, GC–MS and LC–MS. The results showed that, in this environment, there are over a hundred organic compounds in microplastics, detected by LC–MS and GC–MS techniques, while ICP–MS detected several toxic metals, such as Cd, Pb and Hg (Elseblani et al. 2023).

The identification of traces of perfluorooctanoic acid and perfluorooctane sulfonate in surface sediments using a liquid chromatograph coupled to a single quadrupole electrospray ionization mass spectrometer (LC-ESI–Q-MS) indicated that the analyzed pollutants were accurately detected, selective and sensitive (Omotola et al. 2024).

2.3.3 Liquid Chromatography Coupled to Tandem Mass Spectrometry (LC/MS/MS)

Liquid Chromatography tandem Mass Spectrometry (LC/MS/MS), for example, is replacing classical methods for the analysis of steroid hormones. This technique has the advantages of requiring a small sample volume, offering greater specificity, and short analysis times (Krone et al. 2010).

The use of this technique has become promising, due to criticism regarding the validity of the results of steroids, such as testosterone, using older techniques. Although this approach is the gold standard for measuring individual steroids and panels of such compounds, LC/MS/MS has limited use in defining novel metabolomes. In contrast, an advantage of using the GC/MS technique is its nonselective nature. A scanned run, for example, will contain all excreted steroids, providing an integrated picture of an individual's metabolome (Krone et al. 2010).

Chiral pharmaceuticals are emerging pollutants detected in contaminated soil and aquatic ecosystems. It is worth noting that more than 50% of pharmaceutical products currently in use are chiral compounds. The determination of chiral pharmaceutical products is generally carried out by LC–MS/MS and, to a lesser extent, by GC–MS (Arenas et al. 2021).

2.4 Spectrophotometry

Spectrophotometry is a method based on measuring how much a chemical substance absorbs light, by measuring the intensity when a beam of light passes through the sample. Spectrophotometry is an optical analysis methodology used in biological and physicochemical investigations (Winiarski et al. 1995; Kassim et al. 2022; Yousefi and Shemirani 2013).

Determining such small quantities of elements using classical analyses such as gravimetry and volumetry is almost impossible. Therefore, for elementary analyses, the use of spectrophotometry is widely adopted due to its precision and reliability of results (Temel and Çöpür 2023; Rodríguez-Solana et al. 2014; Santos et al. 2006).

2.4.1 Ultraviolet to Visible Spectrophotometry (UV/Vis)

Spectrophotometric methods are considered one of the oldest methods in analytical chemistry. The absorption of visible light by certain chemical substances has long been used to visually determine their concentration. Spectrophotometry and spectrophotometric methods: refer to measurements of radiation intensities using photoelectric transducers or other types of electronic devices (Marczenko and Balcerzak 2000). Ultraviolet to Visible (UV–vis) spectrophotometry played a prominent role in the analysis of environmental pollutants. This technique has gained popularity over the years (Kaur et al. 2021).

Another class of Disinfection Byproducts (DBPs) corresponds to those resulting from the secondary oxidant formed from the reaction of the disinfectant with the aqueous matrix. These DBPs include trichloramine, bromamines, bromochloramine, and organic chloramines. Traditionally, these compounds are analyzed by UV spectroscopy due to ease of operation, low cost, and because it is an instrument that is widely available in laboratories. Concentrations can be determined based on the absorbance of the analyte at a specific wavelength λmax (Allard and Lauritsen 2023; Luh and Mariñas 2014).

However, this method does not present high sensitivity for use in environmentally relevant concentrations, in addition to presenting questionable precision when there are several species present in a mixture, resulting in peak overlap and making result analysis challenging (Allard and Lauritsen 2023).

Hospital activities generate a large volume of toxic solids and liquids containing various inorganic, organic, and microbial residues that, if released without prior treatment into the ecosystem, cause serious environmental damage (Ajala et al. 2022).

The management of hospital wastewater, in particular, has been a major concern due to the presence of unregulated emerging micro pollutants in concentrations in the range of ng L^{-1} to μg L^{-1}. Various emerging pollutants present in hospital waste, such as ciprofloxacin, ofloxacin, norfloxacin, and moxifloxacin can be detected using a UV–visible spectrometer. However, this technique has the disadvantage that only one wavelength can be studied at a time, and spectra must be constructed from data points collected at multiple intervals (Ajala et al. 2022).

However, among the most sensitive detection techniques, such as ICP-OES, spectrophotometry in the UV–visible region is the most used methodology in developing countries, due to its low cost, ease of operation, accuracy and precision, good selectivity, and sensitivity, especially when an appropriate chelating agent is used (Gürkan et al. 2015).

2.4.2 Atomic Absorption Spectrophotometry (AAS)

Atomic Absorption Spectrophotometry (AAS) is a widely technique used for determination of elements at low concentrations, operating across various types of matrices, such as liquid, solid, suspended, and even gaseous samples. This technique can be associated with flow analysis systems, allowing for the study of chemical speciation (Amorim et al. 2008).

The (ASS) can be performed by three different atomization processes, which are: Flame Atomic Absorption Spectrophotometry (FAAS), Electrothermal Atomization Atomic Absorption Spectrophotometry (ETAAS), and Chemical Vapor Generation Atomic Absorption Spectrophotometry (CVG-AAS). The choice of the best technique according to the atomization process depends on the chemical nature of the sample, sample content, chemical composition of the sample, and physical state. (FAAS) is the simplest and most widely used approach, although it has the lowest sensitivity when compared to the others (Ferreira et al. 2018).

One disadvantage of this technique is the that it evaluates only one metal at a time, since for its operation it is necessary to insert the hollow cathode lamp of the element to be analyzed. However, in the early 2000s, a spectrophotometer equipped with software that allowed quick selection of hollow cathode lamps was commercially launched (Ferreira et al. 2018).

The determination of thallium in water samples via solid sampling by HR-CS GF AAS using chromatographic filter paper as a solid support is efficient for detecting thallium in water samples. Through this methodology, the authors eliminated the analyte elution step, necessary for its pre-concentration, thus demonstrating that it is possible to introduce more simplicity by using this technique (Oliveira et al. 2024).

This equipment was capable of operating in sequential mode, allowing the determination of up to 16 elements. In the literature, many studies reported its performance

(Ferreira et al. 2018). Several studies that evaluated environmental pollution quantified the content of potentially toxic metals in effluents and contaminated waters using this technique (Ferreira et al. 2018; Okewale and Grobler 2023; Ozbek and Baysal 2017; Kori et al. 2023; Abebe et al. 2023; Lemessa et al. 2022; Jabbo et al. 2022).

2.4.3 Inductively Coupled Plasma Optical Emission Spectrophotometry (ICP-OES)

Inductively Coupled Plasma Optical Emission Spectrophotometry (ICP-OES) analysis is based on the process of atomic excitation and relaxation. The sample in solution is introduced into the plasma in aerosol form with argon gas acting as a carrier gas. Thermal energy within the plasma excites the atomic and ionic species. Because it is unstable in this state, the atom tends to return to their ground state, emitting energy in the form of light. As each element has different energy levels, emissions will also be different, making it possible to quantify the species under analysis (Brito 2010).

Comparisons between ICP OES and FAAS show that both techniques have similar sensitivity, but FAAS is a cheaper piece of equipment and with lower gas consumption during analysis. Another relevant fact is that both techniques allow for coupling with Flow Injection Analysis (FIA) and Sequential Injection Analysis (SIA) systems (Ferreira et al. 2018).

However, ICP-OES is a technique that offers the advantage of carrying out multi-element analysis, in addition to having the ability to determine metals and non-metals such as chlorine, bromine, iodine, and sulfur in trace quantities in matrices such as food, geological materials, soils, cosmetics and others (Brito 2010; Chi et al. 2023; Douvris et al. 2023; Chand and Prasad 2013; Jarić et al. 2011).

The development of a sulfonated and aminated poly(ethylene–vinyl alcohol) nanofibrous membrane for use as a solid phase extraction matrix, aimed at pre-concentrating trace metal ions from dairy wastewater for direct ICP-OES revealed that the developed membrane presented maximum removal capacities for Al (350 mg g^{-1}), Cu (180 mg g^{-1}), Ni (140 mg g^{-1}) and Cd of 125 mg g^{-1}(Amaly et al. 2023).

ICP-OES will celebrate 50 years on the market in 2024. Since its first appearance on the market (1974), this analytical technique has gained its space and has become one of the most widely used in the world (Luh and Mariñas 2014; Douvris et al. 2023) (Table 1).

Table 1 Applications of water pollutant detection techniques

Technique	Analyte	References
UV/Vis	Ciprofloxacin, ofloxacin, norfloxacin, and moxifloxacin	Ajala et al. (2022)
UV/Vis	Trichloramine, bromamines, bromochloramine, and organic chloramines	Allard and Lauritsen (2023)
Electrochemical sensor based on strontium-doped lanthanum trisulfide	Mesotrione (MTO)	Rajaji et al. (2022)
Synergetic and fast 2-electron water/oxygen cycle reaction based dual-photoelectrode fuel cell (DPFC)	Atrazine	Chen et al. (2024)
Methoxy based covalent organic frameworks (COF–OMe)	Dyes and phenolic pollutants	Zhou et al. (2023)
Electrochemical sensor based on neodymium molybdate/reduced graphene oxide (Nd2Mo3O12/RGO)	Carcinogenic photographic developing agent metol (ML)	Shanlee et al. (2023)
Layered double hydroxides/conductive polymer nanocomposites for electrochemical detection	Pb	Kamel et al. (2024)
GC–MS	Quercetin 7,3′,4′-trimethoxy, octadecadienoic acid, propanoic acid, glycocholic acid methyl ester, cantaxanthin	Chowdhary et al. (2022)
DLLME-GC–MS	Phenols, NPs, CPs, PAEs, PCBs, PAHs, aromatic, heterocyclic and aliphatic compounds	Peñalver et al. (2022)
LC–MS and GC–MS	Chlorpyrifos, phenacetin, minoxidil and etc.	Elseblani et al. (2023)
(LC-ESI–Q-MS)	Perfluorooctanoic acid and perfluorooctane sulfonate	Omotola et al. (2024)
HPLC	Polycyclic aromatic hydrocarbons (PAHs)	Gefenienė et al. (2024)
GC/FID	Volatile organic compounds (BTEX and TMB)	Gefenienė et al. (2024)
GC/FID	Gasoline range organics (GRO)	Gefenienė et al. (2024)
GC/MS	Organic pollutants	Gefenienė et al. (2024)
Gas chromatography electron ionisation-quadrupole mass spectrometer	PAHs, PCBs and OCPs	Leung et al. (2024)

(continued)

Table 1 (continued)

Technique	Analyte	References
ICP–MS	Cd, Pb, Bi and Hg	Elseblani et al. (2023)
ICP-OES	Cu and Zn	Leung et al. (2024)
ICP–MS	Cd, Cr and Pb	Leung et al. (2024)
ICP-OES	Al, Cu, Ni and Cd	Amaly et al. (2023)
ICP-OES	Co, Sn, Al, Ni, Cr, Cd, Pb, Cu, Mn, and Zn	Gefenienė et al. (2024)
AAS	Si, Fe and Al	Okewale and Grobler (2023)
HR-CS GF AAS	Tl	Oliveira et al. (2024)
HR-CS-FAAS	Cu	Kori et al. (2023)
ICP-OES	Co, Sn, Al, Ni, Cr, Cd, Pb, Cu, Mn, and Zn	Gefenienė et al. (2024)

3 Characterization of Water Pollutants

3.1 Potentially Toxic Metals

In the literature, many authors discuss the classification of which elements should be classified as toxic, beneficial, or essential for living organisms (Zoroddu et al. 2019; Maret 2016; Williams 1991).

In general, the classifications divide the elements into organic and bulkcategories, represented by (H—hydrogen), (C—carbon), (N—nitrogen), and (O—oxygen). In addition to these, seven macro-minerals are essential (Na—sodium), (K—potassium), (Mg—magnesium), (Ca—calcium), (Cl—chlorine), (P—phosphorus), and (S—sulfur). In trace elements that are defined as essential, being in this group (Mn—manganese), (Fe—iron), (Cu—copper), (Zn—zinc), (Se—selenium), (Co—cobalt), (Mo—molybdenum) and (I—iodine) (Zoroddu et al. 2019).

The metallic elements considered essential to humans include (Na—sodium), (K—potassium), (Mg—magnesium), (Ca—calcium), (Mn—manganese), (Fe—iron), (Co—cobalt), (Cu—copper), (Zn—zinc) and (Mo—molybdenum) while elements such as (V—vanadium), (Ni—nickel) and (Sn—tin) are presumed to be essential, their status as such has not yet been determined (Zoroddu et al. 2019; Bharat et al. 2024).

There is an insignificant amount of non-essential metals, such as (Li—lithium), (Rb—rubidium), (Sr—strontium), (Zr—zirconium), (Pb—lead), (Au—gold), (U—uranium) and some others that may be present in the human body if they are ingested through food and water, due to their chemical similarity with essential metals, as well as their affinities with the same donor atoms and chemical sites as essential metals (Zoroddu et al. 2019).

Metals also play an important role in non-living systems. An example is humic acid, which originates from the decomposition of organic matter. Humic substances can be defined as condensed organic compounds produced by microbial action, and differ from biopolymers due to their molecular structure and high persistence in the soil (Baldotto and Baldotto 2014). The complexation of humic acids with metal ions is related to biological, biochemical, and geochemical processes (Wöhrle and Pomogailo 2004).

In industrial sectors, several metals are economically exploited. Such as titanium and its alloys with aluminum are used in the aeronautical industry, in the manufacture of turbines and engines. Vanadium through iron-vanadium alloys are used in the manufacture of special steels. Chromium is also used in alloys with iron and non-ferrous metals, and in coatings, copper is used in alloys, wires, pipes, in agriculture as a fungicide, among other applications (Lee 2003).

Potentially toxic metals are designated as such because they occur naturally in the Earth's crust, and their toxicity becomes evident at certain concentrations. At low concentrations, metals are essential nutrients, due to their participation in a range of biological activities, as cofactors of enzymes and proteins (Singh et al. 2018). However, the constant generation of waste and effluents increases their availability in the environment, therefore increasing their toxic nature (Massoud et al. 2019).

The increase in the amount of metals released into the environment will directly influence their bioavailability in the environment, exposing all levels of the food chain to high concentrations of toxic metals, which over time can influence the reduction of biodiversity, and in general cause serious illnesses in humans (Fig. 3) (Baltazar et al. 2018).

Metals with a specific mass greater than 5 g cm^3 are designated as heavy metals and classified into three distinct classes: I. Toxic metals, such as mercury, chromium, lead, zinc, copper, nickel, cadmium, arsenic, cobalt, and tin, among others; II. Precious metals, palladium, platinum, silver, gold, and ruthenium, among others; and III. Radionuclides, e.g. uranium, thorium, radium, and americium (Singh et al. 2018).

High concentrations of the potentially toxic metal Pb can cause neurotoxicity, nephrotoxicity, and other health disorders in humans. Meanwhile, excess Cu can

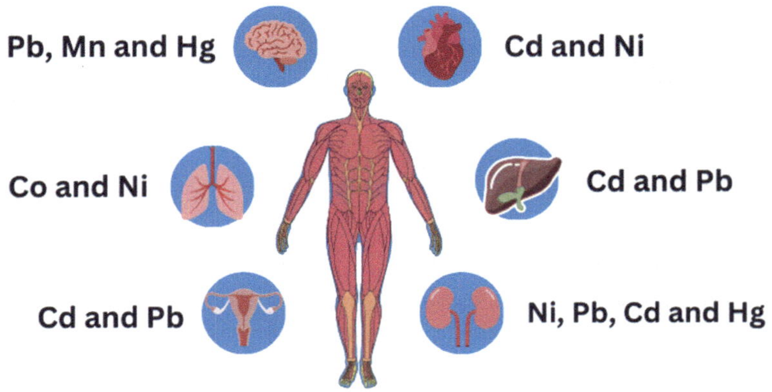

Fig. 3 Main organs of the human body that are affected by contamination by toxic metals

cause toxic effects in fish, invertebrates and amphibians. In fish, excessive intake of Se can be toxic, triggering tissue pathology and reproductive dysfunction (Mutlu 2024).

However, several potentially toxic metals have toxic effects on living beings, such as negative effects on kidney (Pb, Cd and Hg) and liver (Pb and Cd) function, decreased cognitive function (Pb and Hg), decreased reproductive capacity (Cd and Pb), hypertension (Cd), neurological changes (Hg and Pb), teratogenic effects (Hg) and cancers (Cd) (Djedjibegovic et al. 2020).

Other toxic metals that deserve attention are Co, which can cause lung disease, asthma, contact allergies and an increased risk of cancer, Mn, which alters the neurological system and is associated with the development of Parkinson's, and Ni, which can cause contact dermatitis, cardiovascular diseases, asthma, pulmonary fibrosis, and respiratory tract cancer (Bharat et al. 2024).

Metallic nanoparticles have gained prominence in the industry due to their wide range of applications, including antibacterial, photothermal, electrically conductive, medical, military, and others. However, the waste generated from these nanomaterial worries several scientists. Studies have shown that metallic nanoparticles can have negative impacts on plants, aquatic life, and animals. They have a negative impact on animals that live in the soil, in addition to interfering with interactions between plants, microbes and biogeochemical cycles (Wang et al. 2024).

In freshwater crustaceans such as *Daphnia magna*, metallic nanoparticles can cause behavioral changes, reproductive problems, and oxidative stress. For example, copper nanoparticles have been shown to negatively effect on olfactory-mediated behaviors in fish ecology. While, silver nanoparticles when exposed in the lung area have demonstrated the ability to induce pulmonary toxicity, cross the alveolar-capillary barrier, and spread to distant organs (Wang et al. 2024).

Toxic metals are not biodegradable and their bioaccumulation in plant and animal tissues can cause undesirable effects in humans, therefore, their quantities in water and food must always be under control (Lemessa et al. 2022; Massoud et al. 2019; Mutlu 2024; Djedjibegovic et al. 2020; Wang et al. 2024; Hossain et al. 2024).

3.2 Dyes

Dyes are known as a group of synthetic aromatic compounds that comprise several functional groups. Among these are azo dyes (Zafar et al. 2022). Azo dyes are aromatic compounds with $-N=N-$ groups. These synthetic organic dyes are used in various industries such as textiles and pharmaceuticals (Chauhan et al. 2022).

Azo class dyes are recognized as some of the most persistent and recalcitrant industrial pollutants. This class of pollutants can be released into natural waterways through the inadequate disposal of untreated effluents from industries such as textile, leather, and fabric printing (Dissanayake et al. 2021).

Natural dyes are less stable as they degrade, but azo dyes are very stable (Zafar et al. 2022). Research shows that up to 30% of azo dyes used in industry are unable to fully adhere to the products to which they are applied, which results in their elimination in industrial effluents (Dissanayake et al. 2021).

Wastewater contaminated with azo dyes is highly colored and can trigger a variety of serious environmental problems if not removed properly during water treatment. One major concernis that azo dyes and their partial conversion products are known to be genotoxic and cytotoxic (Dissanayake et al. 2021). Additionally, they have the capacity to alter the physical and chemical characteristics of water, such as increasing Biochemical Oxygen Demand (BOD), Chemical Oxygen Demand (COD), causing variations in pH, and organic/inorganic content of pollutants (Ravindiran et al. 2023).

The biological and toxicological activities of each dye vary, even among those belonging to the same class with similar functional groups. Extreme water pollution, which impacts the aquatic ecosystem, is the concerning consequence of this pollutant. Water bodies contaminated by dyes present several environmental problems such as reduced light absorption by organisms (change in watercolor), and the production of different amines under anaerobic conditions make this contaminant cause eutrophication, in addition to high mutagenic and carcinogenic power (Zafar et al. 2022).

Damage in humans is associated with bladder cancer, with a high risk in workers directly involved in the textile industry. There's also a probability of inducing the appearance of tumors and intestinal cancer. In plants, azo dyes reduce photosynthetic activity, and plant growth and reduce soil quality (Zafar et al. 2022).

Azo dyes are the most common dyes, representing more than 80% of all dyes produced annually. Despite having low dyeing efficiency causing damage to several living organisms, and directly impacting water quality, this class of dyes continues to be widely used industrially (Ali et al. 2022).

3.3 Emerging Pollutants

Emerging Pollutants (EPs) have become concerning contaminants due to the widespread consumption of medications, personal care products, antibiotics, and hormones for cosmetic and health reasons. EPs originate from human or animal sources, and have the ability to permeate streams instantly or gradually through the soil. In this way, drinking water sources will be contaminated, and water quality will suffer great damage (Mishra et al. 2023).

Emerging Pollutants (EPs) are also referred to as Contaminants of Emerging Concern (CECs) (Aguilar-Aguilar et al. 2023). These pollutants are defined as synthetic or natural chemicals that are currently not monitored by national or international environmental legislation. They can be found in low concentrations ranging from ng L^{-1} to μ g L^{-1}. These species have low degradation rates, long lifespans, and wide distribution (Martínez-Zamudio et al. 2023).

In recent years, EPs have been a topic of great interest due to their potential pollution, leading to the development of more sensitive analytical techniques that present trace-scale detection (in the range of parts per trillion, ng L^{-1}, to parts per billion, $\mu g\ L^{-1}$) in various environmental matrices, especially in water (Aguilar-Aguilar et al. 2023).

EPs are a group of pollutants that include chemical compounds from various sources, found in pharmaceuticals (including antibiotics), personal care products, pesticides, nutrients, industrial, and household products (e.g., dyes, additives, solvents, endocrine disruptors, hormones, artificial sweeteners, brominated products, and chlorinated flame retardants, among others), illegal drugs, surfactants, microplastics and nanomaterials (Fig. 4) (Aguilar-Aguilar et al. 2023).

The presence of EPs in aquatic environments (drinking water, groundwater, surface water, tributaries, and effluents from wastewater treatment systems), as well as in the atmosphere, marine organisms tissues, and sediments, make these pollutants a category that is receiving more attention by researchers (Aguilar-Aguilar et al. 2023). The main sources of EPs are industrial and domestic effluents, agriculture, urban runoff, and pharmaceuticals, that are discarded in the sewage system (Aguilar-Aguilar et al. 2023).

However, these pollutants are increasingly found in leachate from landfills, groundwater, and surface water systems, and due to their persistence, potential for bioaccumulation, and toxicity, even at low concentrations, they pose a significant environmental and human health concern (Martínez-Zamudio et al. 2023).

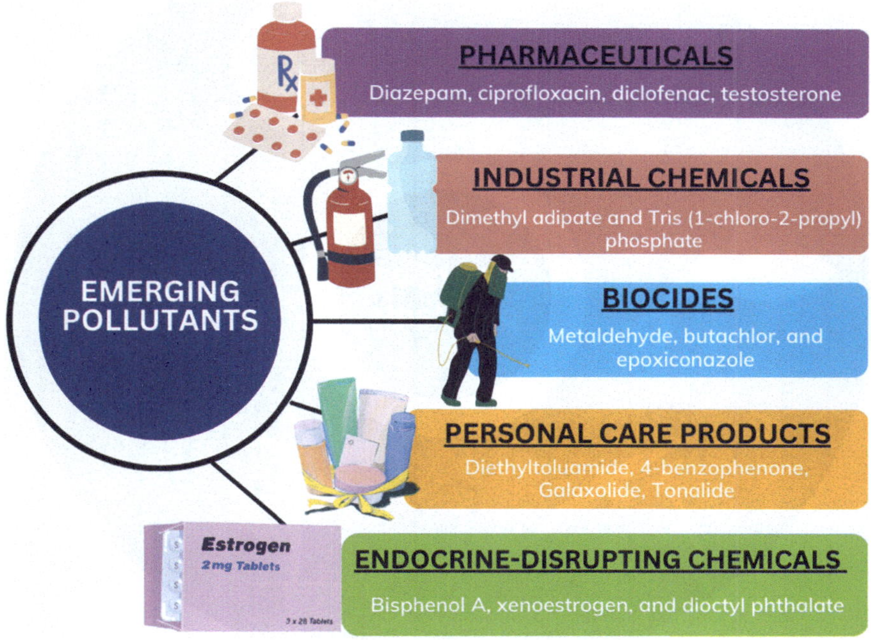

Fig. 4 Main contaminants present in emerging pollutants

These contaminants cause numerous damages to the ecosystem, such as changes in the growth and health of plants, animals, and microorganisms. In humans, the ingestion of plants, animals, or microorganisms contaminated by the bioaccumulation of this pollutants can result in the development of cancer, neuronal damage, stomach ulcers, mitochondrial damage, development of multi-resistant microorganisms (to different medications) that can seriously affect human health (Fig. 5) (Martínez-Zamudio et al. 2023).

In humans and animals, negative effects have been observed after exposure to concentrations of EPs as low as $\mu g\ L^{-1}$. The main reported effects include hormonal interference, genotoxicity, carcinogenicity, endocrine disruption, and immunological toxicity (Coronado-Apodaca et al. 2023).

According to sources and physicochemical characteristics, EPs are grouped into six main classes: (1) Personal Care Products (PCPs), (2) Endocrine Disrupting Chemicals (EDCs), (3) Pharmaceutical Pollutants (PPs), (4) Persistent Organic Pollutants (POPs), (5) Artificial Sweeteners (ASs) and (6) Microplastics (MPs) (Ranjan et al. 2022).

Fig. 5 Main living organisms that are harmed by emerging pollutants

3.3.1 Personal Care Products

The main sources of contaminants in the group of pollutants from personal care products include toothpaste, mouthwash, cosmetics, soaps, among others. Residues of these chemicals reach the soil, subsoil, and groundwater when they are dumped directly into domestic sewage and released from treatment plants (Mishra et al. 2023).

Triclosan is the main antiseptic/antibacterial ingredient in soaps, detergents, shampoos, and toothpaste. This pollutant is also present in many plastic toys, cutlery, cutting boards, and other products. Studies show that it has not havetoxicity potential for humans, however, it is dangerous to aquatic life. In addition to being related to many degradation products that pose serious risks to the health of people and/or animals (Mishra et al. 2023).

3.3.2 Endocrine Disrupting Chemicals

Hormones such as estrogens and androgens affect the functions and bodily mechanisms of the consumer, and activate important administrative processes in the body. Hormones are frequently used in humans and animals. In living things, the adrenal cortex and other parts of the body secrete steroid hormones, which are estrogens created intentionally or naturally (Mishra et al. 2023).

Studies show that pollutants composed of Endocrine Disruptors (EDCs) can interfere with the growth and reproduction of animals, at low levels of contact. By mimicking naturally occurring hormones, causing adhesion to a receptor and blocking endogenous hormones, or causing interruption of routine activity (Mishra et al. 2023).

Normal estrogen hormones identified in wastewater, originating from elimination via human urine, include the hormones estradiol, estriol, and estrone, concerning aspect is that, even without the use of hormonal medications, the body regularly excretes estrogen. Therefore, domestic wastewater entering Wastewater Treatment Plants (WWTPs) through sewer lines is expected to contain a wide range of hormone concentrations. As a result, fish and downstream fish are exposed to more estrogen compounds, which negatively affects their reproduction (Mishra et al. 2023).

Agricultural residues are potential sources of this pollutant. Data shows that this contaminant, like estrogen, has been detected in soil, dairy waste treatment washes, groundwater, beneath untreated runoff ponds, and in rivers draining cattle grazing areas. Cattle, for example, release estrogen in relatively high amounts. Researchers believe that natural estrogens emitted by animal and human waste pose a significant threat to the environment. This ecological catastrophe is particularly worrying because organic farming often uses animal feces and other biosolids as fertilizers (Mishra et al. 2023).

3.3.3 Pharmaceutical Pollutants

Pharmaceutical Pollutants (PPs) are the substances used by a living being, with the aim of personal care aimed at health, as well as agribusiness products aimed at promoting health, production, and animal welfare. PPs include prescription drugs, non-prescription drugs, and veterinary drugs. These substances from human and veterinary medicines are classified as antibiotics, steroids, analgesics, beta-adrenergic blockers, and others (Mishra et al. 2023). Antibiotics and analgesics are one of the main classes of drugs currently consumed by the world population. Among the most commercially available antibiotics are amoxicillin and ciprofloxacin (Ranjan et al. 2022).

The widespread consumption of medications, such as antibiotics found in human and veterinary medicines, has negative consequences for the environment. It promotes bacterial resistance to antibiotics, making this pollutant a major problem for human and animal health (Ranjan et al. 2022).

Pharmaceutical Pollutants (PPs) can indirectly impact the environment through the use of manure and purines as fertilizers in agriculture. This practice can transfer drug residues that were administered to the animal to the soil. Consequently, these pollutant residues can be bio accumulated by plants, potentially contaminating humans via food consumption. There can also be direct impacts when pharmaceuticals are administered to fish (fish farming), leading to contamination of surface waters (Mishra et al. 2023).

3.3.4 Persistent Organic Pollutants

Pesticides can be classified as fungicides, herbicides, or insecticides, depending on their use, and all of them are potential sources of EPs in the aquatic environment. Fungicides are divided into halogenated compounds, conazoles, chloro-alkyl thiocarboxamide, anilinopyrimidine, benzimidazoles, carbamic acid derivatives, morpholines, amides, and some other chemicals according to their chemical constitution. Similarly, herbicides and insecticides are divided into additional classes depending on their purpose and chemical properties (Mishra et al. 2023).

Although they have advantages that justify their use, pesticides have a long history of environmental damage. Due to their widespread use increasing concentrations of pesticide-derived chemicals have been found in both surface water and groundwater. Regarding the damage, pesticides harm the diversity of macroinvertebrates in the marine ecosystem (Mishra et al. 2023).

Scientific reports have shown that conventional treatment plants, which mainly rely on the use of microorganisms fail to effectively eliminate these types of organic chemicals due to their complex molecular structure, low resistance, and elimination rates of only 10%. Drugs such as diclofenac, antacids, steroids, antidepressants, ciprofloxacin, analgesics, salicylic acid, anti-inflammatories, propranolol, antipyretics, beta-blockers, fluoxetine, tranquilizers, and stimulants are among the pharmaceutical products that are most detected in wastewater and drinking water (Mishra et al. 2023).

3.3.5 Artificial Sweeteners

As an emerging pollutant, artificial sweeteners (ASs) are gradually detected in the environment. (ASs) constitute a class of synthetic or semi-synthetic organic compounds that replace sucrose. They act as non-nutritive for sugar substitutes and have high sweetness, and solubility in water (Lin et al. 2024). However, the majority of (ASs) are not completely converted in the body, which is why they are called calorie-free sugars. Consequently, these compounds are generally used by people who need to control calorie intake, such as diabetics and obese patients. In industry, several foods are added with (ASs) such as beverages, low-calorie yogurts, pharmaceutical products, personal hygiene products like toothpaste, and animal feed (Wang et al. 2023; Yue et al. 2023; Praveena et al. 2019).

The first (ASs) consumed by the population were saccharine in 1897. With the development of socioeconomic status, and the increase in obesity cases, the creation of new sweeteners accelerated. Artificial sweeteners can be divided into two categories: the first generation of ASs, which mainly includes saccharin (SAC), cyclamate (CYC), and aspartame (ASP); and the new ASs, which include acesulfame (ACE), sucralose (SUC), neotame (NEO) and neohesperidin dihydrochalcone (NHDC) (Baldotto and Baldotto 2014). In general, the most consumed sweeteners by the world's population are acesulfame (ACE), saccharin (SAC), cyclamate (CYC), sucralose (SUC), aspartame (ASP), alitame (ALI), neotame (NEO) and advantame (ADV) (Yue et al. 2023).

After the consumption of artificial sweeteners, these compounds are metabolized in the human body to a very limited extent or are not metabolized at all, and are subsequently excreted through feces and urine. They are then transported to Wastewater Treatment Plants (STPs) through the sewage system (Yue et al. 2023).

Studies have shown that (ASs) have been detected in various environmental matrices such as tributaries, effluents, and sludges of (STPs), surface waters, groundwater, drinking water, the air, leachate waste, and biosolids. This class of contaminant, upon entering the environment, undergoes migration, transformation, and degradation processes, is deposited in sediments, the oceans, and/or the human body (Wang et al. 2023). Some (ASs) are relatively stable in the environment. The presence of this emerging pollutant has been detected around the world after decades of use of these compounds (Yue et al. 2023).

The negative effects of this contaminant on the environment can especially harm the aquatic ecosystem and impact human health. Some studies have shown that this class of pollutant in the environment is being detected in low concentrations, considered safe for the environment, but in the long term, there may be consequences of these concentrations for living organisms. Toxic organic compounds (ASs) are considered toxic in certain concentrations to aquatic organisms, animals, and plants (Wang et al. 2023).

3.3.6 Microplastics

Plastic is extensively utilized across in various sectors, such as food packaging, industry, agriculture, and medical treatment due to its lightness, durability, corrosion resistance, and low cost. However, the overuse of this material contributes to the direct increase of microplastics in the environment, which are often disposed of in rivers through the domestic sewage system, and the release of industrial effluents (Zhang et al. 2023).

Microplastics (MPs) are particles smaller than 5 mm in size and are widely distributed throughout the world. The migration and accumulation of this pollutant in the environment have caused concerns among scientists (Singh et al. 2018). The main MPs that are found in the environment are polyethylene (PE), polystyrene (PS), polypropylene (PP), polyester (PES), polyamide (PA), polyethylene terephthalate (PET), nylon, styrene copolymer-acrylonitrile (SAN), polyurethane (PU), polyvinyl chloride (PVC), polyacrylate (PEA), alkyd resin (ALK), polyphenylene oxide (PPO), ethylene vinyl acetate (EVA) and acrylamide (Zhang et al. 2023).

The fragmentation of plastics (primary and secondary) results in the formation of micro and nanoplastics (MNPs) (Kumar et al. 2023). Microplastics (MPs) have a diameter of < 5 mm, while nanoplastics (NPs) have a diameter of < 100 nm, Both pose a potential risk to the environment due to their low degradation rate (Kumar et al. 2023). These pollutants can adsorb and transport other pollutants such as potentially toxic metals, emerging pollutants, and organics. These contaminants retained by adsorption can be released into the environment under the influence of UV radiation, temperature, oxygen, soil acidity, and soluble organic matter, and consequently enter the ecosystem through leaching (Zhang et al. 2023).

The ingestion of this pollutant by living organisms can cause physical damage, and trigger biological behavior, and physiological and molecular reactions. Additionally, it can contribute to compound pollution with other pollutants and produce even more harmful toxic substances. Although many researchers have achieved discoveries about MPs, there are still some limitations in their determination, quantification, characterization, and the understanding of the involved interaction mechanisms (Zhang et al. 2023).

3.3.7 Nanopollutants

Pollutants present in the nanometric range are among the contaminants that pose the greatest environmental concerns for regulatory bodies. This is mainly due to the challenges associated with quantifying and subsequently removing them from wastewater. (Lu et al. 2024). As water quality and meteorological-hydrological conditions undergo continuous changes, inorganic pollutants can be transported and enriched in the aquatic system, existing in the form of nanoparticles (NPs, 1–100 nm) (Lu et al. 2024).

NPs of natural origin are of great importance in maintaining the mobility and bioavailability of metals, mediated by complex formation and functional groups.

However, NPs of anthropogenic origin, such as nanomaterials that are produced and discarded into the environment without any prior treatment, can negatively affect biotic and abiotic reactions (Lu et al. 2024) (Table 2).

4 Future Perspectives

The increase in research related to the detection of water pollutants is linked to the technological advances in analytical instrumentation. And, only with the improvement of analytical techniques to be more sensitive and selective, is it possible to detect compounds at trace levels in complex matrices such as water samples. Many contaminants, their byproducts, and metabolites are present in the environment at concentrations on the order of picogram to nanogram per liter (pg L^{-1} a ng L^{-1}, respectively) and require analytical methods with very low detection limits to be detected. It is also important to highlight that the contamination scenario is being increasingly aggravated by a combination of precarious sanitary conditions—such as low investment in sewage treatment plants, which has led to significant contamination of the sewage water receiving channel—and the current level of consumption of goods.

In this direction, science needs to precede regulatory processes, driving research to guarantee the quality of water bodies and understand the risks associated with modern living standards and their consequences for environment health. Several substances have been considered emerging water pollutants, such as Personal Care Products, Endocrine Disrupting Chemicals, Pharmaceutical Pollutants, Persistent Organic Pollutants, Artificial Sweeteners, Micro plastics, and Nano pollutants, among others. In addition to these, some microorganisms and algae toxins are also considered contaminants.

Therefore, research to understand issues related to water pollutants and methods of detection and characterization make it possible to study pollutants from an environmental and public health point of view. As well, studies involving conventional and complementary treatment technologies seek to offer the most efficient way to remove organic contaminants and minimize the formation of transformation products, originated after treatments, especially chlorination, used in practically all treatments as a final disinfection step. Therefore, works like this are very important to support future decision-making involving regulatory processes.

Table 2 Main damages caused by potentially toxic metals and emerging pollutants

Pollutants	Adverse effects	References
AgNPs	In plants, they alter the nitrogen cycle in artificial wetlands	Wang et al. (2024)
CuNPs	High concentrations in food additives, fertilizers and insecticides pollute the environment	Wang et al. (2024)
ZnONPs	Damage to the DNA of aquatic animals	Wang et al. (2024)
ZnONPs	In rats they caused neurobehavioral abnormalities such as impaired learning, memory, depressive behavior and motor problems	Wang et al. (2024)
AuNPs	In beetles, they disrupted ovarian cell structure and microRNA expression, affecting fertility and population dynamics	Wang et al. (2024)
Pb	Neurotoxicity, nephrotoxicity and other health disorders in humans	Mutlu (2024)
Cu	Excess can cause toxic effects on fish, invertebrates, and amphibians	Mutlu (2024)
Se	In fish, excessive intake can be toxic, triggering tissue pathology and reproductive dysfunction	Mutlu (2024)
Pb, Cd and Hg	Impaired renal	Djedjibegovic et al. (2020)
Pb and Cd	Liver function	Djedjibegovic et al. (2020)
Pb and Hg	Decreased cognitive function; neurological changes	Djedjibegovic et al. (2020)
Cd and Pb	Impaired reproductive capacity	Djedjibegovic et al. (2020)
Cd	Hypertension and cancers	Djedjibegovic et al. (2020)
Hg	Teratogenic effects	Djedjibegovic et al. (2020)
Ni	It can cause lung and nasal cancer in humans, damage kidneys and lungs, increase allergies and skin infections in children, and cause congenital heart defects	Hossain et al. (2024)
Co	Cause lung disease, asthma, contact allergy and increased risk of cancer	Hossain et al. (2024)
Mn	Neurological changes and trigger Parkinson's disease	Hossain et al. (2024)

(continued)

Table 2 (continued)

Pollutants	Adverse effects	References
Azo dyes	Bladder cancer and inducing the appearance of tumors and intestinal cancer in humans	Zafar et al. (2022)
Azo dyes	Reduce photosynthetic activity, and plant growth and reduce soil quality	Zafar et al. (2022)
Emerging pollutants	Changes in the growth and health of plants, animals, and microorganisms	Martínez-Zamudio et al. (2023)
Emerging pollutants	In humans and animals, negative effects are hormonal interference, Geno toxicity, carcinogenicity, endocrine disruption, and immunological toxicity	Coronado-Apodaca et al. (2023)
Endocrine Disruptors (EDCs)	Animal growth and reproduction	Mishra et al. (2023)
Pharmaceutical Pollutants (PPs)	Promoting the resistance of bacteria to antibiotics	Ranjan et al. (2022)
Pesticides	Harm the variety of macroinvertebrates in the marine ecosystem	Mishra et al. (2023)
Artificial sweeteners (ASs)	It are considered toxic in certain concentrations to aquatic organisms, animals, and plants	Wang et al. (2023)
Microplastics	The ingestion of this pollutant by living organisms can cause physical damage, and trigger biological behavior, and physiological and molecular reactions	Zhang et al. (2023)

References

Abebe Y, Alamirew T, Whitehead P, Charles K, Alemayehu E (2023) Spatio-temporal variability and potential health risks assessment of heavy metals in the surface water of Awash basin. Ethiopia. Heliyon 9(5):e15832

Abouelsaad O, Matta E, Omar M, Hinkelmann R (2022) Numerical simulation of dissolved oxygen as a water quality indicator in artificial lagoons—case study El Gouna, Egypt. Region Stud Mar Sci 56:102697

Aguilar-Aguilar A, de León-Martínez LD, Forgionny A, Acelas Soto NY, Mendoza SR, Zárate-Guzmán AI (2023) A systematic review on the current situation of emerging pollutants in Mexico: a perspective on policies, regulation, detection, and elimination in water and wastewater. Sci Total Environ 905:167426

Ajala OJ, Tijani JO, Salau RB, Abdulkareem AS, Aremu OS (2022) A review of emerging micro-pollutants in hospital wastewater: environmental fate and remediation options. Results Eng 16:100671

Ali SS, Al-Tohamy R, Mahmoud YA-G, Kornaros M, Sun S, Sun J (2022) Recent advances in the life cycle assessment of biodiesel production linked to azo dye degradation using yeast symbionts of termite guts: a critical review. Energy Rep 8:7557–7581

Allard S, Lauritsen FR (2023) Membrane inlet mass spectrometry (MIMS) for analysis of water disinfection byproducts: recent advances. TrAC Trends Anal Chem 165:117141

Amaly N, El-Moghazy AY, Eweida B, Pandey PK, Sun G (2023) Development of sulfonated and aminated poly(ethylene-vinyl alcohol) nanofibrous membrane as a solid phase extraction matrix for trace metal ions preconcentration from dairy wastewater for direct ICP-OES analyses. React Funct Polym 184:105519

Amorim FAC, Lobo IP, Santos VLCS, Ferreira SLC (2008) Espectrometria de absorção atômica: o caminho para determinações multi-elementares. Quim Nova 31(7):1784–1790

Arenas M, Martín J, Santos JL, Aparicio I, Alonso E (2021) An overview of analytical methods for enantiomeric determination of chiral pollutants in environmental samples and biota. TrAC Trends Anal Chem 143:116370

Ayala-Cabrera JF, Santos FJ, Moyano E (2021) Recent advances in analytical methodologies based on mass spectrometry for the environmental analysis of halogenated organic contaminants. Trends Environ Anal Chem 30:e00122

Bailey RT, Ahmadi M (2014) Spatial and temporal variability of in-stream water quality parameter influence on dissolved oxygen and nitrate within a regional stream network. Ecol Model 277:87–96

Balaram V, Copia L, Kumar US, Miller J, Chidambaram S (2023) Pollution of water resources and application of ICP–MS techniques for monitoring and management—a comprehensive review. Geosyst Geoenviron 2(4):100210

Baldotto MA, Baldotto LEB (2014) Ácidos húmicos. Rev Ceres 61:856–881

Bharat AP, Singh AK, Mahato MK (2024) Heavy metal geochemistry and toxicity assessment of water environment from Ib valley coalfield, India: implications to contaminant source apportionment and human health risks. Chemosphere 352:141452

Chand V, Prasad S (2013) ICP-OES assessment of heavy metal contamination in tropical marine sediments: a comparative study of two digestion techniques. Microchem J 111:53–61

Chauhan V, Gautam P, Kanwar SS (2022) Azo dyes. Develop Wastew Treat Res Process 2022:433–448

Chen Q, Zhang Z, Hu X, Liu M (2024) Novel water/oxygen cycle reaction photo fuel cell for solar conversion and pollutant detection: based on selective 2-electron collaborative catalysis process. Chem Eng J 486:150259

Chi H, Liu X, Xia T, Yang X, He D, Li Z, Liu Y, Lu N (2023) Determination of metal ions in *Paris polyphylla* var. Yunnanensis by ICP-OES and its influence on hemostasis. Microchem J 194:109342

Chowdhary P, Singh A, Chandra R, Kumar PS, Raj A, Bharagava RN (2022) Detection and identification of hazardous organic pollutants from distillery wastewater by GC–MS analysis and its phytotoxicity and genotoxicity evaluation by using *Allium cepa* and *Cicer arietinum* L. Chemosphere 297:134123

Coronado-Apodaca KG, Rodríguez-De Luna SE, Araújo RG, Oyervides-Muñoz MA, González-Meza GM, Parra-Arroyo L, Sosa-Hernandez JE, Iqbal HMN (2023) Occurrence, transport, and detection techniques of emerging pollutants in groundwater. MethodsX 10:102160

de Brito GQ (2010) Espectrometria de emissão óptica com plasma indutivamente acoplado aliada a quimiometria na determinação de íons metálicos no molusco *Mytella falcata* para discriminação de estuários potiguares. Universidade Federal do Rio Grande do Norte

Dessie BK, Aschale M, Polaine X, Melaku S, Alamirew T, Walsh CL, Werner D, Zeleke G (2024) An integrated approach for water quality assessment in African catchments based on physico-chemical and biological indicators. Sci Total Environ 908:168326

Dissanayake M, Liyanage N, Herath C, Rathnayake S, Fernando EY (2021) Mineralization of persistent azo dye pollutants by a microaerophilic tropical lake sediment mixed bacterial consortium. Environ Adv 3:100038

Djedjibegovic J, Marjanovic A, Tahirovic D, Caklovica K, Turalic A, Lugusic A, Omeragic E, Sober M (2020) Heavy metals in commercial fish and seafood products and risk assessment in adult population in Bosnia and Herzegovina. Sci Rep 10(1):13238

dos Baltazar MPG, Gracioso LH, Avanzi IR, Karolski B, Tenório JAS, do Nascimento CAO, Perpetuo EA (2018) Copper biosorption by *Rhodococcus erythropolis* isolated from the Sossego Mine—PA—Brazil. J Mater Res Technol 2018(20):2–10

dos Santos WNL, Costa JLO, Araujo RGO, de Jesus DS, Costa ACS (2006) An on-line pre-concentration system for determination of cadmium in drinking water using FAAS. J Hazardous Mater 137(3):1357–1361

Douvris C, Vaughan T, Bussan D, Bartzas G, Thomas R (2023) How ICP-OES changed the face of trace element analysis: review of the global application landscape. Sci Total Environ 905:167242

Elseblani R, Cobo-Golpe M, Godin S, Jimenez-Lamana J, Fakhri M, Rodríguez I, Szpunar J (2023) Study of metal and organic contaminants transported by microplastics in the Lebanese coastal environment using ICP MS, GC–MS, and LC–MS. Sci Total Environ 887:164111

Ferreira SLC, Bezerra MA, Santos AS, dos Santos WNL, Novaes CG, de Oliveira OMC, Oliveira ML, Garcia RL (2018) Atomic absorption spectrometry—a multi element technique. TrAC Trends Anal Chem 100:1–6

Gefenienė A, Zubrytė E, Kaušpėdienė D, Ramanauskas R, Ragauskas R (2024) Firefighting wastewater from a tire recycling plant: chemical characterization and simultaneous removal of multiple pollutants. J Environ Chem Eng 12(2):112148

Gürkan R, Kır U, Altunay N (2015) Development of a simple, sensitive and inexpensive ion-pairing cloud point extraction approach for the determination of trace inorganic arsenic species in spring water, beverage and rice samples by UV–Vis spectrophotometry. Food Chem 180:32–41

Herawati N, Dahlan MH, Yusuf M, Iqbal MM, Ahmad RK, Nasir S (2023) Removal of total dissolved solids from oil-field-produced water using ceramic adsorbents integrated with reverse osmosis. Mater Today Proc 87:360–365

Honikel KO, Puolanne E (2024) PH measurement. Encyclop Meat Sci 2024:471–477

Hossain MN, Howladar MF, Khan MI, Abu Bakar SM (2024) Appraisal of trace metals toxicity and human health risk using a novel approach in wastewater of four gas fields, Bangladesh. Groundw Sustain Develop 25:101080

Islam MM, Kashem MA, Alyami SA, Moni MA (2023) Monitoring water quality metrics of ponds with IoT sensors and machine learning to predict fish species survival. Microprocessors Microsystems 102:104930

Jabbo JN, Isa NM, Aris AZ, Ramli MF, Abubakar MB (2022) Geochemometric approach to groundwater quality and health risk assessment of heavy metals of Yankari Game Reserve and its environs, Northeast Nigeria. J Cleaner Prod 330:129916

Jarić I, Višnjić-Jeftić Ž, Cvijanović G, Gačić Z, Jovanović L, Skorić S, Lenhardt M (2011) Determination of differential heavy metal and trace element accumulation in liver, gills, intestine and muscle of sterlet (*Acipenser ruthenus*) from the Danube River in Serbia by ICP-OES. Microchem J 98(1):77–81

Kamel M, Abd El-Fatah G, Zaher A, Farghali AA, Othman SI, Allam AA, Rudayni HA, Salah AM (2024) Cost-effective layered double hydroxides/conductive polymer nanocomposites for electrochemical detection of wastewater pollutants. Chin J Anal Chem 52(3):100368

Kassim NSA, Ghazali SAISM, Bohari FL, Abidin AZN (2022) Assessment of heavy metals in wastewater plant effluent and lake water by using atomic absorption spectrophotometry. Mater Today Proc 66:3961–3964

Kaur G, Singh H, Singh J (2021) UV–vis spectrophotometry for environmental and industrial analysis. Green Sustain Process Chem Environ Eng Sci 2021:49–68

Kaur I, Batra V, Bogireddy NKR, Landa SDT, Agarwal V (2023) Detection of organic pollutants, food additives and antibiotics using sustainable carbon dots. Food Chem 406:135029

Kori AH, Uzcan F, Soylak M (2023) $BaTiO_3$ is a novel adsorbent for solid-phase extraction of copper at trace levels in food and water samples before HR-CS-FAAS detection. J Food Compos Anal 122:105474

Krone N, Hughes BA, Lavery GG, Stewart PM, Arlt W, Shackleton CHL (2010) Gas chromatography/mass spectrometry (GC/MS) remains a pre-eminent discovery tool in clinical steroid investigations even in the era of fast liquid chromatography tandem mass spectrometry (LC/MS/MS). J Steroid Biochem Mol Biol 121(3–5):496–504

Kumar V, Singh E, Singh S, Pandey A, Bhargava PC (2023) Micro- and nano-plastics (MNPs) as emerging pollutant in ground water: environmental impact, potential risks, limitations and way forward towards sustainable management. Chem Eng J 459:141568

Lee JD (2003) Química Inorgânica não tão Concisa, 5th edn. Blucher, Rio de Janeiro

Lemessa F, Simane B, Seyoum A, Gebresenbet G (2022) Analysis of the concentration of heavy metals in soil, vegetables and water around the bole Lemi industry park, Ethiopia. Heliyon 8(12):e12429

Leung RK-L, Chui APY, Liu X, Lee H-W, Leung MM-L, Wang Y, Hu M, Kwok KWH (2024) Bioaccumulation of pollutants in the green-lipped mussel *Perna viridis*: assessing pollution abatement in Victoria Harbour and its adjacent aquaculture area, Hong Kong, and the minimal human health risks from mussel consumption. Mar Pollut Bull 201:116086

Lin X, Liu Z, Wang W, Duan G, Zhu Y (2024) Effects of artificial sweetener acesulfame on soil-dwelling earthworms (*Eisenia fetida*) and its gut microbiota. Sci Total Environ 907:167641

Lu Y, Ma T, Lan Q, Liu B, Liang X (2024) Single entity collision for inorganic water pollutants measurements: insights and prospects. Water Res 248(1):120874

Luh J, Mariñas BJ (2014) Kinetics of bromochloramine formation and decomposition. Environ Sci Technol 48(5):2843–2852

Manninen N, Kanerva S, Lemola R, Turtola E, Soinne H (2023) Contribution of water erosion to organic carbon and total nitrogen loads in agricultural discharge from boreal mineral soils. Sci Total Environ 905:167300

Marczenko Z, Balcerzak M (2000) Principles of spectrophotometry. Anal Spectrosc Lib 10:26–38

Maret W (2016) The metals metals in in the the biological biological periodic periodic system system of of the the the elements: concepts concepts and and conjectures conjectures elements. Int J Mol Sci 17:66

Martínez-Zamudio LY, González-González RB, Araújo RG, Rodríguez HJA, Flores-Contreras EA, Melchor-Martínez EM, Parra-Saldívar R, Iqbal HMN (2023) Emerging pollutants removal from leachates and water bodies by nanozyme-based approaches. Curr Opin Environ Sci Health 2023:100522

Massoud R, Hadiani MR, Hamzehlou P, Khosravi-Darani K (2019) Bioremediation of heavy metals in food industry: application of Saccharomyces cerevisiae. Electron J Biotechnol 37, Elsevier España, S.L.U. (2019), pp 56–60—ISBN 07173458

Mishra RK, Mentha SS, Misra Y, Dwivedi N (2023) Emerging pollutants of severe environmental concern in water and wastewater: a comprehensive review on current developments and future research. Water Energy Nexus 6:74–95

Mramba RP, Kahindi EJ (2023) Pond water quality and its relation to fish yield and disease occurrence in small-scale aquaculture in arid areas. Heliyon 9(6):e16753

Mutlu T (2024) Distribution of toxic and trace metals in fish from the Black Sea: implications for human health risks. Emerg Contam 10(2):100295

Nguyen LAT, Ward AJ, Lewis D (2014) Utilisation of turbidity as an indicator for biochemical and chemical oxygen demand. J Water Process Eng 4:137–142

Okewale IA, Grobler H (2023) Assessment of heavy metals in tailings and their implications on human health. Geosyst Geoenviron 2(4):100203

Oliveira LA, Santos JLO, Teixeira LSG (2024) Determination of thallium in water samples via solid sampling HR-CS GF AAS after preconcentration on chromatographic paper. Talanta 266:124945

Omotola EO, Olatunji OS, Moodley B (2024) Trace detection of perfluorooctanoic acid and perfluorooctane sulfonate in surface sediments using a liquid chromatograph coupled to an electrospray ionization single quadrupole mass spectrometer (LC-ESI–Q-MS). Microchem J 199:109928

Oubahmane M, Mihucz VG, Vasanits A (2023) Recent trends in the determination of organic UV filters by gas chromatography–mass spectrometry in environmental samples. TrAC Trends Anal Chem 161:116995

Ozbek N, Baysal A (2017) Determination of sulfur by high-resolution continuum source atomic absorption spectrometry: review of studies over the last 10 years. TrAC Trends Anal Chem 88:62–76

Parron LM, de Muniz DHF, Pereira CM (2011) Manual de procedimentos de amostragem e análise físico-química de água. ISBN: 1980-3958

Patra AK, dos Santos-Ribeiro LP, Yirga H, Sonibare AO, Askar AR, Hussein AH, Puchala R, Goetsch AL (2023) Effects of the concentration and nature of total dissolved solids in drinking water on feed intake, nutrient digestion, energy balance, methane emission, ruminal fermentation, and blood constituents in different breeds of young goats and hair sheep. Anim Nutr 2023:1

Patra AK, dos Santos Ribeiro LP, Yirga H, Puchala R, Goetsch AL (2024) Influence of the concentration and nature of total dissolved solids in brackish groundwater on water intake, nutrient utilization, energy metabolism, ruminal fermentation, and blood constituents in different breeds of mature goats and sheep. Sci Total Environ 907:167949

Peñalver R, Ortiz A, Arroyo-Manzanares N, Campillo N, López-García I, Viñas P (2022) Non-targeted analysis by DLLME–GC–MS for the monitoring of pollutants in the Mar Menor lagoon. Chemosphere 286:131588

Praveena SM, Cheema MS, Guo H-R (2019) Non-nutritive artificial sweeteners as an emerging contaminant in environment: a global review and risks perspectives. Ecotoxicol Environ Saf 170:699–707

Qian Q, He M, Sun F, Liu X (2023) Monitoring and evaluation of the water quality of the Lower Neches River, Texas, USA. Water Sci Eng 2023:1

Rajaji U, Yogesh Kumar K, Arumugam R, Alothman AA, Ouladsmane M, Chung RJ, Liu TY (2023) Sonochemical construction of hierarchical strontium doped lanthanum trisulfide electrocatalyst: an efficient electrode for highly sensitive detection of ecological pollutant in food and water. Ultrasonics Sonochem 92(October 2022):106251

Ramírez SB, Van Meerveld I, Seibert J (2023) Citizen science approaches for water quality measurements. Sci Total Environ 897:165436

Ranjan N, Singh PK, Maurya NS (2022) Pharmaceuticals in water as emerging pollutants for river health: a critical review under Indian conditions. Ecotoxicol Environ Saf 247:114220

Ravindiran G, Sundaram H, Rajendran EM, Ramasamy S, Al-Zaqri N, Ahmed B (2023) Removal of azo dyes from synthetic wastewater using biochar derived from sewage sludge to prevent groundwater contamination. Urban Clim 49:101502

Rho H, Lee Y, Cho J, Woo YC, Chon K (2023) A novel approach to measure carbonate alkalinity in aqueous solutions using a total organic carbon analyzer. Sci Total Environ 903:166127

Rodríguez-Solana R, Salgado JM, Domínguez JM, Cortés S (2014) Assessment of minerals in aged grape marc distillates by FAAS/FAES and ICP–MS. Characterization and safety evaluation. Food Control 35(1):49–55

Sánchez E, Colmenarejo MF, Vicente J, Rubio A, García MG, Travieso L, Borja R (2007) Use of the water quality index and dissolved oxygen deficit as simple indicators of watersheds pollution. Ecol Indic 7(2):315–328

Shanlee SSR, Sundaresan R, Chen S-M, Balaji R, Jeyapragasam T, Peng J-Y, Jothi AI (2023) High-performance electrochemical sensor based on neodymium molybdate/reduced graphene oxide (Nd_2Mo_3O12/RGO) for rapid detection of carcinogenic organic pollutants in water samples. Surf Interfaces 40:103020

Shetty A, Goyal A (2022) Total organic carbon analysis in water—a review of current methods. Mater Today Proc 65:3881–3886

Singh SK, Keshri J, Singh PP, Gupta A, Singh AK (2018) Tolerance of heavy metal toxicity using PGPR strains of *Pseudomonas species*. In: Singh AK, Kumar A, Singh PK (Hrsg.) PGPR amelioration in sustainable agriculture. 1. Aufl. Amsterdã: Elsevier Inc.,—ISBN 9780128158791, pp 239–252

Singh S, Singh J, Singh H (2021) Chemical oxygen demand and biochemical oxygen demand. Green Sustain Process Chem Environ Eng Sci 2021:69–83

Temel NK, Çöpür M (2023) Determination of trace cobalt(II) in spices samples by ultrasonic assisted cloud point extraction with spectrophotometry. J Mol Struct 1284:135433

Wang X, Liang X, Guo X (2023) Global distribution and potential risks of artificial sweeteners (ASs) with widespread contaminant in the environment: the latest advancements and future development. TrAC Trends Anal Chem 159:116915

Wang Y-L, Lee Y-H, Chou C-L, Chang Y-S, Liu W-C, Chiu H-W (2024) Oxidative stress and potential effects of metal nanoparticles: a review of biocompatibility and toxicity concerns. Environ Pollut 346:123617

Wieczorek K, Turek A, Szczesio M, Wolf WM (2024) A holistic approach to the spatio-temporal variability investigation of the main river water quality—the importance of tributaries. Sci Total Environ 906:167588

Williams RJP (1991) The chemical elements of life. J Chem Soc Dalton Trans 1991:539–546

Winiarski T, Thomas O, Charrier C (1995) Analysis of the spatial and temporal variations in the water quality of a karstic aquifer using UV spectrophotometry. J Contam Hydrol 19(4):307–320

Wöhrle D, Pomogailo AD (2004) Metal complexes and metals in macromolecules: synthesis, structures and properties. Bd. 126. 1. Aufl. Weinheim : Wiley—VCH Verlag GmbH & Co. KGaA,—ISBN 9783527304

Young SL, Frongillo EA, Jamaluddine Z, Melgar-Quiñonez H, Pérez-Escamilla R, Ringler C, Rosinger AY (2021) Perspective: the importance of water security for ensuring food security, good nutrition, and well-being. Adv Nutr 12(4):1058–1073

Yousefi SM, Shemirani F (2013) Selective and sensitive speciation analysis of Cr(VI) and Cr(III) in water samples by fiber optic-linear array detection spectrophotometry after ion pair based-surfactant assisted dispersive liquid–liquid microextraction. J Hazardous Mater 254–255:134–140

Yue Y, Li L, Qu B, Liu Y, Wang X, Wang H, Chen S (2023) Levels, consumption, and variations of eight artificial sweeteners in the wastewater treatment plants of Dalian City, China. Sci Total Environ 892:163867

Zafar S, Bukhari DA, Rehman A (2022) Azo dyes degradation by microorganisms—an efficient and sustainable approach. Saudi J Biol Sci 29(12):103437

Zhang Z, Wu X, Liu H, Huang X, Chen Q, Guo X, Zhang J (2023) A systematic review of microplastics in the environment: sampling, separation, characterization and coexistence mechanisms with pollutants. Sci Total Environ 859:160151

Zhou Z-W, Wu Z-N, Cai C-X, Lin R-D, Yu X-Q, Li K, Wang N (2023) Smart bifunctional covalent organic frameworks for water treatment: reversible colorimetric pH detection and photocatalytic removal of pollutants. Microporous Mesoporous Mater 360:112689

Zoroddu MA, Aaseth J, Crisponi G, Medici S, Peana M, Nurchi VM (2019) The essential metals for humans: a brief overview. J Inorgan Biochem 195(February):120–129

Microbial Electrochemical Technologies: Different Configurations, Electrodes and Working Principle/Microbial Electrochemical Technologies—Working Principle and Design of Biocathode

Vita Meylani, Didi Jaya Santri, and Endang Surahman

Abstract Microbial electrochemical system technology integrates microbial-based electrodes and anodes for application in environmental remediation and energy security using household and industrial waste as an action to adapt and mitigate climate change. Basically, the way MESs work is by utilizing interactions between microbes and electrodes to produce electrical energy which can be seen from the density of microbes as well as various internal and environmental factors that influence them. Of the various MESs technologies, the most prominent is the use of microbial fuel cells (MFC) which convert organic matter in wastewater into electrical energy which of course still does not have sufficient practical application in terms of power levels. In minimizing, an effective method to improve its performance is to integrate it with electrochemical capacitors, which are usually used for the purpose of storing electrical charge. This chapter describes the latest developments in integrating capacitors with microbial fuel cells (MFC), starting from explaining the theoretical and operational principles of electrochemical capacitors. In addition, the authors explore various electrochemical metrics that can be used to determine electrochemical (bio)capacitance and offer approaches to analyze such measurement data. In addition, examples of the integration of electrochemical capacitors, both internal and external, that have been used to improve the performance of microbial fuel cells (MFCs) are also described. In this chapter we will evaluate the main benefits and obstacles of implementing capacitive microbial fuel cells (MFC).

V. Meylani (✉)
Department of Biology Education, Faculty of Education and Teacher Training, Universitas Siliwangi, Tasikmalaya, Indonesia
e-mail: vibriovita@unsil.ac.id

D. J. Santri
Department of Biology Education, Faculty of Education and Teacher Training, Universitas Sriwijaya, Indralaya, Indonesia

E. Surahman
Department of Physics Education, Faculty of Education and Teacher Training, Universitas Siliwangi, Tasikmalaya, Indonesia

R. Selvasembian et al. (eds.), *Emerging Trends in Microbial Electrochemical Technologies for Sustainable Mitigation of Water Resources Contamination*,
https://doi.org/10.1007/978-3-031-74636-9_3

Keywords Bioenergy · Microbial electrochemical · Bioelectrochemical · Electrode materials · Advanced applications

1 Introduction

The escalating pollution and global warming are significant issues given the ongoing rise in energy demand and rapid consumption of fossil fuels (Roy and Pandit 2019). This has an impact on the society which is confronted with significant difficulties, including the need for clean water, electricity, and energy security (Roy and Pandit 2019). A large increase in energy supply is needed to meet sustainable and environmentally friendly energy needs and support anthropogenic activities. In this case, technology is needed that comes from waste or garbage considering that the earth's temperature is getting hotter from year to year and carbon emissions are high. The earth's temperature is getting hotter from year to year and high carbon emissions indicate the need for technology that comes from waste. Renewable bioenergy utilization is one approach to resolving the current global warming crisis. Significant interest is currently being devoted to alternative methods of electricity generation (Meylani et al. 2023). With the large potential that bioenergy has, it utilizes microbial agents as natural electricity-producing electrodes known as Microbial Electrochemical Systems (MESs).

Energy recuperation from diverse waste and wastewater sources offers a potential resolution for waste management and the energy shortage. Wastewater has the potential to be used as a renewable source to produce several types of bioenergy, in addition to its role in the remediation process through the regulation of biological processes (Caplan 1993). Lately, the application of microbial electrochemical systems (MESs) has become more significant, offering a highly promising approach for both bioenergy generation and wastewater treatment (Pant et al. 2012). Conventional wastewater treatment technologies require a significant amount of energy.

The concept of harnessing microbes for power generation was initially observed in the early twentieth century During the early 1990s, fuel cells gained significant popularity, leading to the recognition of microbial fuel cells (MFCs) as a potential technology (Bullen et al. 2006). However, a significant breakthrough occurred in the late 1990s with the development and operation of MFC (Microbial Fuel Cells), which utilized bacteria capable of directly transferring electrons to the anode while oxidizing waste, eliminating the requirement for a mediator. The electricity produced in MFCs is approximately 102 times less than e103. The emergence of various Microbial Electrochemical System (MES) designs was driven by reduced power generation and thermodynamic limitations. Over recent decades, there has been a shift from MFCs to exploring alternative MES approaches. Nonetheless, the data gathered from MFCs remains highly valuable, forming the basis for future advancements in MES research (Kim et al. 2007). Various microbial electrochemical technologies (METs) have arisen by incorporating electrochemically active (EA) microorganisms into traditional electrochemical procedures.

In contrast, MESs offers the potential to recover valuable chemicals, produce fuels, or even generate energy while simultaneously treating garbage. It's crucial to understand that Microbial Electrochemical Systems (MESs) represent a rapidly evolving technology leveraging microorganisms to extract chemical energy from bioorganic materials for electrical power generation. MESs offer the dual advantage of microbial-mediated oxidation and reduction mechanisms. They typically feature an anode where electroactive biofilm oxidizes organic substances, creating an electric potential difference between the anode and cathode. The cathode can then perform various functions utilizing in situ electricity, such as hydrogen production in microbial electrolysis cells (MECs), water desalination in microbial desalination cells (MDCs), and the synthesis of valuable fuels in microbial electrosynthesis. This chapter primarily explores the principles of diverse microbial electrochemical systems (MESs), with a specific focus on foundational research concerning microbial fuel cells (MFCs).

2 Fundamental Tenets of Microbial Electrochemical Systems

Microbial Electrochemical Systems (MESs) offer a feasible substitute for traditional anaerobic wastewater treatment and bioenergy extraction methods. In microbial electrochemical systems (MESs), microorganisms play a key role in the oxidation of organic waste through metabolic processes on the anode, resulting in the creation of biofuel or the recovery of valuable chemicals on the cathode by reduction (Wang and Ren 2013). The operational mechanism of MESs relies on the integration of microbial physiology and electrochemistry. The research on MESs originated from thorough examination of MFCs. The MFC primarily adheres to a conventional chemical fuel cell technique. MESs have a variety of applications, including bioremediation via sediment/benthic MFCs, desalination with microbial desalination cells (MDCs), CO_2 sequestration with microbial carbon capture cells (MCCs) utilizing algae, chemical synthesis via microbial electrosynthesis, and biosensors employing microfluidic MFCs.

The MESs typically consists of an anode, a cathode, and one or more electrolytes. The cathodic and anodic chambers can be separated by an ion exchange membrane (IEM). Within the anodic chamber, bacteria exist either in a planktonic state or as a biofilm. These microorganisms oxidize substrates, resulting in the generation of electrons, protons, and other metabolites as end products. The released electrons are collected by the anode and transferred to the cathode via an external circuit. Meanwhile, the protons diffuse through the IEM or directly to the cathode, where they combine with incoming electrons, thus completing the circuit. This electron flow through the external load generates an electric current. The movement of electrons can be utilized for power generation in MFCs or for the production of hydrogen and other valuable molecules in MECs. The voltage produced across the electrodes can also be utilized to desalinate water, particularly through the application of microbial desalination cells (MDCs) (Scott and Yu 2015). However, BES can also be utilized for free or immobilized enzyme-based fuel or electrolysis cell (Sammes 2006).

Bioelectrochemical systems (BES) or microbial electrochemical tools (MXC) have been utilized by several research communities.

2.1 Various Categories of Microbial Electrochemical Systems

The previous publications referred to the many types of MES technology. MFC are the most prevalent and widely used MESs. MEC, or Microbial Electrolysis Cell, is a device that incorporates an external power source to enhance the reduction capacity of the abiotic cathode in anaerobic settings. Hydrogen and various by-products can be produced in MEC. The cathode, which had an abundance of electrons, had several functions, including the reduction of oxidizing pollutants like perchlorate and chlorinated solvents. These systems were referred to as microbial electro-remediation cells (MRC). The utilization of microbial-catalyzed cathodic reduction for the production of value-added compounds has gained popularity in recent years (Hamelers et al. 2010). These systems are known as microbial electrosynthesis. An MDC, or microbial desalination cell, is a bioelectrochemical apparatus that utilizes the electric potential gradient generated by exoelectrogenic bacteria to aid ion transport through a series of ion exchange membranes (IEMs) for water desalination. The gas produced at the anode is directed towards the cathode to capture carbon dioxide in a biocathode within a microbial carbon capture cell (MCC). The aim is to establish a sustainable system that is both carbon-negative and energy-positive. Essentially, the microbial anode in MESs technology derives from microbial species (pure cultures) or microbial communities adhering to the electrode surface, forming a biofilm that oxidizes organic compounds such as sugar, alcohol, acetate, volatile fatty acids, and complex organic materials, and subsequently transfers generated electrons to the electrode through various electron transfer (ET) pathways (Rimboud et al. 2014).

Different types of MESs can be classified into several subsystems based on the utilization of diverse substrates for anodic oxidation. MFCs, in terms of size, can vary from large-scale (measured in liters) to small-scale (employed in biosensor and microfluidic studies). Micro-MFCs are employed in developing biochemical oxygen demand (BOD) biosensors, toxicity sensors, and comparable applications. These devices are normally categorized as either microliter scale or milliliter scale. Photosynthetic bacteria were employed as an anodic biocatalyst in photosynthetic microbial fuel cells (PMFC) to directly convert solar energy into electricity. In a limited number of instances, MESs were combined with pre-existing technology. MFC were combined with membrane bioreactors (MFC-MBRs) and anaerobic digesters (ADs-MFC) to improve the efficiency of treatment and the recovery of bioenergy, respectively (Zhang and Angelidaki 2016). The integration of reverse electrodialysis (RED) with MEC technology has been implemented to enhance hydrogen production in the microbial reverse-electrodialysis electrolysis cell (MREC) without the need for an additional power source (PS). The combination of capacitive deionization (CDI) and microbial desalination cell (MDC) was employed to provide enhanced desalination performance in the microbial capacitive desalination cell (MCDC) (Forrestal et al. 2012). Additional designations may arise from the integration of multiple functionalities within a single system, and they are typically self-explanatory. Instances encompass the microbial electrolysis desalination cell (MEDC), microbial electrolysis desalination and chemical production cell, osmotic microbial fuel cell (MFC), and microbial electrolysis struvite precipitation cell, among additional examples (Chen et al. 2012).

2.2 Shared Characteristics of Microbial Electrochemical Systems

The main components of MESs include a cathode, an anode, and membranes or separators. The anode should be composed of a conductive substance that is resistant to corrosion and facilitates the formation of a biofilm by bacteria. The anode can be made from different materials such as carbon paper, graphite plates, rods, carbon foam, carbon cloth, glassy carbon, carbon felt, stainless steel mesh, and reticulated vitreous carbon. In contrast, the cathode usually consists of carbon felt, graphite, carbon cloth, glassy carbon, carbon paper, titanium mesh, or stainless steel. The choice of cathode in a microelectromechanical system (MESs)iscontingent upon the specific application requirements. Different materials like Pt, MnO2, Pt black, polyaniline, and others were used in MFCs to improve the efficiency of the oxygen reduction reaction (ORR). Conversely, nickel foam, stainless steel, palladium nanoparticles, and similar materials were applied in the cathode to aid the hydrogen evolution reaction (HER). The inclusion of a membrane or separator is unnecessary (Harnisch and Schröder 2010). MFC technology is one of the most prominent and widely applied MES technologies today.

2.2.1 Factors Related to the Operation of Microbial Electrochemical Systems

The effectiveness of MES was evaluated based on anodic and cathodic parameters, design elements, and operational factors. In terms of anodic variables, the selected material must exhibit biocompatibility, a substantial specific surface area, resistance to corrosion and toxicity, good conductivity, stability in wastewater, and be economically viable. Readily accessible within the immediate vicinity. The primary constraints that limit the product production from microbial electrolysis systems (MES) are the cathodic restrictions, specifically the sluggish reduction kinetics and slow rate of electron acceptor reduction processes at the cathode. The essential elements needed to build MESs include bacterial inoculum, wastewater used as anolyte, electrode catalyst, and catholyte. These components play a vital role in shaping the performance of MES (Gil et al. 2003).

2.3 Factors Influencing Design

The performance of a Microbial Electrolysis System (MES) is influenced by various architectural factors, including the shape and type of the MES, the material and spacing of the electrodes, the characteristics and specific surface area of the electrodes, and the distribution of wastewater flow. The electrode material needs to demonstrate biocompatibility, conductivity, and a significant specific surface area, along with low resistance and chemical stability in electrolyte solutions. Different design approaches, such as single-chambered MFC, have been proposed to reduce the internal resistance of MFCs. In this arrangement, the cathode was exposed to air rather than being immersed in the cathodic chamber (Pandit et al. 2012). Furthermore, the arrangement of many MFCs in a series or parallel connection has been devised to enhance the electrical output of the MFCs (Rinaldi et al. 2008). In maximizing overall power density and COD (Chemical Oxygen Demand) removal, appropriate optimization of hydraulic retention time (HRT) is required to maintain sufficient COD concentration for the MESs module as well as balance waste COD and anode power output along the flow path (He et al. 2021).

2.3.1 Chemical Factors

The proliferation and metabolic processes of microorganisms impact the resulting yield. To promote microbial growth kinetics, it is essential to operate the MFC system under optimal operating conditions. The performance of MESs is limited by the supply of protons at the cathode. Therefore, employing NaCl or buffers helps to counteract the restricted movement of protons through the membrane. Temperature has a similar impact on microbial activity and speeds of redox reactions (Li et al. 2010). Substrates that can be used in utilizing MESs technology are liquid waste and hydrogen ion producers as carbon sources (such as acetic acid) where these substrates play an important role in influencing the growth and microbial community in the biocathode development process (Jafary et al. 2015). In addition, environmental controls such as pH (generally close to neutral), salinity and oxygen levels are carried out to optimize ionic interactions between microorganisms and electrodes.

2.3.2 Biological Aspects

The microbial aspect, encompassing characteristics of the inoculum, its capacity to form biofilms, electron transport, and kinetics of substrate use, plays a crucial role in Microbial Electrochemical Systems (MESs). When dealing with intricate wastewater, biological optimization involves choosing appropriate bacterial groups and implementing ways to acclimate bacterial communities to the optimal conditions of the reactor. Efficient exploitation of a substrate for capturing the electric charge and facilitating the movement of electrons results in enhanced current densities at reduced anodic overpotential (Beyenal and Babauta 2015).

2.4 Operating Conditions

The microbial electron transport chain (METC) is a crucial component within living systems, as it serves as the primary source of energy through the oxidation of organic substrates. Evaluating the effectiveness of microbial respiration and the electron transport chain's activity is crucial for comprehending cellular function since these processes are fundamental for the replication and growth od aerobic organisms. Initially, attention was directed towards quantifying the consumption of dissolved oxygen by living cells as a direct measure of microbial viability. This led to the exploitation of electron transfer processes from living microorganisms to electrodes in Microbial Electrochemical Systems (MESs), utilized in microbial fuel cells or as diagnostic tools for rapidly assessing microbial activity. Numerous MESs approaches have been developed and tested for various biological applications. The electrical current generated by MESs is directly proportional to the quantity of viable microbial cells, with non-viable or non-cultivable cells not affecting electrochemical signals. Consequently, bioelectrochemical reactions indicate the extent of anodic respiration, intracellular redox reactions (like intracellular enzyme activities), and other biological interactions. This capability to correlate bioelectrochemical responses with microbial processes has sparked growing enthusiasm for designing advanced MESs, given their myriad potential applications in environmental, energy, and biomedical domains.

The performance of MFC is mostly influenced by the anodic and cathodic working conditions. The performance of a microbial fuel cell (MFC) in terms of energy generation and treatment efficiency is influenced by several operating variables. This encompasses factors such as external resistance, characteristics of the substrate, loading rate, temperature, pH, and retention time. A diverse array of wastewater types, spanning from simple sugars to more intricate compositions, were efficiently employed as a substrate and successfully processed within the anodic chamber of the Microbial Fuel Cell (MFC). The organic loading rate (OLR) is the ratio between the current generation from the microbial fuel cell (MFC) and the degradation of the substrate (Varanasi et al. 2015). At higher external resistance, the reduction in anode potential and the flow of electrons to the cathode are constrained by the load of the external circuit. However, at low external resistance, these processes are limited by either kinetic or mass transfer restrictions. The presence of increased internal resistance led to a decrease in both power density and current density in the microbial fuel cell (MFC) (Li et al. 2010).

2.5 *The Principles of Biochemistry in Microbial Electrochemical Systems*

The idea of microbial-assisted current production differs from standard biogeochemical procedures such as the reduction of Fe^{3+} or Mn^{4+}. This is because it entails employing electrons produced during microbial respiration to lessen the anode, rather than relying on a naturally occurring electron acceptor (Stams et al. 2006). Dissimilatory metal-reducing bacteria and sulfate-reducing bacteria play crucial roles in the bioanode, exerting considerable influence (Logan 2009). The similarity between the mechanisms of extracellular electron transfer (EET) activity for metal reduction and anode reduction is likely because of this reason: Electrochemically Active Bacteria (EAB) can carry out anode reduction by oxidizing the substrate, even when their usual electron acceptors, like O_2 and SO_4, are absent in the anode chamber (Yang et al. 2012). The EAB generate electricity through their metabolism, producing ATP and facilitating their growth. Nevertheless, the energy gain is restricted by the anodic potential, resulting in much lower growth yields compared to those reported in aerobic metabolism. Furthermore, the electric current generated by MESs is directly correlated with the quantity of microbial cells inhabiting the MESs, as they are the source of electrochemical signals (Hassan et al. 2021). The use of smaller growth yields results in reduced waste sludge and increased coulombic efficiency, making it an appealing method for wastewater treatment. The treatment capability of MFCs exhibited a significant rise, reaching as high as 7.1 kg of chemical oxygen demand (COD) per cubic meter (m^3) of reactor volume per day. This level of capacity is comparable to that of an anaerobic digester.

Basically, the degradation process of substrates (such as acetate) by living microorganisms produces protons and carbon dioxide as products, while anolytes and electrons are channeled to the anode as oxidation half-reactions in a two-chamber MFC system. In the conventional aerobic respiration pathway, two electrons are freed through enzymatic oxidation of NADH or NADPH. This process is facilitated by a sequence of enzymatic complexes, commencing with NADH dehydrogenase. Subsequently, the electrons are conveyed to coenzyme Q (CoQ), then to ubisemiquinone, and finally to ubiquinol, which serves as a natural electron carrier. Electrons originating from ubiquinol proceed along the electron transport chain (ETC) through complex III (also known as complex bc1), cytochrome c, complex IV (cytochrome c oxidase), and ultimately to oxygen, which serves as the final electron acceptor. This process culminates in the production of water (H_2O) (Pant et al. 2012; Biffinger et al. 2007). The energy liberated as electrons move along the electron transport chain is harnessed to propel protons outward through complexes I, III, and IV, creating an electrochemical gradient of protons. It is noted that since the electron transport chain is isolated from the external environment by a cytoplasmic membrane, which includes additional layers like the cell wall, peptidoglycan, or outer membrane, the direct transfer of released or biologically produced electrons to the electrode surface becomes crucial.

2.6 Anodic Microorganisms Employed in Microbial Electrochemical Systems

The microorganisms in MESs are of utmost significance in ongoing study. There are essentially three sorts. The first category consists of bacteria that cannot directly transfer electrons to an anode and instead rely on external mediators, such as *E. coli* and *Bacillus* sp. Microorganisms of this kind were employed to oxidize substrates within the anode chamber of the indirect microbial fuel cell (MFC). Another category consists of bacteria capable of directly transmitting electrons to the anode. These bacteria are termed electroactive bacteria (EAB), also recognized as "electricigens," "exoelectrogens," "electrogens," "anodophiles," or "anode-respiring bacteria." Shewanella putrefaciens and *Geobacter sulfurreducens* are commonly acknowledged as examples of electroactive bacteria (EAB) (Stams et al. 2006). They are commonly referred to as anodophiles or exoelectrigens due to their ability to create a conductive biofilm on the anode. The consortia of microorganism act as a biocatalyst within the anode chamber of a microbial fuel cell (MFC) while treating wastewater.

Various phylogenetic groupings are observed to develop the biofilm on the surface of the anode. Firmicutes, Acidobacteria, and a select few groups of Proteobacteria were recognized as effective electron donors at the anode surface, leading to the generation of electrical current. Launching a MES system for practical use in fields like concurrent energy generation and wastewater treatment necessitates employing a mix of bacterial cultures, such as activated sludge or anaerobic digester sludge. Usually a number of metal reducing bacteria such as *Geobacter sulfurreducens* and *Shewanella oneidensis* act as catalysts in transferring electrons from the reduced electron donor to a solid electrode material, which is usually referred to as an anode (generally in the form of graphite), which acts as an electron acceptor (Pant et al. 2012). Under suitable acclimation conditions, it is expected that the MES system will independently select and improve its own electrochemically active communities from the mixed culture gradually, leading to a gradual onset of electrogenesis. The microbial communities differ from one another based on factors such as the source of the inoculum, operational conditions, substrate availability, and MES system architecture.

There are reports that provide direct isolation of EAB from the anode of MFC. A wide variety of electroactive bacteria (EAB) and non-electroactive bacteria (non-EAB) are identified on the anode surface in microbial fuel cell (MFC) systems. Jiang et al. (2010) utilized biological laser printing (BioLP) to isolate and identify a solitary EAB. In their study, Wang et al. (2010) documented the successful isolation and cultivation of a *Bacteroides* sp. W7 that is capable of reducing Fe(III) from the anolyte suspension (Wang et al. 2010). Zuo et al. (2008) describe that *Ochrobactrum anthropi* YZ as exoelectrogenic got by selectively enriching and diluting the anode of a U-tube shaped microbial fuel cell (MFC). Pham et al. (2003) reported the identification of a bacterium with the ability to reduce Fe(III) from the anolyte containing acetate. This bacterium shows close phylogenetic association with *Aeromonas hydrophila*. Another obligate anaerobic microorganism, *Clostridium butyricum* EG3, was found in the anolyte of a microbial fuel cell (MFC) fed with wastewater from starch production. However, it has been observed that not all microorganisms within the biofilm have direct connections to the anode; some may indirectly participate by utilizing electron mediators produced by other anode-respiring microbes. For example, despite its inability to produce energy independently. The main species found in the anode biofilm was *Brevibacillus* sp. PTH1, likely facilitated by the electron mediator pyocyanin or a phenoxazine complex, excreted by *Pseudomonas* sp. This compound facilitates the transfer of electrons to the anode by *Brevibacillus* sp. PTH1 when it is co-cultured with Pseudomonas sp. Interestingly, the current produced was notably decreased when *Brevibacillus* sp. PTH1 was not co-cultured with or exposed to the liquid fraction of a microbial fuel cell (MFC) that had previously operated with the presence of *Pseudomonas* sp. (Pham et al. 2008). Additionally, *Geobacter anodireducens* bacterium, capable of reducing Fe(III), was recently discovered in a microbial electrolysis system (MES), with its closest known relative being *Geobacter sulfurreducens* PCA (Sun et al. 2014).

3 Various Mode of Electron Transfer Modes from Microbes to the Anode

Microorganisms can transport electrons to the electrodes through three main methods. In an indirect MFC or mediator MFC, fermentative microorganisms produce hydrogen, carbon dioxide, alcohols, or ammonia as the ultimate byproducts. These products are formed through the reduction of intermediary substances, such as protons or acids, using electrons created during the breakdown of the substrate. This reduction process results in the formation of hydrogen or alcohol, specifically in anaerobic environments. As result, this bacterium, which undergoes fermentation, lacks the ability to directly transfer electrons to the anode. To utilize these bacteria in the anode, an external mediator is required to shuttle between the cell membrane and the anode. Common synthetic electron carriers like benzylviologen, 2,6-dichlorophenolindophenol, thionine, and 2-hydroxy-1,4-naphthoquinone are often employed for this purpose, facilitating electron transport between the anode and the interior cell membrane of fermentative bacteria (Logan 2009).

3.1 Mediatorless Mechanism of Electron Transfer

In a mediator-less microbial fuel cell (MFC), the electron transport mechanism depends on the ability of electrochemically active bacteria (EAB) to directly convey the resulting electrons from the oxidation of organic materials to the anode (Bretschger et al. 2007). The electron transfer from microorganisms to the electrode can be categorized into two types. In the first method, exoelectrogens release internal mediators to enable electron transportation between the bacteria and the electrode. Some microorganisms such as *Pseudomonas* sp., *Lactobacillus* sp., and *Enterococcus* sp. have been observed to produce electron carrier molecules like phenazines, as seen in *Pseudomonas aeruginosa.* These carriers assist in electron transfer to the anode by utilizing internally generated soluble mediators, a process termed mediator-driven electron transfer (MET). The second approach involves bacteria forming a biofilm on the electrode, facilitating direct electron transfer to the electrode via pili or "nanowires," or through the outermost cytochrome C (Yang et al. 2012).

3.2 Electron Transport via c-Type Cytochromes

The c-type cytochromes known as CTCs are crucial for facilitating electron transport in the process of harvesting electricity by electrochemically active bacteria (EAB). CTC refers to heme-containing proteins that are commonly found in a wide range of archaea and eubacteria. The species *S. oneidensis* MR-1 possesses 42 potential CTCs, with 80% of them found on the outer membrane, covering 8% to 34% of the entire cell surface. CymA is a tetraheme-CTC protein that is anchored in the inner membrane by its N-terminal region, while its C-terminal region is exposed to the periplasm. CymA plays a crucial role in numerous anaerobic respiration activities of *Shewanella.*

The removal of a cymA gene resulted in a decrease of over 80% in the production of electrical current during electrode reduction. CymA has the ability to directly interact with multiple terminal reductases found in the periplasm, such as fumarate reductase (FR) and nitrate reductase. Acting as the primary conduit for electron transfer to the periplasmic region, CymA forms temporary protein complexes with various periplasmic redox proteins, as shown by bacterial two-hybrid assays. Experimental evidence, including electrode-mediated reduction of fumarate by *S. oneidensis* MR-1, suggests that 85% of electrons are transferred to FR through CymA, with the remaining 15% via MtrA. This indicates the existence of a CymA-eFReMtrA complex in the periplasm. The MtrABC conduit, an extension of this complex, serves as a pathway for intracellular electron flow to the anode. MtrA facilitates electron transfer from CymA to MtrC within this complex. The significance of MtrC is highlighted by a reduction of over 90% in power generation in microbial fuel cells upon its removal. Additionally, an alternative electron transfer pathway is proposed by partially replacing the IS-insertion-induced SirCD expression with CymA, indicating the intricate interdependence and functional linkage among the MtrABC proteins. Based on a comparative investigation of genomes, it was found that each *Geobacter* genome has an average of 79 potential CTCs across six *Geobacter* species. Merely 14% of them are discovered to be preserved in every genome (Nevin 2009). The MacA protein facilitated the movement of electrons from the inner membrane to PpcA in the periplasm. Subsequently, PpcA sent the electrons to the OMCs.

4 Microbial Nanowire

The microbial or bacterial nanowire technology is regarded as a novel method for electron transport to the electrode. The bacterial nanowire, identified during the reduction of Fe(III) oxide by *G. sulfurreducens*, is a conductive pilus. These extensions have also been observed in *S. oneidensis* MR-1 and numerous other bacteria, suggesting widespread occurrence of these bacterial structures in diverse environments. *S. oneidensis* MR-1 exhibits multiple conductive outgrowths resembling pilus structures, each with a diameter ranging from 1.5 to 2.5 nm. These outgrowths are observed in conditions where the electron acceptor is limited, or in conditions where diffusion is limited or agitation is low. Recent findings reveal that the nanowires observed in Shewanella oneidensis MR-1 are extensions of the outer membrane and periplasmic components involved in transporting electrons beyond the cell (Pirbadian et al. 2014). The nanowires were split by the pores in the membrane. The expression of *G. sulfurreducens* nanowires occurs during the utilization of metal oxides as electron acceptors or during the reduction of fumarate at low temperatures. Two different PilA isoforms exhibiting unique characteristics were identified in the nanowires of *G. sulfurreducens* (Steidl et al. 2016).

The electrical conductivity of *G. sulfurreducens* nanowires is 6 millisiemens per centimeter (mS/cm), a value that closely resembles that of fabricated metallic nanostructures. The nano-wires of *G. sulfurreducens* exhibit a notably high electronic conductivity, above the suggested threshold of 10^{-3} mS/cm. This characteristic could potentially explain the generation of high volumetric power. Unlike the OMC-dependent conductivity found in *Shewanella* nanowires, *Geobacter* nanowires demonstrate distinct metallic electronic properties.

5 Electron Transporters or Mediators

Mediators or electron shuttles in MFCs are predominantly released by gram-negative bacteria. These mediators are expected to possess specific properties, including solubility, stability, reusability, eco-friendliness, and a redox potential range compatible with the bacterial membrane protein and anode material. The secretion of mediators plays a crucial role in stimulating power production in MFCs. Among the most well-documented electron shuttles in MFCs are flavins, such as riboflavin (RF) and flavin mononucleotide (FMN), which are naturally produced by *Shewanella* species (Pandit et al. 2014). FMN can penetrate the outer membrane and convert to RF in the extracellular space, while RF cannot be transported back to the *Shewanella* periplasm. Without a mediator, the electron transfer rate from pure *S. oneidensis* MR-1 OMCs to insoluble iron ions (Fe + 3) is 100–1000 times slower compared to the whole cell. Phenazines, inherent electron shuttles in MFCs, are produced by various bacterial species, primarily *Pseudomonas*sp. (Pandit et al. 2015). The power generation capacity of a non-electroactive gram-positive bacterium, *Brevibacillus* sp., is enhanced by phenazines. This suggests that the presence of soluble mediators in MFCs provides a synergistic effect for other non-electroactive bacteria to harvest electricity. Besides the biologically produced organic compounds, H_2 generated during fermentation or other chemical reactions can also function as electron shuttles or donors for *Shewanella* sp. and *Geobacter* sp. in MFCs (Stams et al. 2006).

5.1 Cathodic Microorganisms Employed in Microbial Electrochemical Systems

Long-term operation frequently resulted in inadequate power generation for abiotic cathodes. The addition of a biocathode to the microbial electrolysis system (MES) was implemented in order to enhance the sustainability of the process. The biocathode offers numerous advantages compared to the typical abiotic cathode, which relies on costly metal catalysts like platinum to efficiently reduce a TEA (mostly oxygen) or another oxidant. Nevertheless, the potential "poisoning" of a metal catalyst or the oxidation of a catalyst limited its application. This issue can be prevented by implementing a "biocathode" where microorganisms can be employed to catalyze the reduction in the cathode. Furthermore, an algal-based biocathode operates autonomously without the need for an external oxygen source. The utilization of a biocathode in an operation can result in the production of valuable substances or aid in the elimination of undesirable molecules. The cathodic biofilm establishes a robust link between the cathode and the microorganisms. It is responsible for oxidizing soluble electron donors and reducing oxygen or other substances like nitrate, sulfate, or fumarate, as well as carbon dioxide (under anaerobic conditions) (He and Angenent 2006).

Biocathodes can be classified into aerobic and anaerobicdepending on the electron acceptors utilized. Hasvold et al. (1997) showed that seawater bacteria formed a slimy biofilm that aids in the reduction of oxygen on the cathode. The inclusion of a biocathode resulted in an enhancement of the the catalytic function of reducing the cathode, leading to a substantial rise in the on-load cell voltage from 1.2 to 1.6 V. Bergel et al. (2005) and Hasvold et al. (1997) demonstrated that the growth of a biofilm in saltwater on a stainless cathode is highly effective for the process of oxygen reduction. A power output of 270 mW/m^2 was attained when a biofilm was present, which represents the highest recorded value. The biocathode demonstrated an oxygen reduction efficiency, as evidenced by the cleaned cathode, resulting in a maximum power reduction of around 2.8 mW/m^2, decreased to a hundredth of its original power. Manganese and iron are occasionally employed as oxygen transfer mediators in aerobic environments. Rhoads et al. (2005) showed that Mn(IV) oxide was reduced and Mn(II) was subsequently oxidized in the cathode. Manganese dioxide (MnO_2) undergoes reduction to become manganese(II) [Mn(II)] through an intermediary phase that results in the production of manganese oxyhydroxide (MnOOH). The general equation for this process is represented by Equation as bellow.

$$MnO_2(s) + 4H + 2é = Mn^{2+} + 2H_2O \quad (1)$$

The reduction of Mn(IV) is non-biological, but to regenerate the reduced Mn(II), bacteria known as *Leptothrix discophora*, capable of oxidizing manganese, were utilized. Employing manganese reduction/oxidation on an aerated biocathode increased the power output by a factor of forty. Fe(III) oxide, like Mn(IV) oxide, was employed as a cathode mediator for oxygen reduction. This choice was made due to the greater abundance of Fe in the environment compared to Mn. In this instance, the reduction of abiotic Fe(III) was succeeded by the oxidation of biotic Fe(II) to maintain the cycle and accomplish indirect oxygen reduction in the cathode. *Thiobacillus ferrooxidans*, an iron-oxidizing bacterium, was utilized to accelerate the oxidation of Fe(II). It was cultured in the cathode chamber of the reactor to regenerate ferric ions by harnessing energy from the reaction

$$Fe^{2+} = Fe^{3+} + é \quad (2)$$

Gregory et al. (2004) showed that the cathode can act as the only source of electrons for the reduction of nitrate to nitrite in a half-cell that is controlled by a potentiostat and the formula look at this bellow.

$$NO_3 + 2H^+ + 2é = NO^{2-} + H_2O \quad (3)$$

The research revealed that nitrate reduction occurred only in the presence of *G. metallireducens* or an adapted enrichment culture, indicating bacterial involvement in the process. This specialized type of biocathode, based on microbial electrolysis cells (MES), proves advantageous for facilitating denitrification. Additionally, the study utilized sulfate as an electron acceptor. However, anaerobic sulfate reduction is not spontaneous or advantageous primarily due to its negative redox potential. Another study introduced the concept of a bio-cathode employing chemicals with low redox potential, such as CO_2 and fumarate. Park and colleagues demonstrated that carbon dioxide and fumarate could be converted into methane and succinate, respectively, using methanogenic granules and *Actinobacillus succinogenes* in the cathode. Both processes involved the utilization of electrons from the cathode surface, with neutral red acting as an intermediary for electron transfer. However, the redox potentials of CO2/CH4 and fumarate/succinate are not conducive to energy generation in MFCs (Jafary et al. 2015).

$$CO_2 + 8H + 8e = CH^{4-} + 2H_2O \text{ Fumarate} + 2H^+ + 2é = \text{Succinate} \quad (4)$$

Several biofilm factors can influence the performance of the biocathode (Aulenta et al. 2010). The ability of microorganisms to transfer electrons into or out of the cell is regulated by the proportion of oxidized to reduced forms of mediators. Erable et al. (2011) found that the mixed culture yielded higher current densities in microbial fuel cells (MFC) compared to biofilms consisting of pure cultures. Possible reasons for the aforementioned observation may include the existence of a more efficient exo-electrotroph, a synergistic effect in a mixed culture, surface changes, or an underdeveloped biofilm in a pure culture. Behera et al. also investigated the influence of biofilm thickness. The study revealed a reduction in power generation as the biofilm thickness increased, which is contrary to what was reported in the anodic biofilm. The heterogeneity of the biofilm can be influenced by surface roughness, hence impacting the properties of mass transfer and biofilm functioning.

The size of the electrode's surface area is a crucial factor that greatly impacts the biocathode's performance (Huang et al. 2011). Expanding the surface area of the cathode is expected to enhance the performance of the biocathode, akin to the abiotic cathode. However, it is only when there is an increase in the surface area that the microbial population will also rise, allowing for the observation of performance enhancement. The primary difficulties associated with biocathode systems include a lengthy initial activation period, which can be addressed by employing an appropriate cathode potential, as well as the expense of the electrode material (Zhang and Angelidaki 2016). Presently, biocathodes are constructed from materials such as graphite granules, carbon felt, carbon paper, stainless steel mesh, graphite fiber, and graphite pads.

6 Conclusions

Uncontrolled use of fossil fuels results in environmental pollution and depletion of limited fuel reserves. Therefore, it is very important to actively seek clean alternative energy sources. Microbial Electrochemical systems (MESs) technology is a multidisciplinary strategy aimed at finding alternative energy sources that has been developed and extensively studied for its performance over the last decade. MESs studies are carried out through interdisciplinary collaboration, integration of the fields of microbiology, molecular biology, electrochemistry, materials science, and engineering platforms. Most research on microbial electrochemical systems (MESs) focuses on the performance of microbial fuel cells (MFCs), which are considered the most fundamental type of MESs. Significant progress has been made in MES research over the past few decades, resulting in significant performance improvements that are visible in reactor design, material components, and operational parameters have resulted in significant increases in volumetric power production, ranging from 1 mW/m^3 to 2.87 kW/m^3. Although there have been several attempts to increase the scale of research, most of them were unsuccessful due to problems such as air leakage, creation of undesirable byproducts, significant radius in electrical output, and variations in influence. In the future, many aspects will require additional examination. The exact nature of the interaction between the electrode and exoelectrogenic microorganisms remains uncertain due to the challenges faced in conducting experiments under anaerobic conditions that promote growth on the interfaces in the anode chamber. Additional information is needed to reflect the specific mechanisms involved in understanding the colonization process by EAB and biofilm production under various operational circumstances. For practical purposes, the culture mixture can be considered as an inoculum in the anode chamber of a microbial electrolysis system (MES).

Implementing MES on a large scale is difficult due to microbiological, technological and economic obstacles that must be overcome. The economic feasibility of MDC for desalination and MEC for chemical production appears promising. However, research on MDC and microbial electrosynthesis has only begun recently. So far, they have only been limited to the scope of laboratory experiments. More extensive studies are needed to stop long-term operational shutdowns. Preliminary investigations on microbial electrosynthesis systems for the extraction of valuable compounds or fuels from wastewater show great potential. However, these experiments were only carried out on a small scale in controlled laboratories and under ideal operating conditions. Therefore, it is important to conduct large-scale studies to determine the feasibility of this commercial method when treating large amounts of waste air in practical situations. In addition, further research is needed to understand the processes of methanogenesis and electrogenesis in utilizing certain substrates which will help identify ideal conditions for maximizing bioenergy production. Additionally, MFCs need to be arranged in a stack configuration to efficiently generate power for the purposes of powering the device. However, the occurrence of the voltage reversal phenomenon has a significant impact on the performance of microbial fuel cells (MFC) which can accumulate. Enhancing power generation in MFC stacks requires a comprehensive exploration of the enduring implications of voltage reversal and the accumulation of additional insights to tackle this matter proficiently. Comparative studies are essential to assess the energy and material aspects across all MES systems. This data will aid the scientific community in comprehending the rationale behind the simultaneous processes of wastewater treatment and bioenergy extraction. If existing obstacles can be overcome, MESs technology has the potential to make a significant contribution to waste processing and energy generation.

References

Aulenta, F., Reale, P., Canosa, A., Rossetti, S., Panero, S., & Majone, M. (2010). Characterization of an electro-active biocathode capable of dechlorinating trichloroethene and cis-dichloroethene to ethene. *Biosensors and Bioelectronics*, *25*(7), 1796–1802. https://doi.org/10.1016/j.bios.2009.12.033

Bergel, A., Feron, D., & Mollica, A. (2005). Catalysis of oxygen reduction in PEM fuel cell by seawater biofilm. *Electrochem Commun*, *7*(9), 900–904. https://doi.org/10.1016/j.elecom.2005.06.006

Beyenal, H., & Babauta, J. T. (2015). *Biofilms in Bioelectrochemical Systems: From Laboratory Practice to Data Interpretation*. New York, USA: John Wiley & Sons.

Biffinger, J. C., Pietron, J., Ray, R., Little, B., & Ringeisen, B. R. (2007). A biofilm enhanced miniature microbial fuel cell using Shewanella oneidensis DSP10 and oxygen reduction cathodes. *Biosensors and Bioelectronics*, *22*(8), 1672–1679. https://doi.org/10.1016/j.bios.2006.07.027

Bretschger, O., Obraztsova, A., Sturm, C. A., In, S. C., Gorby, Y. A., Reed, S. B., Culley, D. E., Reardon, C. L., Barua, S. R., Margaret F. Z., Jizhong, B., Alexander, S. B., Rachida Saffarini, D., Mansfeld, F., Kim, B. H., Fredrickson, J. K.,Nealson, K. H. (2007). Current production and metal oxide reduction by Shewanella oneidensis MR-1 wild type and mutants. *Applied and Environmental Microbiology*, *73*(21), 7003–7012. https://doi.org/10.1128/AEM.01087-07

Bullen, R. A., Arnot, T. C., Lakeman, J. B., & Walsh, F. C. (2006). Biofuel cells and their development. *Biosens Bioelectron*, *21*(11), 2015–2045. https://doi.org/c

Caplan, J. A. (1993). The worldwide bioremediation industry: prospects for profit. *Trends in Biotechnology*, *11*(8), 320–323. https://doi.org/10.1016/0167-7799(93)90153-Z

Chen, S., Liu, G., Zhang, R., Qin, B., & Luo, Y. (2012). Development of the microbial electrolysis desalination and chemical-production cell for desalination as well as acid and alkali productions. *Environmental Science and Technology*, *46*(4), 2467–2472. https://doi.org/10.1021/es203332g

Erable, B., Etcheverry, L., & Bergel, A. (2011). From microbial fuel cell (MFC) to microbial electrochemical snorkel (MES): maximizing chemical oxygen demand (COD) removal from wastewater. *Biofouling*, *27*(3), 319–326. https://doi.org/10.1080/08927014.2011.564615

Forrestal, C., Xu, P., Jenkins, P. E., & Ren, Z. (2012). Microbial desalination cell with capacitive adsorption for ion migration control. *Bioresource Technology*, *120*(September), 332–336. https://doi.org/10.1016/j.biortech.2012.06.044

Gil, G. C., Chang, I. S., Byung, S. C., Hong, J. K., A, K. M., Kyung, K. J., & Hyung, J. S. (2003). Operational parameters affecting the performance of a mediator-less microbial fuel cell. *Biosensors Bioelectron*, *18*(4), 327–334. https://doi.org/10.1016/S0956-5663(02)00110-0

Gregory, K. B., Bond, D. R., & Lovley, D. R. (2004). Graphite electrodes as electron donor for anaerobic respiration. *Environ Microbiol*, *6*(6), 596–604. https://doi.org/10.1111/j.1462-2920.2004.00593.x

Hamelers, H. V. M., Ter Heijne, A., Sleutels, T. H. J. A., Jeremiasse, A. W., Strik, D. P. B. T. B., & Buisman, C. J. N. (2010). New applications and performance of bioelectrochemical systems. *Applied Microbiology and Biotechnology*, *85*(6), 1673–1685. https://doi.org/10.1007/s00253-009-2357-1

Harnisch, F., & Schröder, U. (2010). From MFC to MXC: Chemical and biological cathodes and their potential for microbial bioelectrochemical systems. *Chemical Society Reviews*, *39*(11), 4433–4448. https://doi.org/10.1039/c003068f

Hassan, R. Y. A., Febbraio, F., & Andreescu, S. (2021). Microbial electrochemical systems: Principles, construction and biosensing applications. *Sensors (Switzerland)*, *21*(4), 1–19. https://doi.org/10.3390/s21041279

Hasvold, O., Henriksen, H., Melvaer, E., Citi, G., Johansen, B. O., Kjonigsen, T., & Galetti, R. (1997). Sea-Water Battery for Subsea Control System. *Power Sources*, *65*, 253–261.

He, W., Jin, W., Wang, Q., & Feng, Y. (2021). Electron flow assisted COD removal in wastewater under continuous flow conditions using microbial electrochemical system. *Science of the Total Environment*, *776*, 145978. https://doi.org/10.1016/j.scitotenv.2021.145978

He, Z., & Angenent, L. T. (2006). Application of bacterial biocathodes in microbial fuel cells. *Electroanalysis*, *18*(19–20), 2009–2015. https://doi.org/10.1002/elan.200603628

Huang, L., Regan, J. M., & Quan, X. (2011). Electron transfer mechanisms, new applications, and performance of biocathode microbial fuel cells. *Bioresource Technology*, *102*(1), 316–323. https://doi.org/10.1016/j.biortech.2010.06.096

Jafary, T., Daud, W. R. W., Ghasemi, M., Kim, B. H., Md Jahim, J., Ismail, M., & Lim, S. S. (2015). Biocathode in microbial electrolysis cell; Present status and future prospects. *Renewable and Sustainable Energy Reviews*, *47*, 23–33. https://doi.org/10.1016/j.rser.2015.03.003

Jiang, X., Hu, J., Fitzgerald, L. A., Biffinger, J. C., Xie, P., Ringeisen, B. R., & Lieber, C. M. (2010). Probing electron transfer mechanisms in Shewanella oneidensis MR-1 using a nanoelectrode platform and single-cell imaging. *Proceedings of the National Academy of Sciences of the United States of America*, *107*(39), 16806–16810. https://doi.org/10.1073/pnas.1011699107

Kim, B. H., Chang, I. S., & Gadd, G. M. (2007). Challenges in microbial fuel cell development and operation. *Applied Microbiology and Biotechnology*, *76*(3), 485–494. https://doi.org/10.1007/s00253-007-1027-4

Li, F., Sharma, Y., Lei, Y., Li, B., & Zhou, Q. (2010). Microbial fuel cells: The effects of configurations, electrolyte solutions, and electrode materials on power generation. *Applied Biochemistry and Biotechnology*, *160*(1), 168–181. https://doi.org/10.1007/s12010-008-8516-5

Logan, B. E. (2009). Exoelectrogenic bacteria that power microbial fuel cells. *Nature Reviews Microbiology*, *7*(5), 375–381. https://doi.org/10.1038/nrmicro2113

Meylani, V., Surahman, E., Fudholi, A., Almalki, W. H., Ilyas, N., & Sayyed, R. Z. (2023). Biodiversity in microbial fuel cells: Review of a promising technology for wastewater treatment. *Journal of Environmental Chemical Engineering*, *11*(2), 109503. https://doi.org/10.1016/j.jece.2023.109503

Nevin, K. P., Kim, B. C., Glaven, R. H., Johnson, J. P., Woodward, T. L., Methé, B. A., … Lovley, D. R. (2009). Anode biofilm transcriptomics reveals outer surface components essential for high density current production in Geobacter sulfurreducens fuel cells. *PLoS ONE*, *4*(5). https://doi.org/10.1371/journal.pone.0005628

Pandit, S., Ghosh, S., Ghangrekar, M. M., & Das, D. (2012). Performance of an anion exchange membrane in association with cathodic parameters in a dual chamber microbial fuel cell. *International Journal of Hydrogen Energy*, *37*(11), 9383–9392. https://doi.org/10.1016/j.ijhydene.2012.03.011

Pandit, S., Khilari, S., Roy, S., Pradhan, D., & Das, D. (2014). Improvement of power generation using Shewanella putrefaciens mediated bioanode in a single chambered microbial fuel cell: Effect of different anodic operating conditions. *Bioresource Technology*, *166*, 451–457. https://doi.org/10.1016/j.biortech.2014.05.075

Pandit, S., Khilari, S., Roy, S., Ghangrekar, M. M., Pradhan, D., & Das, D. (2015). Reduction of start-up time through bioaugmentation process in microbial fuel cells using an isolate from dark fermentative spent media fed anode. *Water Science and Technology*, *72*(1), 106–115.

Pant, D., Singh, A., Van Bogaert, G., Irving Olsen, S., Singh Nigam, P., Diels, L., & Vanbroekhoven, K. (2012). Bioelectrochemical systems (BES) for sustainable energy production and product recovery from organic wastes and industrial wastewaters. *RSC Advances*, *2*(4), 1248–1263. https://doi.org/10.1039/c1ra00839k

Pham, C. A., Jung, S. J., Phung, N. T., Lee, J., Chang, I. S., Kim, B. H., … Chun, J. (2003). A novel electrochemically active and Fe(III)-reducing bacterium phylogenetically related to Aeromonas hydrophila, isolated from a microbial fuel cell. *FEMS Microbiology Letters*, *223*(1), 129–134. https://doi.org/10.1016/S0378-1097(03)00354-9

Pham, T. H., Boon, N., De Maeyer, K., Höfte, M., Rabaey, K., & Verstraete, W. (2008). Use of Pseudomonas species producing phenazine-based metabolites in the anodes of microbial fuel cells to improve electricity generation. *Applied Microbiology and Biotechnology*, *80*(6), 985–993. https://doi.org/10.1007/s00253-008-1619-7

Pirbadian, S., Barchinger, S. E., Leung, K. M., Byun, H. S., Jangir, Y., Bouhenni, R. A., … El-Naggar, M. Y. (2014). Shewanella oneidensis MR-1 nanowires are outer membrane and periplasmic extensions of the extracellular electron transport components. *Proceedings of the National Academy of Sciences of the United States of America*, *111*(35), 12883–12888. https://doi.org/10.1073/pnas.1410551111

Rhoads, A., Beyenal, H., & Lewandowshi, Z. (2005). Microbial fuel cell using anaerobic respiration as an anodic reaction and biomineralized manganese as a cathodic reactant. *Environment ScienceTechnol Ogy*, *39*(12), 4666–4671. https://doi.org/10.1021/es048386r

Rimboud, M., Pocaznoi, D., Erable, B., & Bergel, A. (2014). Electroanalysis of microbial anodes for bioelectrochemical systems: Basics, progress and perspectives. *Physical Chemistry Chemical Physics*, *16*(31), 16349–16366. https://doi.org/10.1039/c4cp01698j

Rinaldi, A., Mecheri, B., Garavaglia, V., Licoccia, S., Di Nardo, P., & Traversa, E. (2008). Engineering materials and biology to boost performance of microbial fuel cells: A critical review. *Energy and Environmental Science*, *1*(4), 417–429. https://doi.org/10.1039/b806498a

Roy, S., & Pandit, S. (2019). Microbial electrochemical system: Principles and application. In S. V. Mohan, S. Varjani, & A. Pandey (Eds.), *Microbial Electrochemical Technology Sustainable*

Platform for Fuels, Chemicals and Remediation (1st ed., pp. 19–48). https://doi.org/10.1016/B978-0-444-64052-9.00002-9

Sammes, N. (2006). *Fuel Cell Technology: Reaching Towards Commercialization*. New York, USA: Springer Science & Business Media, Salmon Tower Building.

Scott, K., & Yu, E. H. (2015). *Microbial Electrochemical and Fuel Cells: Fundamentals and Applications*. Cambridge, USA: Woodhead Publishing.

Stams, A. J. M., De Bok, F. A. M., Plugge, C. M., Van Eekert, M. H. A., Dolfing, J., & Schraa, G. (2006). Exocellular electron transfer in anaerobic microbial communities. *Environmental Microbiology*, *8*(3), 371–382. https://doi.org/10.1111/j.1462-2920.2006.00989.x

Steidl, R. J., Lampa-Pastirk, S., & Reguera, G. (2016). Mechanistic stratification in electroactive biofilms of Geobacter sulfurreducens mediated by pilus nanowires. *Nature Communications*, *7*. https://doi.org/10.1038/ncomms12217

Sun, D., Wang, A., Cheng, S., Yates, M., & Logan, B. E. (2014). Geobacter anodireducens sp. nov., an exoelectrogenic microbe in bioelectrochemical systems. *International Journal of Systematic and Evolutionary Microbiology*, *64*, 3485–3491. https://doi.org/10.1099/ijs.0.061598-0

Varanasi, J. L., Roy, S., Pandit, S., & Das, D. (2015). Improvement of energy recovery from cellobiose by thermophillic dark fermentative hydrogen production followed by microbial fuel cell. *International Journal of Hydrogen Energy*, *40*(26), 8311–8321. https://doi.org/10.1016/j.ijhydene.2015.04.124

Wang, A., Liu, L., Sun, D., Ren, N., & Lee, D. J. (2010). Isolation of Fe(III)-reducing fermentative bacterium Bacteroides sp. W7 in the anode suspension of a microbial electrolysis cell (MEC). *International Journal of Hydrogen Energy*, *35*(7), 3178–3182. https://doi.org/10.1016/j.ijhydene.2009.12.154

Wang, H., & Ren, Z. J. (2013). A comprehensive review of microbial electrochemical systems as a platform technology. *Biotechnology Advances*, *31*(8), 1796–1807. https://doi.org/10.1016/j.biotechadv.2013.10.001

Yang, Y., Xu, M., Guo, J., & Sun, G. (2012). Bacterial extracellular electron transfer in bioelectrochemical systems. *Process Biochemistry*, *47*(12), 1707–1714. https://doi.org/10.1016/j.procbio.2012.07.032

Zhang, Y., & Angelidaki, I. (2016). Microbial Electrochemical Systems and Technologies: It is Time to Report the Capital Costs. *Environmental Science and Technology*, *50*(11), 5432–5433. https://doi.org/10.1021/acs.est.6b01601

Zuo, Y., Xing, D., Regan, J. M., & Logan, B. E. (2008). Isolation of the exoelectrogenic bacterium Ochrobactrum anthropi YZ-1 by using a U-tube microbial fuel cell. *Applied and Environmental Microbiology*, *74*(10), 3130–3137. https://doi.org/10.1128/AEM.02732-07

Application of Microbial Electrochemical Technologies as Biosensors for the Detection of Inorganic Water Pollutants

Javad B. M. Parambath, Ahmad A. L. Ahmad, and Ahmed A. Mohamed

Abstract The study of direct electron transfer (DET) via electroactive microbes extends beyond energy applications and delves into environmental bioremediation. Microbes with DET capabilities play a role in the natural attenuation of pollutants in environments such as contaminated soils, sediments, and contaminated water bodies. The direct interaction with solid surfaces enables these microbes to participate in reducing various contaminants, showcasing the potential of direct electron transfer in sustainable water pollution remediation strategies. Research efforts in microbial electrochemical cells (MECs) are geared toward improving their efficiency, scalability, and economic viability. Advancements in electrode materials, microbial catalysts, and system designs contribute to optimizing MEC performance. The exploration of diverse electroactive microbes and the development of durable materials are paramount in augmenting the capabilities of MECs for inorganic pollutant detection and remediation in water systems. This chapter comprehensively addresses the integration of advanced sensing technologies into MECs, enhancing their ability to detect and quantify specific inorganic pollutants with high sensitivity and accuracy. Additionally, it emphasizes the importance of understanding the complex interplay between electroactive microbes and pollutant substrates at the molecular level for devising tailored bioremediation strategies to effectively address diverse water quality challenges.

Keywords Biosensors · Microbial electrochemical cells · Water quality · Inorganic pollutants · Microbial electron transfer

---

J. B. M. Parambath · A. A. Mohamed (✉)
Center for Advanced Materials Research, Research Institute of Sciences and Engineering, University of Sharjah, 27272 Sharjah, UAE
e-mail: amohamed61@gmail.com

A. A. L. Ahmad
Department of Chemistry, University of Maine, Orono, ME 04469, USA

Advanced Structures and Composites Center, University of Maine, Orono, ME 04469, USA

R. Selvasembian et al. (eds.), *Emerging Trends in Microbial Electrochemical Technologies for Sustainable Mitigation of Water Resources Contamination*,
https://doi.org/10.1007/978-3-031-74636-9_4

Abbreviations

DET	Direct electron transfer
MET	Microbial electrochemical technology
MESs	Microbial electrochemical systems
MECs	Microbial electrochemical cells
MPCs	Microbial phototrophic cells
CV	Cyclic voltammetry
EIS	Electrochemical impedance spectroscopy
BOD	Biological oxygen demand
COD	Chemical oxygen demand
EET	Extracellular electron transfer
METC	Microbial electron transport chain
WHO	World Health Organization
MDCs	Microbial desalination cells
MFCs	Microbial fuel cells
DO	Dissolved oxygen
TDS	Total dissolved solids
OCP	Open circuit potential
OCV	Open circuit voltage
PCB	Polychlorinated biphenyls

1 Introduction

Microbial electrochemical technology (METs) mediated electron transfer biosensors are used to explore the relationship between the output signal (such as voltage or current) and variations in the levels of analytes (e.g., water quality parameters like BOD, COD, and toxicity) (Zhang et al. 2016; Ramírez-Vargas et al. 2018; Do et al. 2022). Achieving highly sensitive detection of these specific compounds stands out as a crucial objective for the biosensor. Within this framework, the sensitivity of MET-based biosensors is defined as the alteration in electrical signal per unit change in the concentration of the analyte. With these setups and electroanalytical techniques, MET biosensors can operate by measuring signals such asOCV, current, or power. These signals dynamically adjust in response to variations in concentrations of the target analyte (Yu et al. 2021; Shanbhag et al. 2023). Also, the advancements in nanotechnology have enabled the fabrication of nanostructured electrodes, enhancing the sensitivity and selectivity of MET biosensors (Ramya et al. 2022; Malik et al. 2023). Integrated data processing algorithms further refine the detection capabilities, enabling real-time monitoring of pollutant levels in water systems. The continual evolution of MET biosensors holds promise for their widespread deployment in environmental monitoring and remediation efforts.

In recent years, significant strides have been made in advancing materials for various components of Microbial electrochemical systems (MESs), including interspecies electron processes, bioelectrodes, engineered microbes, and substrate types, all directed towards enhancing interfacial ET and refining the microbial/electrode interface (Mahmoud et al. 2021). The progress underscores a commitment to overcoming challenges and improving the overall efficiency of METs. Further, emerging trends are in the miniaturization and portability of MET-based biosensors, paving the way for on-site and real-time water quality assessment, crucial for rapid response to pollution incidents (Redha et al. 2017; Liu et al. 2020; Ren and Li 2022). This chapter is focused on microbial electrochemical technologies and their roles as biosensors. Applications of METs for water quality monitoring and inorganic water quality monitoring are highlighted.

2 Microbial Electrochemical Technologies

Microbial electrochemical technology (MET) represents a rapidly advancing system with significant promise, seamlessly combining microbial systems, electrochemistry, and materials science to create energy, environmental, and sensing devices (Hernandez and Osma 2020; Li et al. 2022). This innovative approach capitalizes on the biocatalytic capabilities of living microbes, enabling the extraction of electrons from biodegradable organic substances. Consequently, MET delves into the intricate interplay between living microbial cells (serving as electron donors) and the electrode surface (acting as the electron acceptor) (Khater et al. 2015). A pivotal prerequisite in developing METs is effective integration, fostering electron transfer between microorganisms (capable of accepting or donating electrons to and from electrodes) and the electrode surface. This seamless integration is fundamental to ensuring the optimal performance of microbial electrochemical systems. Utilizing naturally existing electroactive bacteria as living catalysts presents power generation, water treatment, and microbial electrosynthesis opportunities. Technological growth has been restricted due to rate limitations and substrate specificity. The application of synthetic biology tools can enhance power generation, regulate, and customize microbial electrosynthesis, and pave the way for novel technologies like sensors and living electronics (Fig. 1) (Bird et al. 2021).

The METs employ two primary processes to harness their potential: firstly, extracellular electron transfer, which operates when the electrode potential aligns with the microorganisms' physiological range, relying predominantly on Faraday processes. Secondly, METs engage in indirect interactions, where electrochemical processes dictate the metabolite concentrations, pH levels, and oxygen pressure (Hassan and Bilitewski 2013). These processes' efficacy hinges on the proximity of the microbial system to the electrochemical system, necessitating integration into a reactor designed for optimal performance with minimal operation and maintenance costs. Critical factors influencing these processes include the electrode surface, microbial

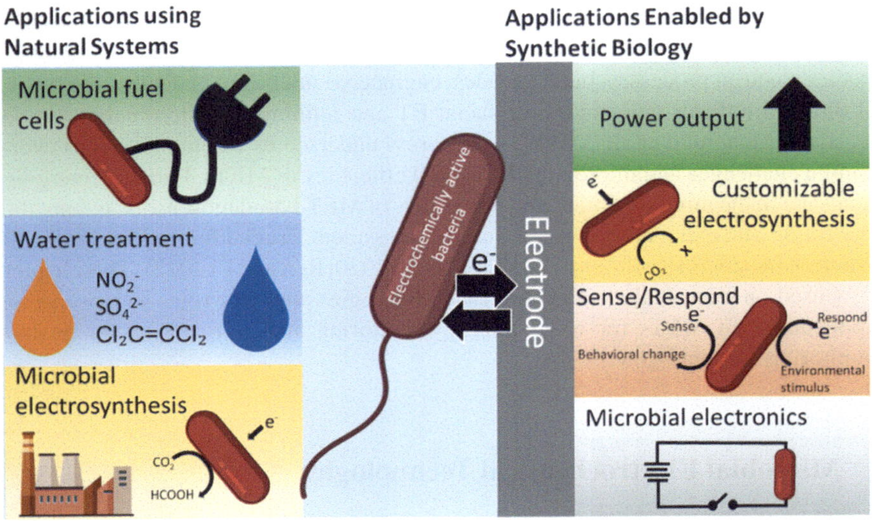

Fig. 1 Existing applications using natural electroactive bacteria and proposed applications enabled by synthetic biology [reprinted with permission from Bird et al. (2021)]

kinetics, and the specific electrogenic microorganisms employed in constructing the device.

The exploration of novel materials contributes to the resilience and longevity of bioelectrodes, ensuring sustained performance across diverse environmental conditions. Furthermore, the focus on engineered microbes seeks to enhance their capabilities for efficient electron transfer. In contrast, the investigation into different substrate types and interspecies electron processes expands the applicability of MESs to a broader array of organic substances.

2.1 *Principles of Microbial Electrochemical Systems*

The Microbial Electron Transport Chain (METC) holds paramount importance within living systems, serving as the primary compartment where degradable organic substrates' oxidation occurs, constituting the primary energy source for microbial cells (Wesolowski et al. 2008; Zhao et al. 2015). Consequently, gauging the efficiency of microbial respiration and the electron transport chain activity stands as crucial indicators of cellular activity, fundamental for the replication and proliferation of aerobic organisms (Kalnenieks et al. 2019; Santoro et al. 2017). Initial attempts focused on assessing microbial survival by measuring dissolved oxygen consumption by living cells (Holder et al. 2017).

This pivotal electron transfer process, wherein living microorganisms transfer electrons to electrodes in MET, has found utility in applications such as microbial fuel cells (Holder et al. 2017; Ng et al. 2017) and diagnostic tools for swiftly evaluating microbial activity (Conners et al. 2022). Numerous MET approaches tailored for biological objectives have been developed and tested (Hernandez and Osma 2020; Li et al. 2022). The electrical current produced by METs directly correlates with the quantity of viable microbial cells incorporated into these systems. This is growing the designing of high-performance METs, driven by their exciting potential in environmental, energy, and biomedical domains.

2.1.1 Extracellular Electron Transfer

Extracellular Electron Transfer (EET) is a pivotal biological process with far-reaching implications, particularly in microbial metabolism and energy generation. At its core, EET involves the transfer of electrons from the interior of microbial cells to external electron acceptors, such as solid surfaces or other microbial cells (Choi and Sang 2016). This phenomenon is of significant interest in developing MET, where microorganisms play a crucial role in catalyzing redox reactions and producing electrical currents. Understanding the mechanisms and regulation of EET is essential for optimizing the efficiency of these systems, which have applications ranging from sustainable energy production in microbial fuel cells to environmental remediation processes (Hernandez and Osma 2020).

The biochemical pathways facilitating EET vary among different microorganisms, and researchers have delved into unraveling the intricate details of these mechanisms. In many cases, bacteria employ specialized proteins and enzymes to mediate the transfer of electrons across their cell membranes. For instance, certain bacteria use outer membrane cytochromes or conductive pili to establish a direct connection between the cell and extracellular electron acceptors (Ishii et al. 2018). This ability of microorganisms to exchange electrons with their surroundings influences their metabolic processes and plays a crucial role in biogeochemical and nutrient cycling in various ecosystems.

Furthermore, EET has garnered attention for its potential in biotechnological applications (Aparici-Carratalá et al. 2023). MET, including microbial fuel cells and bioelectrochemical sensors, leverages the principles of EET for practical purposes. Researchers are exploring ways to enhance EET rates and efficiency, aiming to improve the performance and scalability of these technologies (Hernandez and Osma 2020). As our understanding of EET continues to deepen, it opens avenues for developing sustainable and innovative energy, environmental monitoring, and bioremediation solutions. The exploration of EET represents a frontier where biology meets technology, holding promise for diverse applications with implications for both environmental and industrial advancements.

2.1.2 Direct Electron Transfer via Electroactive Microbes

Direct Electron Transfer (DET) via electroactive microbes is a fascinating phenomenon with profound implications for MET. Electroactive microbes are capable of directly transferring electrons to solid surfaces, such as electrodes, without intermediary electron shuttles (Suzuki et al. 2023). The direct interface linking microbial cells with electrodes plays a vital role in numerous applications, such as MFCs, biosensors, and bioremediation processes (Chen et al. 2023; Falk et al. 2012; Schachinger et al. 2021).

One prominent example of DET occurs in certain strains of bacteria that produce electrically conductive pili, known as bacterial nanowires (Myers et al. 2023). These nanowires facilitate the direct transfer of electrons from the microbial cells to the electrode surface. The discovery of such conductive appendages has opened new avenues for harnessing electroactive microbes in bioenergy applications. Researchers are exploring ways to engineer and enhance the capabilities of these electroactive microbes to improve electron transfer rates, ultimately leading to more efficient energy conversion in microbial fuel cells.

The study of direct electron transfer via electroactive microbes extends beyond energy applications and delves into environmental bioremediation. Microbes with DET capabilities play a role in the natural attenuation of pollutants in environments such as contaminated soils and sediments. The direct interaction with solid surfaces enables these microbes to participate in reducing various contaminants, showcasing the potential of DET in sustainable environmental remediation strategies (Fenner et al. 2021). Moreover, the exploration of DET mechanisms has implications for the development of biosensors. Electroactive microbes that can transfer electrons directly to electrodes are living sensors capable of detecting specific compounds or environmental conditions. Integrating microbial physiology with electronic sensing offers a unique and sensitive approach for monitoring environmental parameters or detecting pollutants in real-time.

As research in this field progresses, manipulating electroactive microbes for enhanced DET is becoming a focus of biotechnological innovation. The genetic engineering of microbial strains to optimize their electron transfer capabilities holds promise for advancing the practical applications of DET in diverse areas. The intersection of microbiology and electrochemistry in understanding and harnessing direct electron transfer via electroactive microbes represents aburgeoning frontier with significant potential for sustainable technologies and environmental solutions.

2.1.3 Mediated Electron Transfer

Mediated electron transfer is a crucial step in microbial electrochemical systems, playing a pivotal role in facilitating the transfer of electrons from microorganisms to solid surfaces or other electron acceptors. In this, electron shuttles or mediators act as intermediaries between the microbial cells and the external electron acceptors, enabling the transport of electrons across the cell membrane (Hernandez and

Osma 2020; Schröder et al. 2015). This process is particularly significant in scenarios where DET is not feasible due to physical barriers or the absence of suitable pathways for electron exchange. Understanding the mechanisms of this transfer has broad applications, ranging from microbial fuel cells to biosensors and environmental remediation.

Various types of mediators are employed, including soluble redox-active molecules and conductive materials. These mediators can effectively shuttle electrons between the microbial cells and the external electron acceptors, allowing for versatile applications in different microbial systems. This field of electron transfer study has led to the development of mediator-based microbial fuel cells, which enhance electron transfer efficiency, expanding the range of microbes that can be utilized in such systems.

The versatility of this mediated electron transfer is evident in its application to biosensors, where the interaction between microorganisms and mediators enables the detection of specific analytes (Wang et al. 2023). The redox properties of the mediators play a crucial role in transducing the biological response into an electrical signal, making it an essential component in developing sensitive and selective biosensing devices. This intersection of biology and electrochemistry holds promise for innovative diagnostic tools with applications in various fields. In environmental bioremediation, mediated electron transfer has proven to be an effective strategy for enhancing the degradation of pollutants. By introducing electron mediators, microbial communities can overcome challenges posed by the redox state of specific contaminants, promoting their transformation and removal (Wang et al. 2020).

2.2 *Types of Microbial Electrochemical Cells*

Microbial Electrochemical Cells (MECs) encompass various types, each designed for specific applications and operating based on distinct principles. The primary types of microbial electrochemical cells include:

2.2.1 Microbial Fuel Cells (MFCs)

Microbial Fuel Cells (MFCs) stand as a remarkable intersection of microbiology and electrochemistry, revolutionizing the landscape of renewable energy technologies. At their core, MFCs capitalize on the metabolic activities of electroactive microorganisms, commonly bacteria, to generate electrical power through a direct conversion process. Microorganisms oxidize organic compounds in the active microbial fuel cell, producing electrons as byproducts. These electrons are then directed through an external circuit, creating an electric current. This unique ability of MFCs to directly convert chemical energy in organic matter into electricity distinguishes them as a promising alternative for sustainable energy production (Vishwanathan 2021).

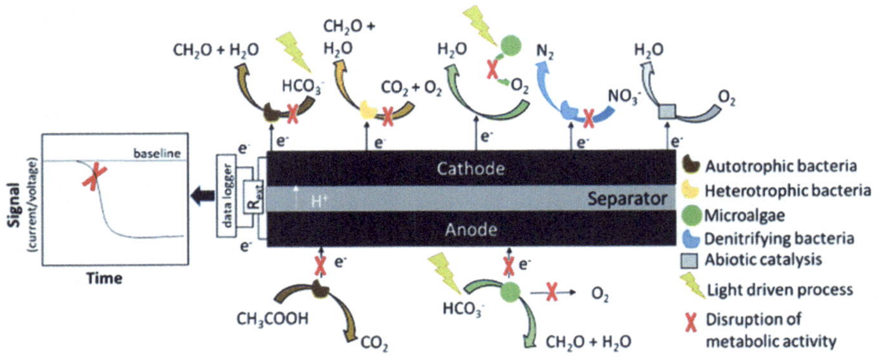

Fig. 2 The functionality of an MFC sensor involves detecting disturbances like bioactive compounds and registering corresponding changes in electrical response [reprinted with permission from Olias and Di Lorenzo (2021)]

One of the critical advantages of MFCs lies in their potential applications in wastewater treatment (Roy et al. 2023). Beyond electricity generation, MFCs play a dual role by aiding the breakdown of organic pollutants in wastewater. As microorganisms engage in the electrochemical process, they simultaneously contribute to the degradation of contaminants, offering a synergistic solution for environmental remediation. Integrating MFCs into wastewater treatment facilities showcases their versatility, providing an innovative approach that produces clean energy and addresses water pollution challenges. Figure 2 illustrates the operational mechanisms of a sensor based on MFCs. The metabolic activity of microorganisms, whether situated at the anode or cathode, can be impacted by a bioactive substance (a toxin) or alterations in operational variables like pH, conductivity, temperature, and organic content (Olias and Di Lorenzo 2021).

The ongoing evolution of MFC technology involves comprehensive research efforts aimed at enhancing efficiency and scalability. Researchers explore a variety of electroactive microbes, optimize electrode materials, and refine system designs to maximize performance. Advances in materials science, bioengineering, and microbial ecology continuously improve MFCs, bringing them closer to practical implementation in various settings. Moreover, researchers investigate potential synergies by integrating MFCs with other renewable energy sources, such as solar or wind, to create hybrid systems that offer more consistent and reliable energy outputs (Vishwanathan 2021).

As MFC technology progresses, its potential applications extend beyond traditional power generation. MFCs have demonstrated their efficacy in powering small electronic devices, remote sensors, and even low-energy demanding applications in remote locations (Ivars-Barceló et al. 2018). Developing portable and scalable MFC systems opens possibilities for decentralized energy production, providing solutions for off-grid communities or remote areas with limited access to conventional power sources. The multifaceted nature of MFC research, spanning microbiology, electrochemistry, and engineering, positions this technology as a frontrunner in pursuing sustainable and environmentally friendly energy solutions. The continued exploration of microbial fuel cells holds promise for addressing global energy challenges and fostering a more sustainable future.

2.2.2 Microbial Electrolysis Cells

Microbial Electrolysis Cells (MECs) represent an innovative approach at the intersection of microbiology and electrochemistry, focusing on the sustainable production of hydrogen gas. The core principle of MECs involves leveraging the metabolic activities of electroactive microorganisms, typically bacteria, to drive water electrolysis and generate hydrogen gas (Logan et al. 2008). In the active microbial electrolysis process, microorganisms mediate the transfer of electrons to the anode, and these electrons are then directed through an external circuit to the cathode, where they react with protons to produce hydrogen gas. MECs hold immense potential as a clean energy solution, addressing the growing demand for hydrogen as a versatile and eco-friendly energy carrier.

One important advantage of MECs is their capability to function under normal atmospheric conditions, including room temperature and pressure, eliminating energy-intensive conditions required in traditional water electrolysis. It makes MECs an attractive option for sustainable and cost-effective hydrogen production. Additionally, MECs can utilize a wide range of organic substrates as feedstocks, making them versatile and adaptable to different environments and waste streams. The integration of MECs with wastewater treatment processes is particularly promising, as it not only produces hydrogen but also aids in the removal of organic pollutants, showcasing the technology's potential for dual environmental and energy benefits.

Research efforts in MECs are geared towards improving their efficiency, scalability, and economic viability. Advancements in electrode materials, microbial catalysts, and system designs contribute to optimizing MEC performance. Exploring diverse electroactive microbes and developing robust, long-lasting materials are critical aspects of enhancing MEC capabilities. Moreover, researchers are investigating the integration of MECs into larger energy systems, such as coupling them with renewable energy sources like solar or wind power. This approach aims to create hybrid systems that offer continuous and reliable hydrogen production, contributing to advancing sustainable energy solutions (Ali et al. 2018).

As MEC technology continues to evolve, its applications extend beyond hydrogen production. MECs have shown promise in bioremediation, utilizing microbial processes to facilitate the elimination of pollutants from water and soil. The potential for MECs to play a role in waste-to-energy conversion is also being explored, demonstrating this technology's versatility in addressing environmental and energy challenges. The ongoing synergy between microbiology and electrochemistry in MEC research positions this technology as a pivotal player in the quest for clean and sustainable energy solutions, with far-reaching implications for a greener and more energy-efficient future.

2.2.3 Microbial Desalination Cells

Microbial Desalination Cells (MDCs) represent an innovative fusion of microbiology and desalination technology, offering a sustainable approach to addressing water scarcity. In the active desalination process, MDCs employ electroactive microorganisms to facilitate the removal of salt from saline water. These microorganisms, often bacteria, actively participate in electrochemical reactions that drive desalination (Odunlami et al. 2023). As the bacteria consume organic matter in the wastewater, they release electrons, creating an electric current that powers ion exchange membranes. This process desalinates water and harnesses energy from microbial metabolism, making MDC an environmentally friendly solution for water treatment and energy generation.

One distinctive feature of MDCs is their capacity to operate with lower energy input than traditional desalination methods. By leveraging the metabolic activities of microorganisms, MDCs reduce the reliance on external energy sources, making them a more sustainable option for communities facing water scarcity. Additionally, MDCs can utilize various organic substrates present in wastewater as a carbon source for microbial metabolism, offering flexibility in their application and contributing to the system's overall efficiency.

Ongoing research in MDCs focuses on optimizing their performance, enhancing desalination efficiency, and expanding their scalability. Advances in microbial electrochemistry, electrode materials, and system designs contribute to the ongoing refinement of MDC technology. The integration of MDCs with other water treatment processes and renewable energy sources holds promise for addressing the global challenge of freshwater scarcity holistically and sustainably (Odunlami et al. 2023).

2.2.4 Microbial Sensor Cells

Microbial Sensor Cells represent a dynamic fusion of microbiology and sensor technology, providing a novel approach to real-time environmental monitoring. These sensor cells harness the unique interactions between microorganisms and electrodes to detect specific analytes or environmental conditions. In active microbial sensing, the redox activity of electroactive microbes generates measurable electrical signals that directly correlate with the presence or concentration of target substances. This intrinsic link between microbial metabolic processes and electronic output makes microbial sensor cells invaluable in various applications, from detecting pollutants in water to monitoring air quality or assessing soil health (Abbasian et al. 2018).

The adaptability of microbial sensor cells is a key feature, allowing their integration into diverse environmental monitoring systems. Researchers focus on optimizing the selectivity and sensitivity of these sensors, tailoring them to detect specific compounds with high accuracy. The design of microbial sensors is often influenced by the unique redox capabilities of microorganisms, including bacteria or algae, which serve as biological indicators. As a result, microbial sensor cells offer a promising avenue for developing rapid and cost-effective biosensing tools, with applications ranging from industrial pollution monitoring to ensuring the safety of water resources.

The development of microbial sensor cells is advancing rapidly, driven by innovations in materials science, synthetic biology, and sensor engineering. Researchers are exploring ways to enhance the stability and longevity of microbial sensors, enabling continuous and reliable monitoring in various environmental conditions. As these technologies progress, microbial sensor cells are poised to play a pivotal role in the transition towards innovative and sustainable environmental monitoring solutions, providing valuable insights for ecological conservation and public health.

2.2.5 Microbial Phototrophic Cells

Microbial Phototrophic Cells (MPCs) represent a fascinating intersection of microbiology and renewable energy technology, harnessing the photosynthetic capabilities of microorganisms for electricity generation (Tanvir et al. 2021). In the active process of microbial phototrophs, photosynthetic microorganisms, such as algae or cyanobacteria, capture light energy and convert it into chemical energy through photosynthesis. In MPCs, this chemical energy is then harnessed to produce electrical power through electrochemical reactions. Using photosynthetic microorganisms distinguishes MPCs as a green and sustainable technology, offering a unique approach to solar energy conversion.

A critical advantage of MPCs lies in their ability to convert sunlight into electricity directly, bypassing the need for external energy inputs, which makes them an attractive option for sustainable power generation, especially in environments with abundant sunlight. The versatility of MPCs extends beyond electricity production; these cells have also shown promise in producing valuable compounds, such as biofuels or biochemicals, through the photosynthetic conversion of carbon dioxide. The integration of MPCs into biotechnological processes highlights their potential role in contributing to both renewable energy and the economy.

Research in MPCs is evolving to enhance efficiency, stability, and scalability. Scientists are exploring various photosynthetic microorganisms, optimizing growth conditions, and developing novel materials to improve electron transfer processes. The potential of MPCs in diverse applications, from decentralized energy production to sustainable biomanufacturing, underscores their importance in the broader context of green technologies. As advancements continue, microbial phototrophic cells are a promising avenue for clean energy solutions and sustainable bioproduction (Knežić et al. 2022; Stephens et al. 2021).

2.3 Mechanisms Involved in MET for Biosensing

Microbial electrochemical technology involves microbial electron transfer in the context of biosensing and represents a cutting-edge field, showcasing the ingenious interplay between microorganisms and electrodes for detecting specific analytes (Hassan et al. 2021). The mechanisms involved in MET for biosensing are diverse and intricate, illustrating the potential for innovative applications in environmental monitoring, healthcare diagnostics, and other fields (Naresh and Lee 2021).

At the heart of MET-based biosensing is DET. In DET, electroactive microorganisms directly exchange electrons with electrodes, creating measurable electrical signals that correlate with specific biological activities or the presence of target substances. This mechanism relies on the inherent redox properties of certain microorganisms, allowing them to transfer electrons from intracellular metabolic processes directly to the electrode surface. The sensitivity and specificity of DET-based biosensors make them valuable tools for real-time monitoring of microbial activities or detecting specific biomolecules (Naresh and Lee 2021).

In addition to DET, MET-based biosensors often utilize mediated electron transfer mechanisms. In this, electron shuttles or mediators act as intermediaries between the microbial cells and the electrode, facilitating the transfer of electrons. These mediators play a crucial role in scenarios where direct electron transfer is challenging due to physical barriers or the absence of suitable pathways (Hassan et al. 2021). The choice of mediator is essential, as it influences the efficiency and selectivity of electron transfer processes. The diverse array of mediators, including redox-active molecules or conductive polymers, offers flexibility in designing biosensors for various applications.

The application of MET in biosensing extends to MFCs and MECs. In MFCs, microorganisms oxidize organic compounds, and the resulting electrons are transferred to an electrode, generating electrical power. This inherent connection between microbial metabolism and electron transfer forms the basis for biosensing applications. MECs, on the other hand, focus on hydrogen production through microbial-mediated water electrolysis, with the potential for biosensing capabilities. They understood and manipulated the mechanisms involved in MET in these bioelectrochemical systems, opening avenues for multifunctional devices that simultaneously produce energy and provide real-time biosensing.

The essential process in MET-based biosensing involves the physical transfer of electrons, liberated or generated biologically, to the electrode surface. This transfer occurs through the separation of the electron transport chain by cellular membranes from the external surroundings, such as cytoplasmic or outer membranes, which poses challenges in efficiently transferring electrons to the electrode surface. Exploring microbial species with natural electron transfer capabilities and developing tailored electrode materials are integral to overcoming these challenges (Hassan et al. 2021). Enhancing the efficiency of electron transfer processes is crucial for optimizing the sensitivity and responsiveness of MET-based biosensors.

Table 1 Overview of METs

METs	Analysis
Overview and principles	MET combines microbial systems, electrochemistry, and materials science to develop energy, environmental, and sensing devices
	It depends on microorganisms' biocatalytic capabilities, extracting electrons from organic substances and transferring them to electrodes
Types of microbial electrochemical cells	MFCs, MECs, MDCs, microbial sensor cells, and MPCs are the primary types of MECs, each serving specific applications in energy production, desalination, sensing, and more
Mechanisms involved in MET for Biosensing	MET for biosensing relies on DET and mediated electron transfer mechanisms
	DET involves direct electron exchange between microorganisms and electrodes, while mediated electron transfer utilizes intermediaries for electron transport

Advances in MET for biosensing have broad implications for environmental monitoring. These biosensors can be tailored to detect specific contaminants, pollutants, or microbial activities indicative of environmental health. Additionally, MET-based biosensors' real-time and in-situ capabilities make them valuable tools for rapid assessment and early warning systems in various environments. The continuous refinement of MET mechanisms and the integrationof these biosensors into monitoring networks contribute to the evolution of environmentally sustainable technologies with wide-ranging applications (Chang et al. 2017) (Table 1).

3 Role of MET as Biosensors

Microbial electrochemical technologies (METs) function as biosensors, requiring effective integration and promotion of electron transfer between microorganisms and the electrode surface. This integration is vital for optimal MET performance, as microorganisms need to transfer electrons to and from electrodes. Essential requirements include achieving proximity between microbial and electrochemical systems and their incorporation into a cost-effective reactor. Success relies on factors like electrode characteristics, microbial kinetics, chamber setups, and the use of appropriate microorganisms in device fabrication (Cui et al. 2019; Do et al. 2020). For example, bacteria possess the ability to detect rapid changes in the concentration of target analytes or their presence. This recognition process triggers a signal that is subsequently recorded as the electric current output in the device (Mustakeem

2015). Recent progress focuses on enhancing the exchange of electrons at interfaces and the interaction between microbes and electrodes. This is achieved through the utilization of innovative materials for bioelectrodes, engineered microorganisms, various substrate varieties, and inter-species electron transfer mechanisms. Techniques such as cyclic voltammetry (CV), electrochemical impedance spectroscopy (EIS), and amperometric measurements contribute to constructing and evaluating METs, providing insights into electron transfer mechanisms and enhancing the understanding of microbe–electrode interactions (Hernández-Flores et al. 2015; Wang et al. 2015; Do et al. 2018; Corbella et al. 2019).

METs are generally explored for electricity generation and wastewater treatment, however, now there is a growing interest in chemical conversions and producing valuable products like hydrogen and organic compounds (Bullen et al. 2006). Meantime, electrochemical biosensors are widely recognized for their high specificity, sensitivity, and user-friendly nature, making them widespread in the online detection of various analytes in complex matrices including wastewater and body fluids (ElMekawy et al. 2018). Here, METs have also demonstrated the versatility to sense such complex analytes in various matrices. This involves a synergistic approach, with electroactive microorganisms aiding in the detection of analytes and electric current output monitoring technology. Additionally, these electroactive microorganisms and biofilms formed actively take part in redox reactions and help in water treatments. Researchers are increasingly exploring METs as biosensors, utilizing their potential (Zhou et al. 2017; Simoska et al. 2021). The cost-effectiveness, scalability, and portability of electrochemical sensors, especially with miniaturized micro/nano-electrodes, make them attractive for high-throughput analysis, as shown in Fig. 3. Within the anode chamber of METs, bacteria with electrochemical capabilities oxidize organic substances. Subsequently, they pass the resulting electrons to the anode, from where they navigate through an external circuit to reach the cathode, generating an electrical current. If a toxic substance from the feed water enters the system, it has the potential to hinder the metabolic activity of these bacteria, leading to a reduction in electron transfer and electrical signal. This implies that the current could serve as a signal to monitor the presence and magnitude of a toxic incident (Stein et al. 2012). Consequently, METs could potentially function as online biosensors for detecting toxic compounds in water. There are a few critical considerations to take into account for constructing feasible MET-based biosensors for water quality checks, including the selection of electrode materials, surface modifications, and the choice of electrochemical methods (potentiometric, voltammetric, amperometric, or impedimetric) to surpass other affordable analytical methods (Do et al. 2018).

3.1 Advantages of Using MET for Biosensing

METs offer distinct advantages when applied to biosensing. Referencing critical studies, METs showcase notable benefits, this involves their capacity for successful integration and enabling electron transfer between microorganisms and

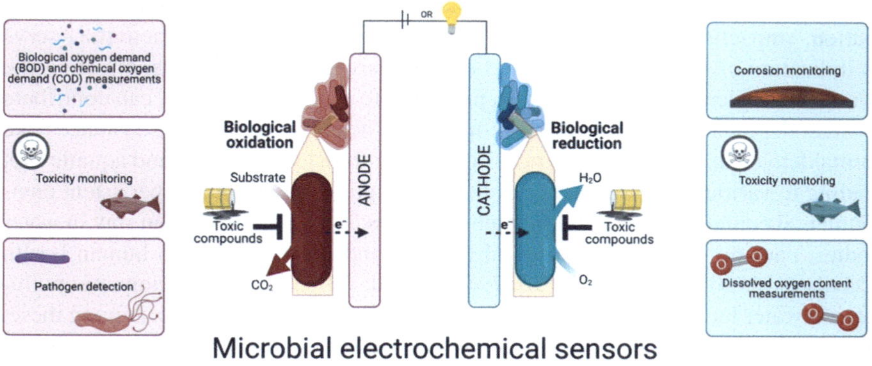

Fig. 3 Schematic representation showing MET-based biosensor designed for monitoring water quality through redox reactions [reprinted with permission from Simoska et al. (2021)]

electrode surfaces. Recent progress in innovating materials for bioelectrodes, modifying microbes, and employing diverse substrate varietieshave enhanced METs' performance in biosensing applications. Microbial Fuel Cells (MFCs) are one such bioelectrochemical devices, where electroactive microorganisms catalytically oxidize organic matter to generate electric current. MFC-based biosensors, owing to their efficiency, ease of application, and economic viability, have found extensive use for monitoring Biological Oxygen Demand (BOD), surpassing conventional methods in adaptability (Bullen et al. 2006; Corbella et al. 2019; Sonawane et al. 2020; Simoska et al. 2021). The need to detect harmful substances in water has propelled the advancement of MFC-based biosensors, showing great potential in analyzing diverse water parameters such as volatile fatty acids (VFA) (Jin et al. 2017), chemical oxygen demand (COD) (Corbella et al. 2019), dissolved oxygen (DO) (Sonawane et al. 2020), biological oxygen demand (BOD) (Zhou et al. 2017; Sonawane et al. 2020) and various toxic components. This contributes significantly to real-time field monitoring, safeguarding public health, and mitigating potential ecological harm resulting from on-site contamination incidents.

3.2 *Common Inorganic Pollutants in Water and Their Hazards*

Common inorganic pollutants in water pose substantial threats to aquatic ecosystems and public health. Monitoring key parameters, including biochemical oxygen demand (BOD), chemical oxygen demand (COD), microbial activities, acetate levels, the presence of toxic substances such as formaldehyde, and the occurrence of heavy metals, is crucial in assessing water quality (ElMekawy et al. 2018). Elevated BOD and COD levels indicate an increased demand for oxygen and potential contamination, impacting aquatic organisms and ecosystems. Microbial activities serve as indicators of biological health and can be disrupted by pollutants, affecting the overall ecological balance. Acetate, a precursor to organic pollutants, can contribute to water degradation if not properly monitored. The presence of toxic substances like formaldehyde and heavy metals poses direct health risks to humans and aquatic life, leading to various illnesses and ecosystem disruption. Additionally, persistent environmental contaminants such as naphthenic acids, if not addressed, can stay in water bodies, causing long-term ecological damage and serious threat to human health (Dávila et al. 2011; Stein et al. 2012; Shen et al. 2013). Common inorganic pollutants in water include arsenic, fluoride, iron, nitrate, and heavy metals. Among these, arsenic, fluoride, and iron have a geological origin, whereas nitrates, phosphates, and heavy metals arise from anthropogenic activities like deficient sewage systems, ineffectively handled agricultural practices, and industrial discharges (Srivastav and Ranjan 2020). These pollutants pose a severe threat, being potentially carcinogenic and detrimental to the health of various living organisms. When their concentration surpasses the levels recommended by the World Health Organization (WHO), it leads to toxic effects on both soil and aquatic systems, circulate through ecosystems via food chains, impacting all living systems (Zamora-Ledezma et al. 2021). Monitoring these parameters using MET-based sensors provides a valuable tool for timely detection and mitigation of these hazards, ensuring the sustainability of water resources and safeguarding environmental and public well-being.

4 Applications of MET in Inorganic Water Pollutant Detection

In the realm of water quality surveillance utilizing biosensors, specific key factors impact the energy production of METs. The fundamental objective of METs-based biosensors is to investigate the correlation between the output signal (like voltage or current) and fluctuations in the concentrations of target substances (for instance, BOD, COD, toxicity). Achieving highly sensitive detection of these specific compounds stands out as a crucial objective for the biosensor. Within this framework, the sensitivity of MET-based biosensors is defined as the alteration in electrical signal per unit change in the concentration of the analyte. With these setups and electroanalytical techniques, MET biosensors can operate by measuring signals in the form of OCV, current, or power. These signals dynamically adjust in response to variations in concentrations of the target analyte.

4.1 Evaluation of Water Quality Indicators

Several significant indicators are employed for evaluating water quality, each providing valuable insights. Dissolved Oxygen (DO) measures the amount of oxygen crucial for aquatic life, while Biochemical Oxygen Demand (BOD) assesses organic pollutant levels. Chemical Oxygen Demand (COD) is another parameter indicating the amount of oxygen needed for chemical oxidation. Total Dissolved Solids (TDS) measure the concentration of dissolved inorganic and organic substances. Amongst, one of the most widely employed and significant indicators for evaluating water quality is biochemical oxygen demand (BOD). Although the traditional BOD test is a universal approach without the need for costly equipment, its drawback lies in a time frame of 5–7 days and reliance on skilled personnel, rendering it unsuitable for online monitoring. An alternative method to overcome these limitations is the utilization of METs for real-time and online BOD monitoring, offering a faster assessment through electrical signals (Pasternak et al. 2017; Do et al. 2020; Sonawane et al. 2020). However, beyond the assessment of water quality parameters with METs, recent studies indicate that cathodic biofilms exhibit higher sensitivity compared to anodic biofilms in response to various inorganic toxicants, including heavy metals.

4.2 Factors Affecting MET Sensitivity as a Toxicity Sensor

Kim et al. successfully used a microbial fuel cell (MFC) to identify Pb(II) and Hg(II) at a concentration of 1 ppm in water (Kim et al. 2007). However, Patil et al. demonstrated that in the presence of Pb(II) and Hg(II) at concentrations ranging from 0.41 to 12.48 and 0.83 to 8.33 ppm, respectively (Patil et al. 2010). The findings from these two investigations indicate that the effectiveness of the electrically conductive biofilm, and thus the sensitivity of METs as a toxicity sensor, can be affected by several factors. Among these, the configuration of the biofilm (comprising its density, permeability, and composition) could be a significant determinant. As the graph shown in Fig. 4A, the impact of diverse toxic compounds on MET-based sensor responses opens avenues for online and early detection of toxic events in water quality control.

4.3 Enhancement of MET Sensitivity for Heavy Metal Detection

Reports have shown that introducing various pollutants to water, includingorganophosphate insecticides, Pb, Hg, and polychlorinated biphenyls (PCBs), led to a substantial reduction in the electrical signal on MET cells. The inhibitory effect

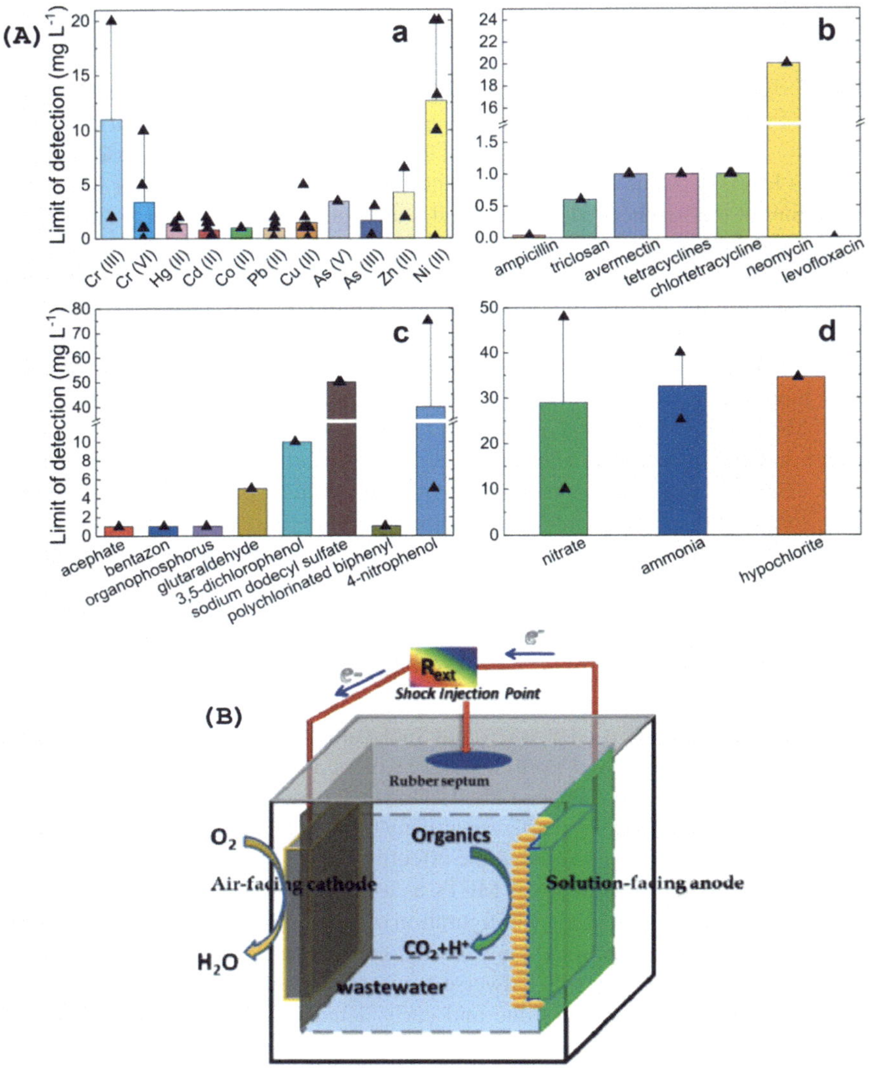

Fig.4 **A** Detection limit compiled in various studies as summarized in Ref. Chu et al. (2021) employing METs to identify **a** metal ions, **b** antibiotics, **c** other organic toxicants, and **d** other inorganic toxicants [reprinted with permission from Simoska et al. (2021)], **B** illustration showing the cube fuel cell functioning as the chemical shock MET biosensor [reprinted with permission from Liu et al. (2014)]

of these pollutants, at a concentration of 1 mg L^{-1}, resulted in significant inhibition relative to the control group. Upon the introduction of actual wastewater, even more substantial declines in current and increased inhibition ratios were observed. In the presence of mixed heavy metals, the inhibition ratio increased to 76% (Kim et al. 2007). Regulating cell over potential involves managing the anode potential, where the concentration and type of substrate, bicarbonate concentration, pH, and temperature play pivotal roles. Monitoring the levels of carbonates and bicarbonates can be effectively achieved using METs (Stein et al. 2010). Yifeng and Irini used adenosine triphosphate (ATP) concentration as an indicator for total living microorganisms. Their study showed a linear relationship between current density and active microorganism concentration (0–6.52 nmol ATP/L), suggesting the potential for detecting phosphate-based pollutants in water (Zhang and Angelidaki 2011). Shen et al. reported a MET-based sensor for quick monitoring of the Cu(II) toxicity in the wastewater. Their study suggests that to enhance the sensitivity of fuel cells as a toxicity sensor for heavy metals, METs should be operated under a low flow rate and intermittent nitrogen sparging (Shen et al. 2013). Stein et al. also explored the potential of MET biosensors to identify the presence of toxic water pollutants, focusing specifically on their sensitivity to nickel. The biosensor exhibited a prompt response with a sensitivity of 0.0027 A/m^2/mg Ni/L at an anode potential of – 0.4 V. Thus a current density demonstrated a correlation with decreasing nickel concentration (Stein et al. 2012). Additionally, according to their findings, the sensitivity of the nickel sensor remains unaffected by the choice of membrane type. Liu et al. devised a cubic shock MET biosensor for evaluating voltage changes and inhibition of output current on Fe^{3+}, NO^{3-}, Na-acetate, and Cr^{6+} chemical shocks applied to the cell, as shown in Fig. 4B. This MET setup has demonstrated the ability to distinguish toxin shocks from non-toxins through voltage signal changes, with anode open circuit potential (OCP) values closely correlating with MET voltage alterations, emphasizing the role of electrogenic bacteria activity on the anode surfaces (Liu et al. 2014). Lorenzo et al. reported when the fuel cell is exposed to higher acetate concentrations, it could detect the presence of cadmium in water from output current changes. Exhibiting a high sensitivity of 0.2 $\mu g\ L^{-1}\ cm^{-2}$ and an LOD of 1 $\mu g\ L^{-1}$, with a dynamic range of 1–25 $\mu g\ L^{-1}$ (Di Lorenzo et al. 2014). Notably, the anode biofilm displayed transient effects from cadmium, fully recovering within 12 min upon reintroduction to pollutant-free wastewater (Table 2).

5 Conclusion and Future Work

Microbial Electrochemical Technologies (MET) represent a cutting-edge and multidisciplinary field at the intersection of microbiology, electrochemistry, and environmental engineering. The principles underlying MET leverage the remarkable ability of certain microorganisms to exchange electrons with solid surfaces, enabling a diverse array of applications. The development of Microbial Fuel Cells (MFCs) is a standout achievement in MET, where electroactive microbes serve as living

Table 2 Overview of MET-based biosensor

Parameters measured	Type of MET	Type of water	Measuring range	References
BOD	Mediator-less MFC	Artificial wastewater	20–200 mg/L	Chang et al. (2017)
BOD	Submersible MFC	Domestic wastewater	17–183 mg/L	Peixoto et al. (2011)
COD	MFC	Wastewater online monitoring	57.7–149.7 mg/L	Grzegorz Pasternak et al. (2017)
COD	3D printed MFC	Wastewater online monitoring	3–164 mg/L	
DO	Submersible MFC	Domestic wastewater	0–8.8 mg/L	Zhang and Angelidaki (2011) and Vishwanathan et al. (2013)
Volatile fatty acid	H-type MFC		0–40 mg/L	Kaur et al. (2013)
Diazinon	Double chamber MFC	Activated sludge	1–10 mg/L	Kim et al. (2007)
Sodium dodecyl sulphate	Double chamber MFC		50 mg/L	Stein et al. (2012) Zeng et al. (2017)
Levofloxacin	Single chamber MFC		0.1–1000 μg/L	
Ni^{2+}	Double chamber MFC		22.7–187.6 mg/L	Stein et al. (2012)
Pb^{2+}	MFC	Wastewater	1–5 mg/L	Kim et al. (2007)
Cu^{2+}	Double chamber MFC	Wastewater	5–7 ppm	Shen et al. (2013)
Hg^{2+}	MFC	Wastewater	1–5 mg/L	Kim et al. (2007)
Cr^{6+}	Cubic microbial fuel cell	Wastewater influent	1–8 mg/L	Liu et al. (2014)
Cd^{2+}	Single chamber MFC	0.1–100 μg/L	0.1–100 μg/L	Lorenzo et al. (2014)

catalysts for the oxidation of organic compounds and the generation of electrical power. This breakthrough technology holds promise for sustainable energy production, offering an environmentally friendly alternative to traditional energy sources. As MET research progresses, there is growing interest in harnessing the metabolic activities of electroactive microbes for applications beyond energy, including bioelectrochemical sensors for environmental monitoring and the bioremediation of contaminated ecosystems. The field's ongoing exploration and innovation in MET continue to unfold new possibilities for addressing global energy sustainability and environmental health challenges.

The commercialization and scaling up of MET applications remain vital challenges. Bridging the gap between laboratory-scale demonstrations and practical, large-scale implementations is crucial for realizing the full potential of MET in real-world scenarios. Additionally, ongoing research in understanding and manipulating microbial communities for enhanced electron transfer efficiency is paving the way for more robust and efficient MET systems. The convergence of technological advancements, improved understanding of microbial physiology, and increased awareness of environmental sustainability positions MET as a pivotal player in the quest for cleaner energy and innovative environmental solutions. The interdisciplinary nature of MET fosters collaboration among scientists, engineers, and environmentalists, fostering a dynamic and evolving landscape that holds great promise for addressing the complex challenges of this century.

References

Abbasian F, Ghafar-Zadeh E, Magierowski S (2018) Microbiological sensing technologies: a review. Bioengineering 5:20. https://doi.org/10.3390/bioengineering5010020

Ali J, Sohail A, Wang L, Rizwan Haider M, Mulk S, Pan G (2018) Electro-microbiology as a promising approach towards renewable energy and environmental sustainability. Energies 11:1822. https://doi.org/10.3390/en11071822

Aparici-Carratalá D, Esclapez J, Bautista V, Bonete M-J, Camacho M (2023) Archaea: current and potential biotechnological applications. Res Microbiol 174:104080. https://doi.org/10.1016/j.resmic.2023.104080

Bird LJ, Kundu BB, Tschirhart T, Corts AD, Su L, Gralnick JA, Ajo-Franklin CM, Glaven SM (2021) Engineering wired life: synthetic biology for electroactive bacteria. ACS Synth Biol 10:2808–2823. https://doi.org/10.1021/acssynbio.1c00335

Bullen RA, Arnot TC, Lakeman JB, Walsh FC (2006) Biofuel cells and their development. Biosens Bioelectron 21:2015–2045

Chang H-J, Voyvodic PL, Zúñiga A, Bonnet J (2017) Microbially derived biosensors for diagnosis, monitoring and epidemiology. Microb Biotechnol 10:1031–1035. https://doi.org/10.1111/1751-7915.12791

Chen S-H, Li Z-T, Zhao H-P (2023) Bioelectrochemical system accelerates reductive dechlorination through extracellular electron transfer networks. Environ Res 235:116645. https://doi.org/10.1016/j.envres.2023.116645

Choi O, Sang B-I (2016) Extracellular electron transfer from cathode to microbes: application for biofuel production. Biotechnol Biofuels 9:11. https://doi.org/10.1186/s13068-016-0426-0

Chu N, Liang Q, Hao W, Jiang Y, Liang P, Zeng RJ (2021) Microbial electrochemical sensor for water biotoxicity monitoring. Chem Eng J 404:127053

Conners EM, Rengasamy K, Bose A (2022) Electroactive biofilms: how microbial electron transfer enables bioelectrochemical applications. J Ind Microbiol Biotechnol 49:kuac012. https://doi.org/10.1093/jimb/kuac012

Corbella C, Hartl M, Fernandez-gatell M, Puigagut J (2019) MFC-based biosensor for domestic wastewater COD assessment in constructed wetlands. Sci Total Environ 660:218–226

Cui Y, Lai B, Tang X (2019) Microbial fuel cell-based biosensors. Biosensors 9:92

Dávila D, Esquivel JP, Sabate N, Mas J (2011) Silicon-based microfabricated microbial fuel cell toxicity sensor. Biosens Bioelectron 26:2426–2430

Di Lorenzo M, Thomson AR, Schneider K, Cameron PJ, Ieropoulos I (2014) A small-scale air-cathode microbial fuel cell for on-line monitoring of water quality. Biosens Bioelectron 62:182–188

Do MH, Ngo HH, Guo WS, Liu Y, Chang SW, Nguyen DD, Nghiem LD, Ni BJ (2018) Challenges in the application of microbial fuel cells to wastewater treatment and energy production: a mini review. Sci Total Environ 639:910–920

Do MH, Ngo HH, Guo W, Chang SW, Nguyen DD, Liu Y, Varjani S, Kumar M (2020) Microbial fuel cell-based biosensor for online monitoring wastewater quality: a critical review. Sci Total Environ 712:135612

Do MH, Ngo HH, Guo W, Chang SW, Nguyen DD, Liu Q, Nghiem DL, Thanh BX, Zhang X, Hoang NB (2022) Performance of a dual-chamber microbial fuel cell as a biosensor for in situ monitoring Bisphenol A in wastewater. Sci Total Environ 845:157125

ElMekawy A, Hegab HM, Pant D, Saint CP (2018) Bio-analytical applications of microbial fuel cell-based biosensors for onsite water quality monitoring. J Appl Microbiol 124:302–313

Falk M, Blum Z, Shleev S (2012) Direct electron transfer based enzymatic fuel cells. Electrochem Front Glob Environ Energy 82:191–202. https://doi.org/10.1016/j.electacta.2011.12.133

Fenner K, Elsner M, Lueders T, McLachlan MS, Wackett LP, Zimmermann M, Drewes JE (2021) Methodological advances to study contaminant biotransformation: New prospects for understanding and reducing environmental persistence? ACS EST Water 1:1541–1554. https://doi.org/10.1021/acsestwater.1c00025

Hassan RYA, Bilitewski U (2013) Direct electrochemical determination of *Candida albicans* activity. Biosens Bioelectron 49:192–198. https://doi.org/10.1016/j.bios.2013.05.015

Hassan RYA, Febbraio F, Andreescu S (2021) Microbial electrochemical systems: principles, construction and biosensing applications. Sensors 21:1279. https://doi.org/10.3390/s21041279

Hernandez CA, Osma JF (2020) Microbial electrochemical systems: deriving future trends from historical perspectives and characterization strategies. Front Environ Sci 8:1

Hernández-Flores G, Poggi-Varaldo HM, Solorza-Feria O, Ponce-Noyola MT, Romero-Castañón T, Rinderknecht-Seijas N, Galíndez-Mayer J (2015) Characteristics of a single chamber microbial fuel cell equipped with a low cost membrane. Int J Hydrog Energy 40:17380–17387

Holder SL, Lee C-H, Popuri SR (2017) Simultaneous wastewater treatment and bioelectricity production in microbial fuel cells using cross-linked chitosan–graphene oxide mixed-matrix membranes. Environ Sci Pollut Res 24:13782–13796. https://doi.org/10.1007/s11356-017-8839-2

Ishii S, Suzuki S, Tenney A, Nealson KH, Bretschger O (2018) Comparative metatranscriptomics reveals extracellular electron transfer pathways conferring microbial adaptivity to surface redox potential changes. ISME J 12:2844–2863. https://doi.org/10.1038/s41396-018-0238-2

Ivars-Barceló F, Zuliani A, Fallah M, Mashkour M, Rahimnejad M, Luque R (2018) Novel applications of microbial fuel cells in sensors and biosensors. Appl Sci 8:1184. https://doi.org/10.3390/app8071184

Jin X, Li X, Zhao N, Angelidaki I, Zhang Y (2017) Bio-electrolytic sensor for rapid monitoring of volatile fatty acids in anaerobic digestion process. Water Res 111:74–80

Kalnenieks U, Balodite E, Rutkis R (2019) Metabolic engineering of bacterial respiration: high vs. low P/O and the case of *Zymomonas mobilis*. Front Bioeng Biotechnol 7:327

Kaur A, Kim JR, Michie I, Dinsdale RM, Guwy AJ, Premier GC, Centre SER (2013) Microbial fuel cell type biosensor for specific volatile fatty acids using acclimated bacterial communities. Biosens Bioelectron 47:50–55

Khater DZ, El-Khatib KM, Hazaa MM, Hassan RYA (2015) Development of bioelectrochemical system for monitoring the biodegradation performance of activated sludge. Appl Biochem Biotechnol 175:3519–3530

Kim M, Hyun MS, Gadd GM, Kim HJ (2007) A novel biomonitoring system using microbial fuel cells. J Environ Monit 9:1323–1328

Knežić T, Janjušević L, Djisalov M, Yodmuang S, Gadjanski I (2022) Using vertebrate stem and progenitor cells for cellular agriculture, state-of-the-art, challenges, and future perspectives. Biomolecules 12:699

Li J, Liu G, Chen D, Li C, Liang D, Wang F, Wu J, He W, Yu Y, Feng Y (2022) Enhanced microbial electrochemical systems performance by optimizing the "anode-collector" collection mode: from enhancement mechanism to construction strategy. ACS EST Eng 2:263–270. https://doi.org/10.1021/acsestengg.1c00396

Liu B, Lei Y, Li B (2014) A batch-mode cube microbial fuel cell based "shock" biosensor for wastewater quality monitoring. Biosens Bioelectron 62:308–314

Liu D, Wang J, Wu L, Huang Y, Zhang Y, Zhu M, Wang Y, Zhu Z, Yang C (2020) Trends in miniaturized biosensors for point-of-care testing. TrAC Trends Anal Chem 122:115701

Logan BE, Call D, Cheng S, Hamelers HVM, Sleutels THJA, Jeremiasse AW, Rozendal RA (2008) Microbial electrolysis cells for high yield hydrogen gas production from organic matter. Environ Sci Technol 42:8630–8640. https://doi.org/10.1021/es801553z

Mahmoud RH, Samhan FA, Ibrahim MK, Ali GH, Hassan RYA (2021) Formation of electroactive biofilms derived by nanostructured anodes surfaces. Bioprocess Biosyst Eng 44:759–768. https://doi.org/10.1007/s00449-020-02485-4

Malik S, Singh J, Goyat R, Saharan Y, Chaudhry V, Umar A, Ibrahim AA, Akbar S, Ameen S, Baskoutas S (2023) Nanomaterials-based biosensor and their applications: a review. Heliyon 2023:1

Mustakeem (2015) Electrode materials for microbial fuel cells: nanomaterial approach. Mater Renew Sustain Energy 4:1–11

Myers B, Catrambone F, Allen S, Hill PJ, Kovacs K, Rawson FJ (2023) Engineering nanowires in bacteria to elucidate electron transport structural–functional relationships. Sci Rep 13:8843. https://doi.org/10.1038/s41598-023-35553-2

Naresh V, Lee N (2021) A review on biosensors and recent development of nanostructured materials-enabled biosensors. Sensors 21:1109. https://doi.org/10.3390/s21041109

Ng I-S, Hsueh C-C, Chen B-Y (2017) Electron transport phenomena of electroactive bacteria in microbial fuel cells: a review of Proteus hauseri. Bioresour Bioprocess 4:53

Odunlami OA, Vershima DA, Tagbo CV, Ogunlade S, Nkongho S (2023) Microbial desalination cell technique—a review. S Afr J Chem Eng 46:312–329

Olias LG, Di Lorenzo M (2021) Microbial fuel cells for in-field water quality monitoring. RSC Adv 11:16307–16317. https://doi.org/10.1039/D1RA01138C

Pasternak G, Greenman J, Ieropoulos I (2017) Self-powered, autonomous biological oxygen demand biosensor for online water quality monitoring. Sensors Actuators B Chem 244:815–822

Patil S, Harnisch F, Schröder U (2010) Toxicity response of electroactive microbial biofilms—a decisive feature for potential biosensor and power source applications. ChemPhysChem 11:2834–2837

Peixoto L, Min B, Martins G, Brito AG, Kroff P, Parpot P, Angelidaki I, Nogueira R (2011) In situ microbial fuel cell-based biosensor for organic carbon. Bioelectrochemistry 81(2):99–103

Ramírez-Vargas CA, Prado A, Arias CA, Carvalho PN, Esteve-Núñez A, Brix H (2018) Microbial electrochemical technologies for wastewater treatment: principles and evolution from microbial fuel cells to bioelectrochemical-based constructed wetlands. Water 10:1128

Ramya M, Kumar PS, Rangasamy G, Rajesh G, Nirmala K, Saravanan A, Krishnapandi A (2022) A recent advancement on the applications of nanomaterials in electrochemical sensors and biosensors. Chemosphere 308:136416

Redha ZM, Yusuf HA, Ahmed HA, Fielden PR, Goddard NJ, Baldock SJ (2017) A miniaturized injection-moulded flow-cell with integrated conducting polymer electrodes for on-line electrochemical degradation of azo dye solutions. Microelectron Eng 169:16–23

Ren Z, Li Y (2022) A miniaturized electrochemical nitrate sensor and the design for its automatic operation based on distributed model. Math Probl Eng 2022:1–9

Roy H, Rahman TU, Tasnim N, Arju J, Rafid MM, Islam MR, Pervez MN, Cai Y, Naddeo V, Islam MS (2023) Microbial fuel cell construction features and application for sustainable wastewater treatment. Membranes 13:1

Santoro C, Arbizzani C, Erable B, Ieropoulos I (2017) Microbial fuel cells: from fundamentals to applications. A review. J Power Sources 356:225–244

Schachinger F, Chang H, Scheiblbrandner S, Ludwig R (2021) Amperometric biosensors based on direct electron transfer enzymes. Molecules 26:4525

Schröder U, Harnisch F, Angenent LT (2015) Microbial electrochemistry and technology: terminology and classification. Energy Environ Sci 8:513–519

Shanbhag MM, Manasa G, Mascarenhas RJ, Mondal K, Shetti NP (2023) Fundamentals of bio-electrochemical sensing. Chem Eng J Adv 16:100516

Shen Y, Wang M, Chang IS, Ng HY (2013) Effect of shear rate on the response of microbial fuel cell toxicity sensor to Cu(II). Bioresour Technol 136:707–710

Simoska O, Gaffney EM, Minteer SD, Franzetti A, Cristiani P, Grattieri M, Santoro C (2021) Recent trends and advances in microbial electrochemical sensing technologies: an overview. Curr Opin Electrochem 30:100762

Sonawane JM, Ezugwu CI, Ghosh PC (2020) Microbial fuel cell-based biological oxygen demand sensors for monitoring wastewater: state-of-the-art and practical applications. ACS Sensors 5:2297–2316

Srivastav AL, Ranjan M (2020) Inorganic water pollutants. In: Inorganic pollutants in water. Elsevier, London, pp 1–15

Stein NE, Hamelers HVM, Buisman CNJ (2010) Stabilizing the baseline current of a microbial fuel cell-based biosensor through overpotential control under non-toxic conditions. Bioelectrochemistry 78:87–91

Stein NE, Hamelers HVM, Buisman CNJ (2012) Influence of membrane type, current and potential on the response to chemical toxicants of a microbial fuel cell based biosensor. Sensors Actuators B Chem 163:1–7

Stephens S, Mahadevan R, Allen DG (2021) Engineering photosynthetic bioprocesses for sustainable chemical production: a review. Front Bioeng Biotechnol 8:610723

Suzuki Y, Makino F, Miyata T, Tanaka H, Namba K, Kano K, Sowa K, Kitazumi Y, Shirai O (2023) Essential insight of direct electron transfer-type bioelectrocatalysis by membrane-bound d-fructose dehydrogenase with structural bioelectrochemistry. ACS Catal 13:13828–13837. https://doi.org/10.1021/acscatal.3c03769

Tanvir RU, Zhang J, Canter T, Chen D, Lu J, Hu Z (2021) Harnessing solar energy using phototrophic microorganisms: a sustainable pathway to bioenergy, biomaterials, and environmental solutions. Renew Sustain Energy Rev 146:111181

Vishwanathan AS (2021) Microbial fuel cells: a comprehensive review for beginners. 3 Biotech 11:248

Vishwanathan AS, Rao G, Siva Sankara Sai S (2013) A novel minimally invasive method for monitoring oxygen in microbial fuel cells. Biotechnol Lett 35:553–558

Wang H, Park J-D, Ren ZJ (2015) Practical energy harvesting for microbial fuel cells: a review. Environ Sci Technol 49:3267–3277

Wang X, Aulenta F, Puig S, Esteve-Núñez A, He Y, Mu Y, Rabaey K (2020) Microbial electrochemistry for bioremediation. Environ Sci Ecotechnology 1:100013. https://doi.org/10.1016/j.ese.2020.100013

Wang Z, Li D, Shi Y, Sun Y, Okeke SI, Yang L, Zhang W, Zhang Z, Shi Y, Xiao L (2023) Recent implementations of hydrogel-based microbial electrochemical technologies (METs) in sensing applications. Sensors 23:641

Wesolowski J, Hassan RYA, Hodde S, Bardroff C, Bilitewski U (2008) Sensing of oxygen in microtiter plates: a novel tool for screening drugs against pathogenic yeasts. Anal Bioanal Chem 391:1731–1737. https://doi.org/10.1007/s00216-008-1947-6

Yu Y-Y, Ding X-L, Quan W-Z, Niu Q, Fang Z, Dapaah MF, You T, Xiao X, Cheng L (2021) Dynamically controlling the electrode potential of a microbial fuel cell-powered biocathode for sensitive quantification of nitrate. Electrochim Acta 369:137661

Zamora-Ledezma C, Negrete-Bolagay D, Figueroa F, Zamora-Ledezma E, Ni M, Alexis F, Guerrero VH (2021) Heavy metal water pollution: a fresh look about hazards, novel and conventional remediation methods. Environ Technol Innov 22:101504

Zeng L, Li X, Shi Y, Qi Y, Huang D, Tadé M, Wang S, Liu S (2017) FePO4 based single chamber air-cathode microbial fuel cell for online monitoring levofloxacin. Biosens Bioelectron 91:367–373

Zhang Y, Angelidaki I (2011) Submersible microbial fuel cell sensor for monitoring microbial activity and BOD in groundwater: focusing on impact of anodic biofilm on sensor applicability. Biotechnol Bioeng 108:2339–2347

Zhang Q, Hu J, Lee D-J (2016) Microbial fuel cells as pollutant treatment units: research updates. Bioresour Technol 217:121–128

Zhao Z, Zhang Y, Wang L, Quan X (2015) Potential for direct interspecies electron transfer in an electric-anaerobic system to increase methane production from sludge digestion. Sci Rep 5:11094

Zhou T, Han H, Liu P, Xiong J, Tian F, Li X (2017) Microbial fuels cell-based biosensor for toxicity detection: a review. Sensors 17:2230

Applications of Microbial Electrochemical Biosensors for the Detection of Organic Pollutants in Water

Umashankar Eshaa, Pasupathi Rathinasabapathi, and Ethiraj Selvarajan

Abstract Water pollution, driven by organic compounds, represents an escalating global concern due to its adverse effects on aquatic ecosystems and human health. This chapter delves into the multifaceted challenges emanating from anthropogenic activities that introduce organic contaminants into aquatic environments. It underscores the imperative need for innovative sensor technologies to combat this issue, focusing on the promising domain of microbial-based electrochemical sensors. The chapter unravels the intricate mechanisms and undeniable advantages of microbial electrochemical biosensors in the meticulous detection and quantification of organic water pollutants. It comprehensively outlines diverse applications of these biosensors, ranging from their pivotal role in wastewater management to their exceptional proficiency in monitoring specific organic pollutants like volatile organic compounds. However, this narrative does not shy away from confronting the limitations faced by microbial-based electrochemical sensors. It places a spotlight on strategies designed to bolster sensitivity, selectivity, and sustainability. Furthermore, the chapter propels readers into the realm of cutting-edge trends within biosensor technology. It elucidates the transformative potential of nanomaterials and data analytics in refining biosensor performance. Moreover, it accentuates the importance of regulatory frameworks and standardization protocols in ensuring the robust and widespread adoption of biosensors within the realm of environmental monitoring. As the global community grapples with the escalating challenges posed by water pollution, this chapter offers an insightful and optimistic perspective on the role of microbial-based electrochemical sensors in mitigating the threat. It invites readers to embark on a journey through the dynamic landscape of biosensor innovation and research, ultimately advocating for a more sustainable and environmentally conscious future.

U. Eshaa · P. Rathinasabapathi (✉) · E. Selvarajan (✉)
Department of Genetic Engineering, School of Bioengineering, College of Engineering and Technology, SRM Institute of Science and Technology, Kattankulathur, Chengalpattu, Tamil Nadu 603203, India
e-mail: biovijay@gmail.com; rathinap1@srmist.edu.in

E. Selvarajan
e-mail: Selvarajanselrajan@gmail.com; selvarae@srmist.edu.in

R. Selvasembian et al. (eds.), *Emerging Trends in Microbial Electrochemical Technologies for Sustainable Mitigation of Water Resources Contamination*,
https://doi.org/10.1007/978-3-031-74636-9_5

Keywords Electrochemical biosensor · Organic pollutants · Microbes · Water pollutants

1 Introduction

Water pollution attributed to organic compounds encompasses chemical substances originating from living organisms or containing carbon atoms within their molecular structures. These contaminants, encompassing pesticides, antibiotics, and detergents, have the capacity to deplete dissolved oxygen levels in industrial effluents, thereby presenting a substantial hazard to aquatic ecosystems and marine life (Ewis et al. 2022). Organic pollutants introduce challenges due to their toxicity, persistence, and the complexities associated with their detection and monitoring. These pollutants emanate from diverse sources, including industrial discharges, agricultural runoff, and improper waste disposal practices. The management of environmental pollution is further complicated by the presence of emerging contaminants and nonpoint source pollution. Nonpoint source pollution, such as agricultural runoff and stormwater, poses a formidable challenge due to its widespread occurrence and decentralized nature. Inadequate treatment methods can also result in the presence of resilient organic pollutants in drinking water supplies (Ewis et al. 2022).

1.1 Challenges Arising from Anthropogenic Activities

A foremost challenge in the realm of environmental monitoring arises from the introduction of organic contaminants through a myriad of anthropogenic activities. These contaminants manifest in various forms, stemming from industrial, residential, and agricultural practices (Saha et al. 2017). For instance, organic herbicides and pesticides, extensively employed in agriculture to combat weeds, pests, and undesirable vegetation, frequently find their way into wastewater streams. Industrial effluent, on the other hand, carries a mixture of hydrocarbons, chlorine compounds, aromatic chemicals, and surfactants Simultaneously, industrial wastewater harbors persistent organic pollutants (POPs) such as polychlorinated biphenyls (PCBs), polybrominated diphenyl ethers (PBDEs), polybrominated biphenyls (PBBs), phthalate esters (PAEs), and various other deleterious substances (Qing et al. 2006). Household waste also contributes to this complex milieu, with compounds like phenyl ether, commonly found in everyday products such as soaps, deodorants, plastics, and cosmetics, infiltrating the environment. Organic water pollutants encompass a diverse array of substances, including heavy metals, pesticides, herbicides, polycyclic aromatic hydrocarbons (PAHs), polychlorinated biphenyls (PCBs), pharmaceuticals, personal care products (Caliman and Gavrilescu 2009), volatile organic compounds (VOCs), and endocrine-disrupting chemicals (EDCs) (Rudel et al. 2003).

Over the last two decades, there has been a growing global concern about the increasing presence of new pollutants such as cosmetics, surfactants, medicines, personal care products, surfactants, and pesticides in aquatic habitats. This is particularly worrying because certain substances among them act as endocrine disruptors, while others have been confirmed as carcinogens and mutagens (Morin et al. 2022). These organic pollutants demand comprehensive investigation and stringent control measures due to their association with severe diseases such as cataracts, kidney and liver damage, jaundice, and cancer. Moreover, they can adversely affect human health by compromising the immune, reproductive, nervous, and endocrine systems (Saha et al. 2017). The types of organic pollutants found in water are discussed below:

(a) *Volatile Organic Compounds*—The wastewater produced during the synthesis of pesticides represents a significant potential source of elevated concentrations of volatile organic compounds (VOCs), posing potential risks to both human health and the environment (Yang et al. 2019). Apart from natural sources, aquatic VOCs can also originate from human activities. One example is the introduction of cyclic volatile methyl siloxanes into Arctic environments, originating from their use in various industrial sectors such as cosmetics, surface treatments, and construction materials. This contamination occurs through the volatilization of these substances into the atmosphere, followed by long-range transport (Warner et al. 2010). Another anthropogenic source of aquatic VOCs is the discharge of wastewater into water bodies, containing pollutants from industrial, domestic, urban, or agricultural origins (Pozzer et al. 2022). This discharge occurs through several mechanisms, such as septic systems, wastewater treatment facilities, surface runoff, and storm drain runoff. Insufficiently treated wastewater can cause eutrophication as a result of an excessive amount of phosphate and nitrogen, leading to the development of algal blooms, algal clumps, and unpleasant smells.

    Akpor and Muchie (2011). Volatile organic compounds (VOCs) encompass a range of substances, including benzene, toluene, ethylbenzene, and xylenes, as well as formaldehyde, acetaldehyde, acrolein, benzaldehyde, and butyraldehyde Long-term exposure to carbonyl and compounds has been associated with a range of enduring health implications on humans, including compromised immune function, kidney and liver damage leading to jaundice, severe respiratory issues, lung function irregularities, tumor development, and cancer. The prolonged presence of these compounds can lead to multifaceted diseases, impacting genetic integrity, clonal expansion, and resistance to programmed cell death (apoptosis) (Soni et al. 2018).

(b) *Polycyclic Aromatic Hydrocarbons*—Polycyclic aromatic hydrocarbons (PAHs) are organic pollutants generated by anthropogenic actions related to industrialization and urbanization, as well as by natural processes such as volcanic eruption, forest fires, the incomplete combustion of fossil fuels (Qiao et al. 2018). PAH surfactants are extensively employed in industrial processes globally, and subsequent discharge into diverse water sources can lead to persistent presence, resulting in significant health and environmental concerns (Zambianchi et al. 2017). Recent research has documented the presence of organic pollutants, including PAHs, in different aquatic environments such as influent and effluent from wastewater treatment facilities, groundwater, surface waters, and seawater (Grandclement et al. 2017).

The Environmental Protection Agency of the United States (USEPA) has identified 16 polycyclic aromatic hydrocarbons (PAHs) as priority contaminants due to their prevalence of occurrence at hazardous disposal facilities, potential for human exposure, toxicity, and the extent of available information (Bojes and Pope 2007). The set of 16 PAHs consists of indeno[1,2,3-cd] pyrene, naphthalene, fluoranthene, fluorene, benzo[ghi]perylene, anthracene, pyrene, chrysene, benz[a]anthracene, phenanthrene, acenaphthene, acenaphthylene, benzo[b]fluoranthene, benzo[k]fluoranthene, dibenz[a,h]anthracene, benzo[a]pyrene (Mojiri et al. 2019). Short-term exposure to PAHs has been documented to induce symptoms such as nausea, vomiting, eye irritation, diarrhea, impaired lung function in individuals with asthma, and thrombotic complications in those with heart ailments. Conversely, prolonged exposure is associated with an elevated risk of developing lung, skin, gastrointestinal, and bladder cancers (Kim et al. 2013).

(c) *Persistent Organic Pollutants*—The compounds encompass a range of substances, including pesticides such as aldrin, heptachlor, dieldrin, chlordane, endrin, dioxins, mirex, furans, and toxaphene. Additionally, industrial compounds such as hexachlorobenzene and polychlorinated biphenyls (PCBs) are included, along with other chemical by-products, flame retardants, specifically polychlorinated dibenzofurans (PCDF) and polychlorinated dibenzodioxins (PCDD), which are collectively referred to as "dioxins" (Qing et al. 2006). The origins of persistent organic pollutants encompass the direct utilization of pesticides in environmental settings, industrial waste, and the byproducts of industrial operations. Furthermore, activities spanning agriculture, forestry, horticulture, industry, municipal practices, and medical processes contribute to the dissemination of these deleterious pollutants into our environment. Moreover, certain natural phenomena, including volcanic eruptions and wildfires, play a role in introducing these pollutants into our ecosystem. Both hydrophobic and hydrophilic POPs are deposited onto freshwater and marine sediment (Alharbi et al. 2018). Polychlorinated biphenyls (PCBs) were primarily employed as additives in compressor oils, electrical oils, hydraulic oils, paint enhancers, impregnating agents, plasticizers for plastics anti-dust agents, heat-carrying agents, and rubber softeners. However, due to their demonstrated harmful effects on living organisms, the production and usage of PCBs were ceased. PCBs, along with chloro-organic pesticides, were categorized as toxic organic compounds (TOCs) (Pawełczyk 2013). POPs are linked to a range of health issues, including diabetes, obesity, disruptions in the endocrine system, cancer, cardiovascular diseases, reproductive problems, and various other health complications (Alharbi et al. 2018). Initial concerns about the harmful effects of POPs exposure stemmed from studies on wildlife communities, revealing endocrine, carcinogenic, reproductive, and developmental impacts, including high rates of malformed genitalia, abnormal mating behavior, infertility, cancer, and disruptions to immune and thyroid function in various species (Qing et al. 2006; Johnson et al. 1998).

(d) *Endocrine Disrupting Compounds*—EDCs, recognized as a group of chemicals, originate from xenobiotic and exogenous sources, while they mimic or impede the natural functions of the endocrine system in both animals and humans. These chemicals can affect processes such as synthesis, secretion, transport, and binding within the endocrine system. Their impact extends to the maintenance of homeostasis, metabolism reproduction, development, and behavior across various organisms (Chang et al. 2009). In water, there exist both natural compounds, such as estrone (E1) and 17β-estradiol (E2), and synthetic compounds, including nonylphenol monoethoxylate carboxylate (NP1EC), diethylstilbestrol (DES), mestranol (MES), nonylphenol (NP), bisphenol-A (BPA), 17α-ethinylestradiol (EE2), and benzophenone (BP), which function as endocrine-disrupting compounds (Pojana et al. 2007). Also, Endocrine-disrupting chemicals (EDCs) disrupt the metabolism of natural hormones. This disruption ultimately leads to the dysregulation of homeostatic mechanisms. Moreover, EDCs have the capability to alter transcriptional signals by either inhibiting or synthesizing new proteins. Endocrine-disrupting compounds (EDCs) that imitate natural steroid hormones can trigger rapid nongenomic reactions by attaching to receptors on the cell membrane and initiating signal pathways via secondary messengers. These alterations result in changes in signaling processes, cellular motility, and the rapid synthesis of hormones (Annamalai et al. 2015). The impacts of EDCs on wildlife have been extensively researched, revealing phenomena such as feminization, reduced fecundity, infertility, shifts in sex ratios, and other forms of reproductive toxicity (Annamalai et al. 2015). Table 1 provides a comprehensive overview of distinct organic pollutants and their associated effects.

2 Biosensor-Based Pollutant Surveillance

Within the realm of environmental monitoring and detection of organic water pollutants, conventional analytical methods often involve labor-intensive and time-consuming laboratory procedures. These methods, while reliable, may need to be better suited for real-time monitoring or the establishment of early warning systems. As a result, there has been an increasing interest in the advancement of innovative sensor technologies that can provide timely and accurate detection of organic contaminants within aquatic environments. Among these technologies, biosensors have emerged as promising tools (Ma et al. 2022).

Biosensors are analytical devices designed to harness biological components such as enzymes, antibodies, or whole cells to detect specific analytes and generate measurable signals (Ma et al. 2022). Their versatility extends to the identification of endocrine-disrupting chemicals (EDCs) and volatile organic compounds (VOCs)

Table 1 Types of organic pollutants and their effects on human health

Types of organic pollutants	Examples	Health effects	References
Volatile organic compounds	Benzene, toluene, formaldehyde, xylene	Damage to liver, kidney, and central nervous system, cancer	Soni et al. (2018)
Polycyclic aromatic hydrocarbons	Benzo[a]pyrene, Naphthalene	Respiratory issues, cancer	Kim et al. (2013)
Chlorinated organic compounds	Polychlorinated biphenyls (PCBs), dioxins	Cancer, cognitive issues	Qing et al. (2006)
Organophosphates	Pesticides, herbicides	Neurological effects	Qing et al. (2006)
Industrial byproduct	Dioxins and furans	Carcinogenic effects, immune system suppression, skin disorders	Alharbi et al. (2018)
Volatile organic solvents	Methanol, ethanol	Neurological effects	Soni et al. (2018)
Flame retardants	Penta-bromo diphenyl ether, hydroxylated metabolites, nitrosamine	Cardiovascular disorders, neurological effects	Alharbi et al. (2018)
Perfluoroalkyl substances (PFAS)	Perfluorooctanoic acid (PFOA), perfluoro-octane sulfonic acid (PFOS)	Developmental effects, disruption of the endocrine system	Annamalai et al. (2015)
Endocrine disrupting compound	Phthalates, bisphenol A (BPA)	Reproductive and developmental issues	Annamalai et al. (2015)

present in water, making them invaluable for a range of applications, including the detection of environmental pollutants, pathogenic microorganisms, substances of health concern, and the assessment of food quality. It's important to note that the choice of microorganism-based biosensors introduces unique advantages and limitations in terms of analyte specificity (Ma et al. 2022). In contrast to traditional physical and chemical sensors, which often struggle with issues of selectivity, microbial sensors offer improved selectivity owing to their engineered specificity (D'souza et al. 2001). These biosensors exhibit a remarkable ability to discriminate between target analytes and potential interfering substances, minimizing the occurrence of false positives and cross-reactivity. Microbial biosensors have been explored for the identification of Halogenated hydrocarbons, which serve various purposes such as insecticides, flame-retardants, foaming agents, pharmaceuticals, and intermediates in polymer production. This group constitutes one of the most substantial categories of environmental contaminants (D'souza et al. 2001).

A significant drawback encountered in utilizing microbial biosensors, particularly in waste monitoring applications, pertains to the toxicity posed by heavy metals. Consequently, organisms resistant to metals assume critical importance for such microbial biosensor implementations. When juxtaposed with enzyme sensors, microbial biosensors exhibit fundamental limitations such as sluggish response times and relatively low sensitivity and detection thresholds. Enhanced enzyme electrode sensitivity and detection limits can be accomplished, according to theoretical frameworks for enzyme-based biosensors that use non-reactive transducers. This may be achieved by either decreasing the Michaelis constant (KM) or increasing the bimolecular rate constant. The progress in enzyme engineering has rendered these objectives achievable through methodologies like site-directed mutagenesis. Furthermore, the reason for the slower reaction time of microbial sensors, as opposed to enzyme sensors, is attributed to the difficulties in diffusion caused by cell membranes (D'souza et al. 2001). However, the integration of microbial electrochemical sensors can mitigate this limitation, as they offer rapid response times and real-time online monitoring capabilities (Prevoteau et al. 2017).

Electrochemical biosensors are highly sensitive and can function in turbid solutions (Palchetti et al. 2009). A typical electrochemical biosensor comprises a bioreceptor that binds precisely to the target, transducer elements, and a data analyzer. When a bioreceptor and a target engage, electrons are created as a signal. This signal is measured electronically and analyzed using a transducer. A conventional electrochemical biosensor requires a bioreceptor and a transducer electrode for analyte detection (Gupta et al. 2021). The microbial-based electrochemical sensors capitalize on the natural interactions between microorganisms and specific organic pollutants. Microbial biosensors that incorporate many types of microorganisms, such as bacteria, microalgae, fungi, and viruses, have emerged as a suitable technology for addressing the limits of traditional analytical methods. Microorganisms possess remarkable adaptability and sensitivity to environmental changes, making them ideal candidates for biosensing applications. When exposed to target organic pollutants, certain microorganisms exhibit distinctive responses, such as changes in metabolism, gene expression, and membrane potential. These microbial responses can be harnessed to create biosensors that transduce their reactions into measurable electrochemical signals, enabling the detection and quantification of pollutants in water samples (Chu et al. 2021). Microbial electrochemical sensors have the following advantages: Direct generation of electrical signals without requiring an additional transducer, self-repair capability of the electrode-biofilm for extended operational lifespan, real-time online monitoring, affordability, self-sustainability akin to a fuel cell, and simplified operation and maintenance (Prevoteau et al. 2017). The types of electrochemical biosensors are discussed below:

(i) *Amperometric biosensor*—An amperometric biosensor represents a self-contained integrated device that relies on the current generated from oxidation to provide precise quantitative analytical data. Typically, these biosensors exhibit dynamic ranges, reaction times, and sensitivities akin to potentiometric biosensors. Within amperometric biosensors, the output current serves as the basis for sensing. Sensitivity in amperometric biosensors is ascertained by comparing the current responses at various analyte concentrations. These biosensors employ only two electrodes: one for applying voltage and the other for measuring current flow through the device. Unlike optical or electrochemical devices, amperometric biosensors rely solely on current measurements for their operation (Singh et al. 2021).

(ii) *Potentiometric biosensor*—A potentiometric biosensor operates on the principle of detecting potential differences between working and reference electrodes. Unlike amperometric biosensors, the measured species are not consumed in the process. The sensor's response is directly proportional to the concentration of the analyte, achieved by comparing its activity to that of the reference electrode. Potentiometric biosensors offer significant advantages in sensitivity and selectivity, particularly when employing a highly stable and accurate reference electrode (Singh et al. 2021).

Potentiometry, a widely utilized electrochemical technique in sensor technology, is known for its cost-effectiveness and applicability across a broad range of ion concentrations. Sensors developed using potentiometric methods are readily available in the market and can be easily fabricated. Moreover, their performance remains unaffected by reductions in size. The integration of potentiometric tools in biosensor technology has greatly expanded possibilities for diagnosis and sensing applications. The primary advantages of potentiometric biosensors lie in their generation of potential-based signals and their utilization of biochemical components as integral parts of the receptor (Konchi et al. 2007; Singh et al. 2021).

(iii) *Immunosensor*—Immunosensors represent solid-state devices wherein the immunochemical reaction interfaces with a transducer. They constitute a pivotal category among affinity biosensors, relying on the specific recognition of antigens by antibodies to form a stable complex, akin to immunoassays. Antibodies (Ab) are the predominant bio-elements utilized in the fabrication of electrochemical immunosensors, followed by aptamers (Apt) and microRNA (miRNA). Kanso's research team (Kanso et al. 2013) conducted a study focusing on the development of electrochemical immunosensors utilizing magnetic beads to detect oestradiol and ethinyl oestradiol in wastewater samples. These compounds, found in the nanomolar range in water sources, are known as endocrine-disrupting chemicals, capable of interfering with human endocrine systems and potentially leading to testicular and breast cancers. The magnetic beads were subjected to chemical modification using four different synthetic estrogen derivatives containing either carboxylic or amino groups. The assay principle was based on the competitive interaction between free estrogen in the sample and immobilized estrogen, employing a primary antibody labeled with alkaline phosphatase (ALP) (Kanso et al. 2013). Additionally, an electrochemical immunosensor capable of simultaneously detecting pesticides, namely endosulfan, and paraoxon, was also reported (Cristea et al. 2015).

(iv) *Enzymatic sensor*—Electrochemical biosensors utilizing immobilized enzymes stand out as one of the most favored and commercially prosperous categories within the biosensor realm. Their popularity stems from several factors, including their cost-effectiveness, expedited and straightforward analytical procedures, and the compactness of their measuring units, often rendering them portable. Furthermore, the advent of screen-printed planar electrodes, known for their affordability, sensitivity, and potential for miniaturization, significantly contributes to the success of biosensors (Tudorache and Bala 2007). Typically, enzymes serve as the sensitive components of electrochemical biosensors due to their swift catalytic reactions, high substrate specificity, and relatively enduring stability. The availability of a wide array of commercially purified enzymes further facilitates the creation of biosensors grounded on diverse biochemical reactions (Kucherenko et al. 2020). Table 2 provides insights into the types of electrochemical biosensors used for detecting organic water pollutants.

Table 2 Types of electrochemical sensors in detecting organic water pollutants

Organic water pollutant category	Examples	Electrochemical sensor type	References
Volatile organic compounds (VOCs)	Benzene, toluene, xylene, trichloroethylene	Amperometry sensors	Nomngongo et al. (2012)
Polycyclic aromatic hydrocarbons (PAHs)	Benzo[a]pyrene, naphthalene, anthracene	Potentiometric sensors, immunosensors	Nigam et al. (2015) and Cristea et al. (2015)
Chlorinated organic compounds	Polychlorinated biphenyls (PCBs), dioxins	Amperometric sensors	Nomngongo et al. (2012)
Organophosphates	Pesticides like malathion, chlorpyrifos	Enzymatic sensors	Baeumner et al. (2003)
Plastic	Bisphenol A (BPA), phthalate	Potentiometric sensors, enzymatic sensors	Nomngongo et al. (2012) and Baeumner et al. (2003)
Aromatic amines	Aniline, benzidine	Amperometric sensors, potentiometric sensors	Nigam et al. (2015)
Byproducts of industrial processes	Dioxins and furans	Potentiometric sensors, amperometric sensors	Nigam et al. (2015) and Baeumner et al. (2003)
Perfluoroalkyl substances (PFAS)	Perfluorooctanesulfonic acid (PFOS)	Enzymatic sensors	Baeumner et al. (2003)
Volatile organic solvents	Methanol, ethanol, acetone	Amperometry sensors, potentiometric sensors	Singh et al. (2021) and Nigam et al. (2015)

3 Components and Mechanisms of Microbial Electrochemical Biosensors

Several components influence the efficacy of microbial electrochemical biosensors, including the working electrode, biofilm, counter electrode, anion exchange membrane, and reference electrode as shown in Fig. 1. Every single one of these components significantly contributes to ensuring the accuracy and reliability of the biosensor's measurements, making them essential considerations in sensor design and optimization (Hill et al. 2020).

(i) *Working Electrode*—The Working Electrode (WE) operates as the location where the desired faradaic reaction occurs. Normally, its electrochemical potential is regulated in relation to a reference electrode positioned below it, with the assistance of a potentiostat. A positive current is observed when electrons are transferred 'to' the electrode due to an oxidation reaction (for instance, acetate oxidation yielding electrons, H+, and HCO3–). In such instances, the working electrode (WE) is referred to as the "anode" electrode. In certain experiments or measurements, the polarity of the working electrode (WE) can be reversed

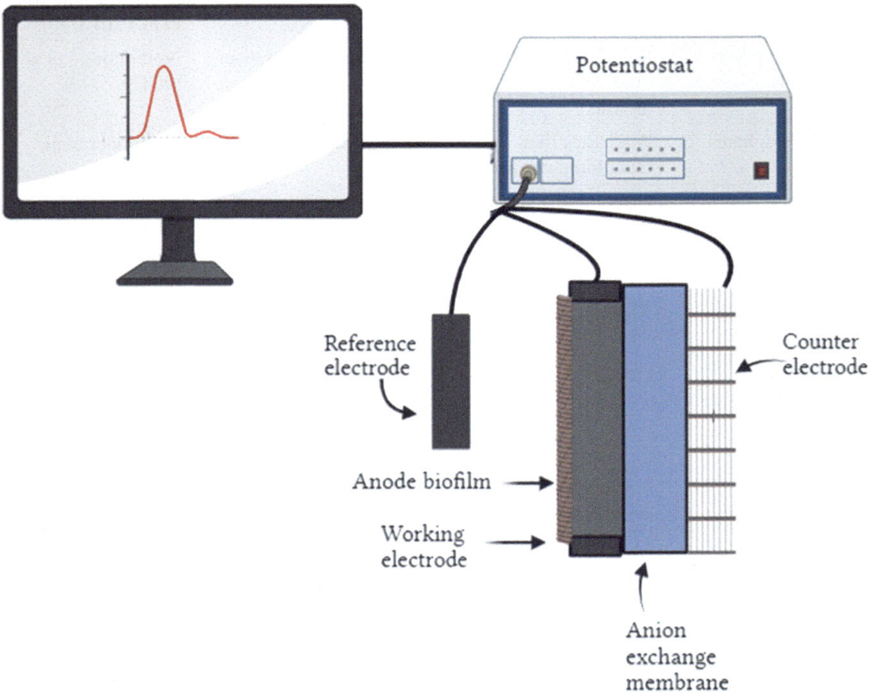

Fig. 1 Microbial electrochemical biosensor (Ref.: Chung et al. 2020)

to enable a reduction process, such as in a cyclic voltammetry (CV) experiment (Hill et al. 2020).

(ii) *Counter Electrode*—The Counter Electrode (CE) facilitates a counter-reaction to equilibrate the electron flow at the Working Electrode (WE) and uphold charge equilibrium within the electrolyte. This reaction is instrumental in consuming the H + ions produced by the anodic oxidation at the working electrode, thereby sustaining charge balance. It achieves this either through the reduction of oxygen or through the evolution of hydrogen (Hill et al. 2020).

(iii) *Reference Electrode*—The Reference Electrode (RE) facilitates the imposition of a predetermined electrochemical potential difference onto the Working Electrode using a potentiostat. For instance, an RE composed of a silver/silver chloride (Ag/AgCl, Sat. KCl) half-cell possesses a potential of 0.197 mV, enabling the potentiostat to sustain this voltage at the working electrode (Hill et al. 2020).

(iv) *Ion Exchange Membrane*—The Ion Exchange Membrane facilitates the diffusion of counter ions across chambers to uphold charge neutrality. Simultaneously, it serves to physically separate electrolytes, thereby safeguarding the biofilm from inhibitory compounds or particulates (Hill et al. 2020).

(v) *Biofilm*: In microbial electrochemical systems that are not mediated by external agents, the direct link between metabolically active microorganisms and electrodes is established by the creation of electroactive biofilms (Mahmoud et al. 2021, 2020, 2018; Cong et al. 2013). These biofilms consist of diverse groups of microbial cells enveloped within a dense layer of exopolysaccharides matrix (EPSs), firmly adhering to living tissues or solid surfaces (Saratale et al. 2017; Flemming et al. 2010). Typically, living cells constitute less than 10% of the total content, with the matrix comprising approximately 90%. The composition of biofilm matrices varies, encompassing polysaccharides, proteins, and extracellular DNA, contingent upon bacterial species and environmental factors (Azeredo et al. 2017). Furthermore, the EPS environment imparts high mechanical stability, facilitating cell–cell adhesion to solid surfaces and creating a cohesive three-dimensional network interconnecting biofilm cells. The mechanism underlying biofilm-assisted bioelectrochemical signals involves physical connections facilitated by bacterial appendages, microbial nanowires, ferric iron, and cyt-c in certain microbes like *Thiobacillus denitrificans, Geobacter-sulfurreducens* and *Shewanella oneidensis* (Reguera et al. 2006; Zhou et al. 2015). O'Toole et al. emphasized the notable role of flagella and motility in the first step of transmitting signals from the cell to the surface (O'Toole et al. 1998). Their research demonstrated that mutants with defective flagella (flgK) failed to form microcolonies on solid substrates throughout the experimental duration, providing evidence for the crucial role of intact flagella in microcolony formation (Hassan et al. 2021).

In Microbial Electrochemical Biosensors, an oxidation reaction transpires at the anode, whereas a reduction reaction transpires at the cathode. Both of these reactions are amenable to catalysis by microbes. These microorganisms, referred to as electroactive microorganisms, possess the capacity to either discharge electrons to the electrode (anode) or receive electrons from the electrode (cathode) via their metabolic processes. The process of electron transfer encompasses a range of mechanisms that can be classified into distinct groups: direct transfer facilitated by outer membrane cytochromes, direct transfer facilitated by nanowires, indirect transfer facilitated by mediators, and indirect transfer facilitated by catabolites. Direct electron transfer is mostly favored in microbial electrochemical technologies due to its lower energy consumption compared to other mechanisms (Simoska et al. 2018).

Most investigations regarding microbial electrochemical systems are designed with the anode serving as the sensing component. This bioanode exhibits the ability to engage in outward Electron Exchange Transfer (EET), where it utilizes organic materials as the source of electrons and the anode as the recipient of electrons (Chu et al. 2021; Jiang et al. 2017). It is predicted that the utilization of a biocathode as the sensing element would result in an increase in the sensitivity of microbial electrochemical sensors. This is because the biocathode operates as an electron donor and oxygen functions as an electron acceptor, hence facilitating inward electron transfer (Jiang et al. 2018). The efficacy of employing a biocathode sensing element for cost-effective and non-selective monitoring has been assessed within the 3-electrode configuration, although this setup does not function in a self-sustaining manner (Liao et al. 2018; Prevoteau et al. 2020). In a self-sustaining microbial electrochemical sensor, the bioanode's reduced redox potential can initiate the oxygen reduction reaction (ORR) at the biocathode. Nevertheless, when employing the effluent from anodic chambers as the inflow for cathodic chambers, the presence of unprocessed organic material might significantly hinder the biocathode's internal Extracellular Electron Transfer (EET) (Jiang et al. 2017). To address this challenge, a sequential flowing biosensor incorporates air sparging with an active oxygen source to diminish the accumulation of unutilized organic material in the cathodic chamber (Chu et al. 2021).

Microbial electrochemical sensors are typically categorized as self-powered electrochemical biosensors based on fuel cells (Ejeian et al. 2018). In these sensors, the electrode-biofilm complex serves as the sensing element (Chen et al. 2019; Hill et al. 2020). Electroactive microorganisms are capable of forming biofilms on electrodes and establishing electrical communication through a phenomenon referred to as extracellular electron transfer (Logan et al. 2019; Chu et al. 2021; Sanche et al. 2020; Jiang et al. 2017). The formation of the electrode-biofilm assembly is often a natural process, where the electrode serves as a solid source or recipient of electrons for the respiratory activity of electroactive bacteria (Oon et al. 2018). Alternatively, constructing an electrode-biofilm involves immobilizing microorganisms that are non-electrogenic or weakly electrogenic, along with the inclusion of a redox mediator or redox polymer (Tucci et al. 2019; Lazzarini et al. 2020). However, the discussion of microbial biosensors utilizing planktonic cells with redox mediators can be found in a comprehensive review article for further exploration.

The electrode-biofilm within a microbial electrochemical sensor serves the dual purpose of acting as both a bioreceptor and a transducer (Jiang et al. 2018). Toxic substances have the ability to reduce the metabolic functions of electroactive microorganisms, which in turn hinders the flow of electrons between these microbes and electrodes. This leads to a decrease in electrical activity (Yu et al. 2017). The whole cellular response is a reflection of the extensive influence of the toxicant on a catalytic system involving several enzymes (Si et al. 2015). In addition, biofilm formations enhance electrical signals by aggregating and communicating across multiple cells (Chu et al. 2021).

Microbial electrochemical sensors, which rely on electrode-biofilm configurations, typically exhibit reduced specificity compared to subcomponent-based biosensors such as electrochemical enzymatic biosensors (Karimian et al. 2020; Karimi-Maleh et al. 2020). Additionally, they generally demonstrate lower sensitivity than electrochemical sensors equipped with chemically modified electrodes (Gupta et al. 2013). However, the advantage of self-powered microbial electrochemical sensors lies in the fact that they do not require the labor-intensive, costly, and time-consuming extraction and purification processes associated with other sensor types (Antonacci et al. 2020).

Microbial electrochemical sensors have successfully achieved targeted monitoring in laboratory-scale research by utilizing modified strains that have been cloned with zinc-sensitive properties (Khan et al. 2020) or arsenic-sensitive promoters (Webster et al. 2014). The lack of specificity in these sensors is not deemed a drawback when monitoring water biotoxicity, as specificity is not a requisite (Lazzarini et al. 2020). Instead, the lack of specificity in microbial electrochemical sensors can be advantageous, particularly in aquatic environments where they can serve as early detection systems for the presence of several toxic substances. Hence, microbial electrochemical sensors hold promise for detecting toxicity shocks, characterized by short periods of significantly elevated pollutant concentrations compared to normal levels (Diaz-Gonzalez et al. 2019).

Microbial electrochemical sensors offer a range of advantages. Firstly, they generate electrical signals directly, eliminating the need for an additional transducer. Secondly, the electrode-biofilm system possesses self-repairing capabilities, ensuring long-term functionality. Thirdly, these sensors enable real-time online monitoring, providing continuous data feedback. Additionally, microbial electrochemical sensors are cost-effective, making them accessible for various applications. Furthermore, their self-sufficiency, akin to fuel cells, adds to their appeal. Lastly, they offer convenient operation and maintenance procedures, enhancing their usability in practical settings (Chu et al. 2021).

4 Applications and Significance of Microbial-Based Electrochemical Sensors

The applications of microbial electrochemical biosensors are diverse and multifaceted. These sensors have found their niche in wastewater management, where they excel in detecting organic pollutants. They are equally adept at monitoring environmental conditions, identifying pathogens, measuring chemical oxygen demand, assessing biological oxygen demand, conducting toxicity monitoring, and even monitoring corrosion (Simoska et al. 2018). This versatility positions them as indispensable tools for safeguarding environmental health.

Within the realm of biosensors, specific applications have emerged to address the detection of VOCs. For instance, sensitive olfactory biosensors have been developed, exhibiting exceptional performance with high sensitivity and selectivity even at deficient concentrations. Protein and peptide biosensors have played a crucial role in realizing these applications (Brito et al. 2020). Moreover, nonspecific biosensors based on microbial electrochemical cell-based (MXC) structures have demonstrated their efficacy in monitoring specific pollutants like toluene (Chung et al. 2021). Engineered whole-cell biosensors, utilizing metabolic genes, have been developed to specifically detect benzene, phenol, toluene, and similar contaminants. This advancement enhances the efficiency of biosensor technology (Roy et al. 2021). Whole-cell luminescence biosensors have also found success in measuring VOCs and polycyclic aromatic hydrocarbons (PAHs) across various environmental media, including wastewater, saltwater, and soil (Voon et al. 2022). However, the advancement of biosensors for specific organic pollutants like newly identified toxins, faces challenges due to a lack of comprehensive knowledge regarding essential sensing components based on microbial metabolic pathways (Kannappan et al. 2022). To overcome this limitation, high-throughput sequencing and synthetic biology methods are being employed to delve deeper into the genetic details of sensing components (Thavarajah et al. 2020).

(a) *Toxicity Monitoring*:

Anodic microbial electrochemical sensors—Reports have surfaced regarding electroactive microorganisms exhibiting resilience against elevated concentrations of diverse heavy metals such as chromium (Cr), nickel (Ni), and cadmium (Cd), thus presenting promising avenues for the advancement of shock sensors. These discoveries lay the groundwork for the advancement of economical and portable microbial sensors. An especially notable finding is the demonstration that the electrical signal can be regenerated despite prolonged exposure to metal concentrations (Xu et al. 2015). It is imperative to acknowledge that microbial electrochemical sensors for toxicity assessment lack significant selectivity (Chu et al. 2021). In contrast to biosensors for biochemical oxygen demand/chemical oxygen demand that emphasize broadening linear ranges, microbial electrochemical sensors at the anode, are specially intended for toxicity monitoring and attempt to improve the detection limits of contaminants. Presently, these toxicity biosensors continue to demonstrate detection limits that surpass the water quality standards established by the World Health Organization (Jiang et al. 2018).

Cathodic microbial electrochemical sensors—Comparable to the observed behavior at the anode, the presence of toxic compounds at the cathode, regardless of bacterial involvement, can impede bacterial activity. Additionally, these bacteria are prone to being affected by other pollutants and ions, which can lead to a decrease in electricity production due to inhibition of the electroactive biofilm. To broaden the usability of conventional heterotrophic anodic sensors to encompass oxygenated water and autotrophic environments, without requiring additional organic carbon supplementation, cathodic biosensors have been suggested (Liao et al. 2018). The microbial populations present at the biocathode aid in the process of reducing several electron acceptors, including oxygen. When there is a sudden release or a significant rise in the concentration of toxic substances, the metabolic rates of electroactive microorganisms decrease, thereby altering the output electrical signal. The application of a biocathode for the oxygen reduction reaction (ORR) as the sensing component in microbial fuel cell-based sensors for monitoring toxicity has been shown by Jiang et al. (2017).

(b) *Pathogen detection*: Microbial electrochemical sensors exhibit considerable potential as tools for the swift detection and surveillance of pathogens, aiding in the diagnosis of bacterial infectious diseases (Amiri et al. 2018; Simoska et al. 2019). These sensing systems offer a quick, highly sensitive, and qualitative method to detect pertinent microbial pathogens, as well as signaling chemicals, metabolites, and hormones produced by cellular activity. Demonstrations of such capabilities include the detection of glucose and L-lactate through microbial electrochemical sensors utilizing *Gluconobacteroxidans* (Cevik et al. 2019) and *Hansenula polymorpha* (Smuto et al. 2007), respectively. In clinical settings, microbial electrochemical sensors have been customized to detect electrochemically active pathogenic microbes present in clinically significant samples (Karbellar et al. 2020). In this context, pathogenic microorganisms that have been detected in clinical samples serve as catalysts, generating redox-active mediators as secondary metabolites. Consequently, they generate a discernible signal exclusively in the presence of the target pathogen. These sensors have the potential to be used as analytical tools for pathogen identification in both environmental surveillance and medical diagnostics. They present appealing attributes such as rapid responsiveness, cost-efficiency, and adaptability (Simoska et al. 2018). However, although many microbial electrochemical sensing systems have been developed for detecting pathogens, only a limited number have been subjected to thorough characterization of analytical performance parameters, such as assessing the speed at which a sensor responds and its ability to accurately measure a wide range of values, as well as evaluating its performance in complex biological environments (Simoska et al. 2018).

(c) *Corrosion sensing*: Biofilms have the remarkable ability to thrive on a wide array of materials and metals, such as copper alloys, despite copper's toxic effects on numerous biological organisms (Carvalho et al. 2014). The settlement of bacteria induces alterations in the electrochemical properties of conductive surfaces. This modification primarily involves changes in corrosion processes that take place at the interface between the metal and the biofilm, notably augmenting the cathodic current, observed in both anaerobic and aerobic environments (Wg 1990). Taking advantage of this phenomenon, simple electrochemical devices known as BIOX (Mollica et al. 2003) have been created.

These devices consist of a cathode made of stainless steel or titanium and an anode made of zinc. The cathode and anode are coupled by a high-value electrical resistance. They are engineered for simultaneous real-time monitoring of biofilm proliferation, microbial corrosion susceptibility, and the efficacy of antifouling interventions such as chlorination (Cristiani 2014). These sensor types marked the initial utilization of a bioelectrochemical system on an industrial scale (Cristiani et al. 2021).

(d) *Environmental monitoring-Dissolved oxygen determination*: Cathodic microbial electrochemical sensors are employed for assessing dissolved oxygen levels. The cathode can operate either in an abiotic manner or with the presence of a biofilm (referred to as a biocathode). Since oxygen acts as the final electron acceptor (oxidant), changes in its concentration, whether dissolved or in a gaseous state, are anticipated to impact both the thermodynamics and kinetics of the oxygen reduction reaction (Chu et al. 2021). Gonzales Olias et al. (2020) established a correlation between dissolved oxygen (DO) concentration and the performance of ceramic-based soil microbial fuel cells. An elevation in dissolved oxygen (DO) concentration within the electrolyte elicited a favorable response from the entire microbial fuel cell, typically noticeable approximately 3.3 min thereafter. It is crucial to note that the presence of oxygen may have adverse effects on the anaerobic/anoxic conditions predominantly favored by the anode biofilm. Notably, the microbial fuel cell design featured a ceramic separator, safeguarding the anaerobic environment prevailing at the anode (Olias et al. 2020). Furthermore, the impact of operational factors such as pH, temperature, and solution conductivity was assessed, revealing negligible effects on sensor responsiveness. Moreover, another investigation documented the utilization of a DO sensor to monitor DO concentration at various depths of a lake over 67 days (Song et al. 2019). This demonstrated significant potential for early detection of environmental hazards in aquatic ecosystems.

(e) *BOD/COD Measurement*: Microorganisms with anodic electroactivity are sensitive to variations in the substrate they utilize for respiration and extracellular electron transfer (ET). Upon achieving steady-state conditions in microbial electrochemical technologies (MET), an increase in the concentration of organic substrate typically results in a corresponding rise in the generated current (Spurr et al. 2018). This holds until a saturation threshold is reached, beyond which further increments in organic concentration fail to enhance the recorded current output. Conversely, a reduction in organic concentration results in a corresponding decline in the signal output. Hence, biosensors utilizing microbial electrochemical technologies have the capacity to operate as real-time detectors for biochemical oxygen demand (BOD) or chemical oxygen demand (COD), substantially decreasing response times relative to conventional colorimetric detection kits. Traditionally, BOD determination requires 5 days, while COD determination typically takes a few hours. Upon appropriate calibration, these biosensors promptly reflect the current output, allowing for immediate correlation with organic matter concentrations in aqueous solutions. However, it is crucial to acknowledge that the selection of a calibration solution profoundly influences the sensor's efficacy for BOD/COD measurements (Simoska et al. 2018).

(f) *Detection of Organic Pollutants:* Microbial biosensors can detect BTEX—benzene, toluene, ethylbenzene, and xylene. These volatile aromatic compounds were commonly found in crude oil, natural gas, and crude oil, as well as being extensively produced by industries for use as solvents and raw materials in chemical synthesis. They were recognized as significant environmental pollutants. Additionally, the microbial biosensor can detect 3-chlorobenzoate, chloromuconic acids, polycyclic aromatic hydrocarbons, biphenyl, 2-chlorobiphenyl and triclosan, monochlorinated biphenyls, and hydroxylated polychlorinated biphenyls. Different phenol derivatives recognized for their toxic properties could also be detected. Examples included 2-hydroxybiphenyl, a prevalent disinfectant and fungicide, alongside 2,4-dichlorophenoxyacetic acid (2,4-D), an extensively employed herbicide (Tecon et al. 2008). Microbial biosensors have been explored for a diverse range of environmental applications, including the detection of polyhalogenated hydrocarbons. These compounds, utilized as insecticides, foaming agents, pharmaceuticals, flame-retardants, and intermediates in polymer production, are among the most extensive categories of environmental contaminants (D'souza et al. 2001) (Fig. 2).

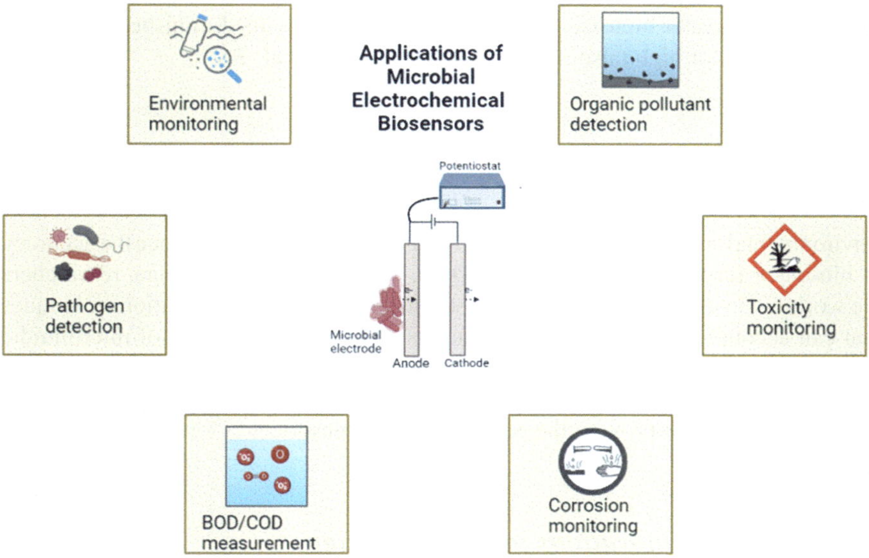

Fig. 2 Applications of microbial electrochemical biosensor (Ref.: Simoska et al. 2018)

5 Overcoming Limitations and Paving the Way Forward

While microbial-based electrochemical sensors have made significant strides in environmental monitoring and organic pollutant detection, they are not without limitations. Addressing these limitations is crucial to further advancing the field and ensuring the effectiveness of these sensors in safeguarding our environment. To further advance biosensor technology, research focused on understanding the metabolic genes associated with persistent organic pollutants (POPs) is essential. In this context, microbial electrochemical biosensor technology holds promise for effectively monitoring challenging-to-degrade organic pollutants like EDCs, PCBs, POPs, TPHs, PAHs, and VOCs. We delve into these challenges and the strategies that pave the way forward.

5.1 Enhancing Sensitivity and Selectivity

One of the primary challenges in the realm of biosensors is enhancing sensitivity and selectivity (Simoska et al. 2018). Sensitivity refers to the ability to detect low concentrations of pollutants, while selectivity pertains to the ability to distinguish between different substances. Improving sensitivity and selectivity is essential for detecting trace amounts of organic pollutants in complex environmental matrices (Jiang et al. 2017). The practical implementation of a bioanode as a sensing element for monitoring water biotoxicity is greatly constrained by its diminished sensitivity and propensity for high false-negative outcomes (Chu et al. 2021).

5.2 Managing Interfering Factors

Environmental conditions, such as humidity and pH levels, can influence the accuracy of biosensor readings (Simoska et al. 2018). To address this challenge, researchers are working on the development of robust sensor coatings and calibration techniques that can account for these interfering factors. Additionally, the use of microfluidic systems diminishes power consumption, reduces response time, eliminates the need for a membrane, decreases the usage of specialized reagents, and ensures heightened sensitivity, further improving the reliability of biosensors.

5.3 Material Innovations in Sensor Design

Materials play a crucial role in sensor design, biofilm development, cost considerations, and reliability. Various alternative materials for electrodes like copper have been recognized as viable options for reducing response times. These materials are capable of generating higher electron flows compared to traditional carbon-based electrodes, as identified in studies such as that by Hill et al. (2020). The development of innovative materials for bioelectrodes, modified microorganisms, substrate types, and interspecies electron transfer processes has seen major advancements in recent years. These advancements have been made possible by scientific advancements. These developments aim to enhance interfacial electron transfer and optimize the interaction between microbes and electrodes (Hassan et al. 2021).

5.4 Nanomaterial Advancements

The development of the chemical sensing layer for an abiotic cathode sensing element mandates the utilization of advanced nanostructured materials to achieve heightened selectivity and sensitivity (Chu et al. 2021). Continued research into novel nanomaterials, such as graphene, carbon nanotubes, and nanocomposites, is expected to lead to significant advancements in biosensor performance. These nanomaterials can be tailored to specific applications, enhancing the capabilities of biosensors in detecting an extensive array of organic pollutants.

5.5 Integration of Advanced Technologies

This enables real-time imaging techniques, structure prediction, and mathematical modeling, allowing for more promising outcomes in environmental monitoring and pollutant management. Standardization efforts are crucial for ensuring the reliability and comparability of biosensor data across different monitoring applications (Patil et al. 2012).

To ensure the widespread adoption of biosensors in environmental monitoring, efforts are underway to establish regulatory frameworks and standardization protocols. This includes defining performance benchmarks, quality control measures, and validation procedures to meet regulatory requirements.

6 Conclusion

Effective monitoring and detection methods are crucial in addressing the significant environmental issue of water pollution caused by organic compounds. This chapter highlights the challenges arising from human activities that contribute to the spread of organic contaminants in aquatic environments. It emphasizes the critical role of microbial-based electrochemical sensors in safeguarding our ecosystems. Remarkably sensitive and selective, microbial electrochemical biosensors are valuable tools for detecting organic water pollutants, with applications in wastewater management and the monitoring of volatile organic compounds. Despite their many benefits, however, biosensors face challenges related to sensitivity, selectivity, and sustainability. To overcome these hurdles, researchers are developing specific bioreceptors, advanced nanomaterials, and calibration techniques to enhance sensor performance. Sustainability considerations, such as the use of renewable materials and optimizing energy efficiency, are also top priorities. Integration with data analytics and adherence to regulatory standards are essential steps toward ensuring the reliability and comparability of biosensor data. In conclusion, microbial-based electrochemical sensors offer promising solutions to the complex issue of organic water pollution. Ongoing research and innovation in this field will contribute to more effective environmental monitoring and pollutant management, ultimately preserving the health of aquatic ecosystems and human well-being.

References

Akpor OB, Muchie B (2011) Environmental and public health implications of wastewater quality. Afr J Biotech 10(13):2379–2387

Alharbi OM, Khattab RA, Ali I (2018) Health and environmental effects of persistent organic pollutants. J Mol Liq 263:442–453

Amiri M, Bezaatpour A, Jafari H, Boukherroub R, Szunerits S (2018) Electrochemical methodologies for the detection of pathogens. ACS Sens 3(6):1069–1086

Annamalai J, Namasivayam V (2015) Endocrine disrupting chemicals in the atmosphere: their effects on humans and wildlife. Environ Int 76:78–97

Antonacci A, Scognamiglio V (2020) Biotechnological advances in the design of algae-based biosensors. Trends Biotechnol 38(3):334–347

Azeredo J, Azevedo NF, Briandet R, Cerca N, Coenye T, Costa AR, Sternberg C (2017) Critical review on biofilm methods. Crit Rev Microbiol 43(3):313–351

Baeumner AJ (2003) Biosensors for environmental pollutants and food contaminants. Anal Bioanal Chem 377:434–445

Bojes HK, Pope PG (2007) Characterization of EPA's 16 priority pollutant polycyclic aromatic hydrocarbons (PAHs) in tank bottom solids and associated contaminated soils at oil exploration and production sites in Texas. Regul Toxicol Pharmacol 47(3):288–295

Brito NF, Oliveira DS, Santos TC, Moreira MF, Melo ACA (2020) Current and potential biotechnological applications of odorant-binding proteins. Appl Microbiol Biotechnol 104:8631–8648

Caliman FA, Gavrilescu M (2009) Pharmaceuticals, personal care products and endocrine disrupting agents in the environment—a review. Clean: Soil, Air, Water 37(4–5):277–303

Carvalho ML, Doma J, Sztyler M, Beech I, Cristiani P (2014) The study of marine corrosion of copper alloys in chlorinated condenser cooling circuits: the role of microbiological components. Bioelectrochemistry 97:2–6

Cevik E, Cerit A, Tombuloglu H, Sabit H, Yildiz HB (2019) Electrochemical glucose biosensors: whole cell microbial and enzymatic determination based on 10-(4*H*-Dithieno [3, 2-b: 2′, 3′-d] Pyrrol-4-yl) Decan-1-amine interfaced glassy carbon electrodes. Anal Lett 52(7):1138–1152

Chang HyunShik CH, Choo KwangHo CK, Lee ByungWhan LB, Choi SangJune CS (2009) The methods of identification, analysis, and removal of endocrine disrupting compounds (EDCs) in water. J Hazardous Mater 172(1):1–12

Chen Y, Ji W, Yan K, Gao J, Zhang J (2019) Fuel cell-based self-powered electrochemical sensors for biochemical detection. Nano Energy 61:173–193

Chu N, Liang Q, Hao W, Jiang Y, Liang P, Zeng RJ (2021) Microbial electrochemical sensor for water biotoxicity monitoring. Chem Eng J 404:127053

Chung TH, Meshref MN, Dhar BR (2020) Microbial electrochemical biosensor for rapid detection of naphthenic acid in aqueous solution. J Electroanal Chem 873:114405

Chung TH, Meshref MN, Dhar BR (2021) A review and roadmap for developing microbial electrochemical cell-based biosensors for recalcitrant environmental contaminants, with emphasis on aromatic compounds. Chem Eng J 424:130245

Cong Y, Xu Q, Feng H, Shen D (2013) Efficient electrochemically active biofilm denitrification and bacteria consortium analysis. Biores Technol 132:24–27

Cristea C, Florea A, Tertiş M, Săndulescu R (2015) Immunosensors. In: Biosensors-micro and nanoscale applications. IntechOpen, London

Cristiani P (2014) Risk assessment of biocorrosion in condensers, pipework and other cooling system components. Understanding Biocorrosion 2014:357–384

Cristiani P, Perboni G (2021) Corrosion monitoring in microbial environments. In: Techniques for corrosion monitoring. Woodhead Publishing, London, pp 335–377

Díaz-González M, Fernández-Sánchez C (2019) Decentralized analysis of water contaminants using compact (bio) electroanalytical tools. Curr Opin Environ Sci Health 10:47–56

D'souza SF (2001) Microbial biosensors. Biosens Bioelectron 16(6):337–353

Ejeian F, Etedali P, Mansouri-Tehrani HA, Soozanipour A, Low ZX, Asadnia M, Razmjou A (2018) Biosensors for wastewater monitoring: a review. Biosens Bioelectron 118:66–79

Ewis D, Ba-Abbad MM, Benamor A, El-Naas MH (2022) Adsorption of organic water pollutants by clays and clay minerals composites: a comprehensive review. Appl Clay Sci 229:106686

Flemming H, Wingender J (2010) The biofilm matrix. Nat Publ Gr 8:623–633

Grandclément C, Seyssiecq I, Piram A, Wong-Wah-Chung P, Vanot G, Tiliacos N, Doumenq P (2017) From the conventional biological wastewater treatment to hybrid processes, the evaluation of organic micropollutant removal: a review. Water Res 111:297–317

Gupta VK, Yola ML, Atar N, Ustundağ Z, Solak AO (2013) A novel sensitive Cu(II) and Cd(II) nanosensor platform: graphene oxide terminated *p*-aminophenyl modified glassy carbon surface. Electrochim Acta 112:541–548

Gupta R, Raza N, Bhardwaj SK, Vikrant K, Kim KH, Bhardwaj N (2021) Advances in nanomaterial-based electrochemical biosensors for the detection of microbial toxins, and pathogenic bacteria in food matrices. J Hazard Mater 401:123379

Hassan RY, Febbraio F, Andreescu S (2021) Microbial electrochemical systems: principles, construction and biosensing applications. Sensors 21(4):1279

Hill A, Tait S, Baillie C, Virdis B, McCabe B (2020) Microbial electrochemical sensors for volatile fatty acid measurement in high strength wastewaters: a review. Biosens Bioelectron 165:112409

Jiang Y, Liang P, Liu P, Wang D, Miao B, Huang X (2017) A novel microbial fuel cell sensor with biocathode sensing element. Biosens Bioelectron 94:344–350

Jiang Y, Yang X, Liang P, Liu P, Huang X (2018) Microbial fuel cell sensors for water quality early warning systems: fundamentals, signal resolution, optimization and future challenges. Renew Sustain Energy Rev 81:292–305

Johnson BL, Hicks HE, Jones DE, Cibulas W, Wargo A, De Rosa CT (1998) Public health implications of persistent toxic substances in the Great Lakes and St. Lawrence basins. J Great Lakes Res 24(3):698–722

Kannappan S, Ramisetty BCM (2022) Engineered whole-cell-based biosensors: sensing environmental heavy metal pollutants in water—a review. Appl Biochem Biotechnol 2022:1–27

Kanso H, Barthelmebs L, Inguimbert N, Noguer T (2013) Immunosensors for estradiol and ethinylestradiol based on new synthetic estrogen derivatives: application to wastewater analysis. Anal Chem 85(4):2397–2404

Karbelkar AA, Furst AL (2020) Electrochemical diagnostics for bacterial infectious diseases. ACS Infect Dis 6(7):1567–1571

Karimian N, Hashemi P, Khanmohammadi A, Afkhami A, Bagheri H (2020) The principles and recent applications of bioelectrocatalysis. Anal Bioanal Chem Res 7(3):281–301

Karimi-Maleh H, Karimi F, Alizadeh M, Sanati AL (2020) Electrochemical sensors, a bright future in the fabrication of portable kits in analytical systems. Chem Rec 20(7):682–692
Khan A, Salama ES, Chen Z, Ni H, Zhao S, Zhou T, Li X (2020) A novel biosensor for zinc detection based on microbial fuel cell system. Biosens Bioelectron 147:111763
Kim KH, Jahan SA, Kabir E, Brown RJ (2013) A review of airborne polycyclic aromatic hydrocarbons (PAHs) and their human health effects. Environ Int 60:71–80
Koncki R (2007) Recent developments in potentiometric biosensors for biomedical analysis. Anal Chim Acta 599(1):7–15
Kucherenko IS, Soldatkin OO, Dzyadevych SV, Soldatkin AP (2020) Electrochemical biosensors based on multienzyme systems: main groups, advantages and limitations–a review. Anal Chim Acta 1111:114–131
Lazzarini Behrmann IC, Grattieri M, Minteer SD, Ramirez SA, Vullo DL (2020) Online self-powered Cr(VI) monitoring with autochthonous Pseudomonas and a bio-inspired redox polymer. Anal Bioanal Chem 412:6449–6457
Liao C, Wu J, Zhou L, Li T, Du Q, An J, Wang X (2018) Optimal set of electrode potential enhances the toxicity response of biocathode to formaldehyde. Sci Total Environ 644:1485–1492
Logan BE, Rossi R, Ragab AA, Saikaly PE (2019) Electroactive microorganisms in bioelectrochemical systems. Nat Rev Microbiol 17(5):307–319
Ma Z, Meliana C, Munawaroh HSH, Karaman C, Karimi-Maleh H, Low SS, Show PL (2022) Recent advances in the analytical strategies of microbial biosensor for detection of pollutants. Chemosphere 306:135515
Mahmoud RH, Samhan FA, Ali GH, Ibrahim MK, Hassan RY (2018) Assisting the biofilm formation of exoelectrogens using nanostructured microbial fuel cells. J Electroanal Chem 824:128–135
Mahmoud RH, Abdo SM, Samhan FA, Ibrahim MK, Ali GH, Hassan RY (2020) Biosensing of algal-photosynthetic productivity using nanostructured bioelectrochemical systems. J Chem Technol Biotechnol 95(4):1028–1037
Mahmoud RH, Samhan FA, Ibrahim MK, Ali GH, Hassan RY (2021) Formation of electroactive biofilms derived by nanostructured anodes surfaces. Bioprocess Biosyst Eng 44:759–768
Mojiri A, Zhou JL, Ohashi A, Ozaki N, Kindaichi T (2019) Comprehensive review of polycyclic aromatic hydrocarbons in water sources, their effects and treatments. Sci Total Environ 696:133971
Mollica A, Cristiani P (2003) On-line biofilm monitoring by" BIOX" electrochemical probe. Water Sci Technol 47(5):45–49
Morin-Crini N, Lichtfouse E, Liu G, Balaram V, Ribeiro ARL, Lu Z, Crini G (2022) Worldwide cases of water pollution by emerging contaminants: a review. Environ Chem Lett 20(4):2311–2338
Nigam VK, Shukla P (2015) Enzyme based biosensors for detection of environmental pollutants—a review. J Microbiol Biotechnol 25(11):1773–1781
Nomngongo PN, Ngila JC, Msagati TA, Gumbi BP, Iwuoha EI (2012) Determination of selected persistent organic pollutants in wastewater from landfill leachates, using an amperometric biosensor. Phys Chem Earth Parts a/b/c 50:252–261
Olias LG, Otero AR, Cameron PJ, Di Lorenzo M (2020) A soil microbial fuel cell-based biosensor for dissolved oxygen monitoring in water. Electrochim Acta 362:137108
Oon YS, Ong SA, Ho LN, Wong YS, Oon YL, Lehl HK, Nordin N (2018) Disclosing the synergistic mechanisms of azo dye degradation and bioelectricity generation in a microbial fuel cell. Chem Eng J 344:236–245
O'Toole GA, Kolter R (1998) Flagellar and twitching motility are necessary for *Pseudomonas aeruginosa* biofilm development. Mol Microbiol 30(2):295–304
Palchetti I, Laschi S, Mascini M (2009) Electrochemical biosensor technology: application to pesticide detection. In: Biosensors and biodetection: methods and protocols: electrochemical and mechanical detectors, lateral flow and ligands for biosensors, pp 115–126
Patil SA, Hägerhäll C, Gorton L (2012) Electron transfer mechanisms between microorganisms and electrodes in bioelectrochemical systems. Bioanal Rev 4:159–192

Pawełczyk A (2013) Assessment of health risk associated with persistent organic pollutants in water. Environ Monit Assess 185:497–508

Pojana G, Gomiero A, Jonkers N, Marcomini A (2007) Natural and synthetic endocrine disrupting compounds (EDCs) in water, sediment and biota of a coastal lagoon. Environ Int 33(7):929–936

Pozzer AC, Gómez PA, Weiss J (2022) Volatile organic compounds in aquatic ecosystems—detection, origin, significance and applications. Sci Total Environ 838:156155

Prévoteau A, Rabaey K (2017) Electroactive biofilms for sensing: reflections and perspectives. ACS Sens 2(8):1072–1085

Prévoteau A, Carvajal-Arroyo JM, Ganigué R, Rabaey K (2020) Microbial electrosynthesis from CO2: Forever a promise? Curr Opin Biotechnol 62:48–57

Qiao M, Bai Y, Cao W, Huo Y, Zhao X, Liu D, Li Z (2018) Impact of secondary effluent from wastewater treatment plants on urban rivers: Polycyclic aromatic hydrocarbons and derivatives. Chemosphere 211:185–191

Qing Li Q, Loganath A, Seng Chong Y, Tan J, Philip Obbard J (2006) Persistent organic pollutants and adverse health effects in humans. J Toxicol Environ Health A 69(21):1987–2005

Reguera G, Nevin KP, Nicoll JS, Covalla SF, Woodard TL, Lovley DR (2006) Biofilm and nanowire production leads to increased current in Geobacter sulfurreducens fuel cells. Appl Environ Microbiol 72(11):7345–7348. https://doi.org/10.1128/AEM.01444-06

Roy R, Ray S, Chowdhury A, Anand R (2021) Tunable multiplexed whole-cell biosensors as environmental diagnostics for ppb-level detection of aromatic pollutants. ACS Sens 6(5):1933–1939

Rudel RA, Camann DE, Spengler JD, Korn LR, Brody JG (2003) Phthalates, alkylphenols, pesticides, polybrominated diphenyl ethers, and other endocrine-disrupting compounds in indoor air and dust. Environ Sci Technol 37(20):4543–4553

Saha JK, Selladurai R, Coumar MV, Dotaniya ML, Kundu S, Patra AK, Patra AK (2017) Organic pollutants. Soil Poll Emerg Threat Agric 2017:105–135

Sánchez C, Dessì P, Duffy M, Lens PN (2020) Microbial electrochemical technologies: electronic circuitry and characterization tools. Biosens Bioelectron 150:111884

Saratale GD, Saratale RG, Shahid MK, Zhen G, Kumar G, Shin HS, Kim SH (2017) A comprehensive overview on electro-active biofilms, role of exo-electrogens and their microbial niches in microbial fuel cells (MFCs). Chemosphere 178:534–547

Si RW, Zhai DD, Liao ZH, Gao L, Yong YC (2015) A whole-cell electrochemical biosensing system based on bacterial inward electron flow for fumarate quantification. Biosens Bioelectron 68:34–40

Simoska O, Stevenson KJ (2019) Electrochemical sensors for rapid diagnosis of pathogens in real time. Analyst 144(22):6461–6478

Simoska O, Sans M, Fitzpatrick MD, Crittenden CM, Eberlin LS, Shear JB, Stevenson KJ (2018) Real-time electrochemical detection of pseudomonas aeruginosa phenazine metabolites using transparent carbon ultramicroelectrode arrays. ACS Sens 4(1):170–179

Singh A, Sharma A, Ahmed A, Sundramoorthy AK, Furukawa H, Arya S, Khosla A (2021) Recent advances in electrochemical biosensors: applications, challenges, and future scope. Biosensors 11(9):336

Smutok O, Dmytruk K, Gonchar M, Sibirny A, Schuhmann W (2007) Permeabilized cells of flavocytochrome b2 over-producing recombinant yeast *Hansenula polymorpha* as biological recognition element in amperometric lactate biosensors. Biosens Bioelectron 23(5):599–605

Song N, Yan Z, Xu H, Yao Z, Wang C, Chen M, Jiang HL (2019) Development of a sediment microbial fuel cell-based biosensor for simultaneous online monitoring of dissolved oxygen concentrations along various depths in lake water. Sci Total Environ 673:272–280

Soni V, Singh P, Shree V, Goel V (2018) Effects of VOCs on human health. Air Pollut Control 2018:119–142

Spurr MW, Eileen HY, Scott K, Head IM (2018) Extending the dynamic range of biochemical oxygen demand sensing with multi-stage microbial fuel cells. Environ Sci Water Res Technol 4(12):2029–2040

Tecon R, Van der Meer JR (2008) Bacterial biosensors for measuring availability of environmental pollutants. Sensors 8(7):4062–4080

Thavarajah W, Verosloff MS, Jung JK, Alam KK, Miller JD, Jewett MC, Lucks JB (2020) A primer on emerging field-deployable synthetic biology tools for global water quality monitoring. NPJ Clean Water 3(1):18

Tucci M, Grattieri M, Schievano A, Cristiani P, Minteer SD (2019) Microbial amperometric biosensor for online herbicide detection: photocurrent inhibition of *Anabaena variabilis*. Electrochim Acta 302:102–108

Tudorache M, Bala C (2007) Biosensors based on screen-printing technology, and their applications in environmental and food analysis. Anal Bioanal Chem 388:565–578

Voon CH, Yusop NM, Khor SM (2022) The state-of-the-art in bioluminescent whole-cell biosensor technology for detecting various organic compounds in oil and grease content in wastewater: from the lab to the field. Talanta 241:123271

Warner NA, Evenset A, Christensen G, Gabrielsen GW, Borgå K, Leknes H (2010) Volatile siloxanes in the European arctic: assessment of sources and spatial distribution. Environ Sci Technol 44(19):7705–7710

Webster DP, TerAvest MA, Doud DF, Chakravorty A, Holmes EC, Radens CM, Angenent LT (2014) An arsenic-specific biosensor with genetically engineered *Shewanella oneidensis* in a bioelectrochemical system. Biosens Bioelectron 62:320–324

Wg C (1990) Transport and interfacial transport phenomena. Biofilms 1990:265–340

Xu Z, Liu B, Dong Q, Lei Y, Li Y, Ren J, Li B (2015) Flat microliter membrane-based microbial fuel cell as "on-line sticker sensor" for self-supported in situ monitoring of wastewater shocks. Biores Technol 197:244–251

Yang K, Wang C, Xue S, Li W, Liu J, Li L (2019) The identification, health risks and olfactory effects assessment of VOCs released from the wastewater storage tank in a pesticide plant. Ecotoxicol Environ Saf 184:109665

Yu D, Zhai J, Liu C, Zhang X, Bai L, Wang Y, Dong S (2017) Small microbial three-electrode cell based biosensor for online detection of acute water toxicity. ACS Sens 2(11):1637–1643

Zambianchi M, Durso M, Liscio A, Treossi E, Bettini C, Capobianco ML, Melucci M (2017) Graphene oxide doped polysulfone membrane adsorbers for the removal of organic contaminants from water. Chem Eng J 326:130–140

Zhou G, Yuan J, Gao H (2015) Regulation of biofilm formation by BpfA, BpfD, and BpfG in Shewanella oneidensis. Front Microbiol 6:790. https://doi.org/10.3389/fmicb.2015.00790

Microbial Fuel Cell for the Treatment of Wastewater Pollutants and Resource Recovery

Deepika Jothinathan, Ashutosh Pandey, Kamlesh Choure, and Lavanyasri Rathinavel

Abstract Microbial fuel cells (MFCs) represent an innovative and sustainable technology for the simultaneous treatment of water pollutants and resource recovery. This study explores the application of MFCs in water treatment, focusing on their capacity to harness the metabolic activities of microorganisms for the degradation of diverse water pollutants. The microbial electrochemical reactions within MFCs facilitate the removal of organic contaminants, inorganic pollutants, and emerging micropollutants from water bodies. Moreover, MFCs play a pivotal role in resource recovery by generating electrical energy during the microbial oxidation of organic matter. This electricity generation not only provides a clean and renewable energy source but also contributes to the overall sustainability of the water treatment process. Additionally, MFCs enable the recovery of valuable resources, including nutrients such as nitrogen and phosphorus, which can be repurposed as fertilizers for agricultural applications. The study highlights the environmental benefits of using MFCs, such as reduced chemical usage, lower energy consumption and the potential for circular economy practices through the recovery of resources. Various case studies and successful applications demonstrate the versatility of MFCs in treating municipal wastewater, remediating industrial effluents, and addressing agricultural runoff. The integration of microbial communities in MFCs contributes to the formation of biofilms, enhancing pollutant degradation efficiency. In conclusion, microbial fuel cells emerge as a promising technology for sustainable water treatment, combining the removal of

D. Jothinathan
Department of Biotechnology, SRM Arts and Science College, Kattankulathur, Chennai, Tamil Nadu, India

A. Pandey (✉)
Institute for Water and Wastewater Technology, Durban University of Technology, Durban, South Africa
e-mail: ashuthoshp@dut.ac.za

K. Choure
Department of Biotechnology, AKS University, Satna, MP, India

L. Rathinavel (✉)
Department of Biotechnology, Mercy College, Palakkad, Kerala 678006, India
e-mail: biotechcreditsu@gmail.com

R. Selvasembian et al. (eds.), *Emerging Trends in Microbial Electrochemical Technologies for Sustainable Mitigation of Water Resources Contamination*,
https://doi.org/10.1007/978-3-031-74636-9_6

water pollutants with the recovery of valuable resources. This study provides insights into the mechanisms, applications, potential environmental and economic benefits associated with the utilization of MFCs in water treatment processes.

Keywords Product recovery · Metals · Organic contaminants · Biofilm formation · Water treatment · Nutrient recovery

1 Introduction

Water pollution is a critical environmental challenge that poses significant threats to ecosystems and human health globally. It is primarily caused by the discharge of pollutants into water bodies, leading to adverse effects on aquatic life and the overall balance of ecosystems (Ekubo and Abowei 2011). Recognizing the importance of addressing water pollution, researchers have been exploring innovative technologies to remediate contaminated water while simultaneously recovering valuable resources. One such promising technology is the utilization of Microbial Fuel Cells (MFCs).

Water pollution has become a pressing concern due to rapid industrialization, urbanization, and agricultural activities. Effluents from industrial processes, untreated sewage and agricultural runoff contribute to the contamination of water bodies with pollutants such as heavy metals, organic compounds, and nutrients (Hampel et al. 2015). These pollutants can have detrimental effects on aquatic ecosystems, disrupting the balance of ecosystems and threatening the availability of clean water resources.

Conventional water treatment methods often prove insufficient in addressing the complex nature of emerging contaminants and persistent pollutants (Bolong et al. 2009). Additionally, these methods are energy-intensive and may result in the generation of sludge, requiring further disposal measures (Klöpffer 1996). Thus, there is a growing need for sustainable and cost-effective technologies that not only remediate water pollution but also contribute to resource recovery.

The significance of addressing water pollution extends beyond environmental concerns to encompass socio-economic and public health aspects. Contaminated water sources directly impact communities relying on these resources for drinking water, agriculture, and industrial processes. Moreover, the economic burden associated with water pollution, including healthcare costs and loss of productivity, emphasizes the urgent need for effective remediation technologies (Coppola and Trevisan 2012). Resource recovery from polluted water sources further enhances the economic feasibility of water treatment processes (Pasqualino et al. 2011). Valuable resources, such as energy and nutrients can be extracted from wastewater, contributing to a circular economy approach. This not only mitigates the environmental impact of pollutants but also creates opportunities for sustainable resource management (Guerra-Rodríguez et al. 2020).

The main objective of this chapter is to underscore the environmental advantages of Microbial Fuel Cells (MFCs) in water treatment, emphasizing their ability to reduce chemical usage, lower energy consumption, and facilitate circular economy practices through resource recovery. Through a presentation of various case studies and successful applications, the versatility of MFCs in treating municipal wastewater, remediating industrial effluents, and mitigating agricultural runoff is showcased. Additionally, the integration of microbial communities within MFCs is highlighted for its role in enhancing pollutant degradation efficiency through biofilm formation. Ultimately, the study positions MFCs as a promising technology for sustainable water treatment, merging pollutant removal with valuable resource recovery. By providing insights into the mechanisms, applications, and potential environmental and economic benefits associated with MFC utilization in water treatment processes, the content aims to offer comprehensive understanding and support for the adoption of MFCs in environmental management practices.

2 Microbial Fuel Cells (MFCs)

Microbial fuel cells (MFCs) have emerged as a promising technology for addressing water pollution and recovering resources simultaneously. MFCs harness the metabolic activity of microorganisms to generate electrical energy from organic matter present in wastewater. The microbial processes involved in MFCs facilitate the degradation of pollutants, offering a dual benefit of remediation and energy production (Kim et al. 2007). The basic principle of MFCs involves the microbial oxidation of organic compounds, generating electrons that flow through an external circuit, producing electricity. The use of MFCs in water treatment not only provides an environmentally friendly means of electricity generation but also promotes the degradation of pollutants, thereby improving water quality (Mink et al. 2012). This chapter aims to explore the potential of Microbial Fuel Cells in treating water pollution while emphasizing the importance of resource recovery. By understanding the microbial processes within MFCs and optimizing their performance, this study seeks to contribute to the development of sustainable and efficient solutions for water pollution remediation and resource management.

2.1 Principle of Microbial Fuel Cells

MFCs operate on the fundamental principle of harnessing the metabolic activities of microorganisms to generate electrical energy. These devices facilitate the electron transfer process from microbial oxidation of organic matter to an electrode, resulting in the production of electricity (Rabaey and Verstraete 2005). The core concept revolves around the utilization of microorganisms as biocatalysts that facilitate the breakdown of organic compounds, releasing electrons in the process. These electrons

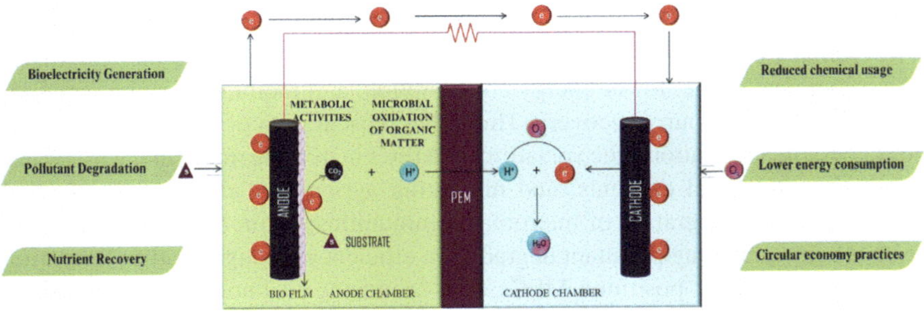

Fig. 1 Schematic diagram of microbial fuel cell (MFC) and its different applications

are then captured by an anode, creating a flow of current through an external circuit (Rabaey and Rozendal 2010).

Microbial fuel cells (MFCs) represent an innovative technology harnessing electricity through microbial metabolism of substances. This process illustrated in Fig. 1 involves oxidation–reduction reactions, releasing electrons that generate electricity. MFCs consist of two chambers—an anode and a cathode—separated by a proton exchange membrane (PEM). Within the anode chamber, microorganisms metabolize organic substances, producing protons and electrons. These electrons travel to the anode and reach the cathode through an external circuit, while protons migrate to the cathode chamber via the PEM.

In the cathode chamber, electrons and protons combine, facilitating the reduction of oxygen to water. MFCs feature multiple gas inflows and outflows, with oxygen inlets in the cathode chamber significantly impacting electricity output. The source of oxygen varies depending on the MFC type, with atmospheric air used for single-cell MFCs and mechanical aeration for dual-cell MFCs. Carbon dioxide is the primary gaseous end product, with glucose, acetate, or wastewater serving as substrates. The cathode chamber maintains an alkaline condition, enhancing carbon dioxide absorption from the anode. This alkaline environment arises from the accumulation of hydroxide ions resulting from oxygen reduction at the cathode. However, practical limitations exist in MFC applications for oxygen gas delivery and carbon dioxide accumulation. These challenges can be addressed through the adoption of efficient and sustainable catalysts for cathode reactions (Rathinavel et al. 2018).

2.2 *Factors Influencing MFC Performance*

The components and design of Microbial Fuel Cells are crucial aspects influencing their efficiency and performance. MFCs consist of an anode, a cathode and a membrane separating the two chambers. The anode serves as the site for microbial oxidation, while the cathode facilitates reduction reactions. The ion-selective membrane ensures the migration of protons while preventing direct contact between the anode and cathode (Oh and Logan 2006). Electrodes, often made of conductive materials like graphite, play a critical role in facilitating electron transfer (Bélafi-Bakó et al. 2014). The design parameters such as surface area, electrode spacing and reactor configuration significantly impact the overall efficiency of MFCs (Hu 2008).

3 Microbial Community in Water Treatment

3.1 Role of Microorganisms in Water Pollutant Degradation

Microorganisms play a pivotal role in the degradation of water pollutants within Microbial Fuel Cells. Various bacterial species possess the metabolic capability to oxidize organic compounds, serving as electron donors in MFCs. Microbial fuel cells show great diversity in bacterial communities, ranging from delta-Proteobacteria in sediment MFCs to alpha-, beta-, gamma-, gamma- or delta-Proteobacteria, Firmicutes, and uncharacterized clones (Logan and Regan 2006). *Enterobacter* sp., *Staphylococcus* sp., and *Pseudomonas* sp. are key microorganisms involved in microbial fuel cell energy production (Das 2013). These microorganisms utilize enzymatic pathways to break down complex organic molecules into simpler components, releasing electrons in the process (Fetzner 2000). The electron transfer to the anode not only generates electrical power but also contributes to the removal of organic pollutants from water, thereby improving water quality (Sirés et al. 2014).

3.2 Diversity and Functionality of Microbial Consortia

The microbial community within MFCs is characterized by a diverse consortium of microorganisms with distinct functionalities. Different species collaborate in a synergistic manner, each contributing to specific stages of pollutant degradation and electron transfer (Logan and Regan 2006). This diversity enhances the system's robustness and adaptability, allowing for the treatment of a broad spectrum of pollutants (Polzin et al. 2017). Understanding the interactions and dynamics of microbial consortia is crucial for optimizing MFC performance in water treatment applications.

3.3 Biofilm Formation in MFCs

Biofilm formation is a key aspect of Microbial Fuel Cells, influencing both the microbial community structure and overall system performance. Microorganisms adhere to the electrode surfaces, forming biofilms that enhance electron transfer efficiency (Saratale et al. 2017). Biofilms serve as conductive matrices, facilitating the direct transfer of electrons from microorganisms to electrodes. *Shewanella decolorationis* S12 biofilms in microbial fuel cells has been reported to generate more current than planktonic cells, with closed-circuit operation maintaining biofilm viability (Yang et al. 2011). However, biofilm formation also poses challenges such as potential clogging of pores and reduced mass transfer rates (Marcus et al. 2007). Balancing biofilm development is essential for maintaining MFC stability and efficiency in water treatment applications.

4 Water Pollutants Targeted by Microbial Fuel Cells

Water pollution is a persistent global issue, necessitating innovative technologies for effective remediation. Microbial Fuel Cells (MFCs) have emerged as promising tools for addressing various water pollutants, demonstrating versatility in targeting a range of contaminants. In this review, we explore the effectiveness of MFCs in targeting specific classes of water pollutants, namely organic contaminants, inorganic contaminants, and emerging pollutants/micropollutants.

4.1 Organic Contaminants

Organic contaminants constitute a significant portion of water pollution, originating from industrial discharges, agricultural runoff, and domestic wastewater. Microbial Fuel Cells exhibit remarkable efficiency in targeting organic contaminants through the metabolic activities of microorganisms (Gude 2016). MFC reactors are an effective way to treat organic pollutants such as synthetic organic dyes, pesticides, pharmaceuticals, phenolic compounds, polyhydrocarbons etc. (Suresh et al. 2022). In 2019, researchers designed a photoelectrocatalytic MFC to remove 95% of organic pollutants such as phenol and aniline from wastewater. This system also resulted with high voltage, maximum power density, low internal resistance and high coloumbic efficiency (Zhang et al. 2019). Likewise, a photocatalytic MFC showed 79.3% of 2,4,6 Trichlorophenol in 10 h. Pseudomonas was the dominating bacterium in the anode and Rhodococcus in cathode also aided in the TCP removal from wastewater (Wang et al. 2019). Bacterial species within the MFCs catalyze the oxidation of organic matter, releasing electrons in the process. This electron flow not only generates electrical power but also facilitates the breakdown of complex organic compounds, leading to the degradation of pollutants (Passos et al. 2016). In another study, benthic MFC was employed with yam waste to remediate heavy metals such as Pb^{2+}, Cd^{2+}, Cr^{3+}. As a result, after 30 days of time period, the maximum remediation efficiency of 90.14%, 88% and 90.34% respectively was achieved (Fadzli et al. 2021). The unique ability of MFCs to harness the microbial-driven degradation of organic contaminants makes them a promising technology for sustainable water treatment.

4.2 Inorganic Contaminants

Inorganic contaminants, including heavy metals, nutrients and salts pose serious threats to water quality and ecosystem health. Microbial Fuel Cells play a crucial role in addressing certain inorganic pollutants through microbial processes (Wehrer et al. 2011). For instance, in 2013 researchers attempted to remove arsenic in MFC coupled with ZVI (Zerovalent Iron) hybrid process. As a result, they could achieve water quality standard as per EPA and WHO (Xue et al. 2013). The redox reactions occurring at the anode can facilitate the reduction and removal of specific metal ions. In another study, Cu removal was successfully performed in metallurgical MFC under aerobic and anaerobic conditions. It was found that in both the conditions, the microbes were able to remove more than 99% (Heijne et al. 2010). In order to convert Cr(VI) into non-toxic Cr(III), Li et al. used three distinct cathode materials: carbon felt, carbon brush, and carbon cloth. The findings shown that, at an ideal pH of 2, 80 mg/L of Cr(VI) could be entirely eliminated by MFC using a carbon cloth cathode in 72 h. In contrast, the removal of Cr(VI) was only achieved with an efficiency of 33.45% and 12.72% when carbon brush and carbon felt were used as the cathode materials, respectively (Li et al. 2018). Additionally, the electrochemical reactions within MFCs influence the chemical conditions, contributing to the precipitation or sorption of inorganic contaminants (Jadhav et al. 2018). While the efficacy may vary depending on the specific contaminants, MFCs offer a potential solution for mitigating the impact of inorganic pollutants on water bodies.

4.3 Emerging Pollutants and Micropollutants

Emerging pollutants and micropollutants present a modern challenge in water pollution, as conventional treatment methods often struggle to remove these persistent compounds (Bolong et al. 2009). Microbial Fuel Cells demonstrate promise in addressing emerging pollutants by harnessing diverse microbial consortia capable of enzymatically breaking down complex compounds. Pharmaceuticals, personal care products, and industrial chemicals fall into this category and MFCs offer a sustainable approach for their removal (Ben et al. 2018). Using a single chambered up-flow MFC, Thung et al. found a color removal efficiency of roughly 80.6% (Acid orange 7; starting concentration of 50 mg L^{-1}) Biofilm attachment growth was boosted on the inert gravel surface of the MFC operating in the up-flow mode with intermediate gravel layers between the electrodes (Thung et al. 2015). Penicillin's impact on power generation in a single chambered MFC fed with synthetic wastewater based on glucose was evaluated by Wen et al. When fed 1000 mg L^{-1} glucose + 50 mg L^{-1} penicillin, the maximum power density value of 101.2 W m^{-3} was achieved (Wen et al. 2011). A comparison of the Congo red degradation produced in a membrane-less MEC and a dual chambered MEC revealed that the internal resistance of the membrane-less MEC was 5.2 Ω, eight times lower than that of the dual chambered

MEC mode. It was found that the addition of catholytes in cathode enhanced congo red removal (Huang et al. 2018). The microbial-driven degradation of emerging pollutants showcases the adaptability and potential of MFCs in addressing contemporary water quality challenges (Suresh et al. 2022).

5 Resource Recovery from Microbial Fuel Cells

Resource recovery is an integral aspect of sustainable water treatment strategies, and Microbial Fuel Cells (MFCs) have emerged as innovative technologies that not only remediate water pollutants but also facilitate the retrieval of valuable resources. This part of the chapter explores the concept of resource recovery from MFCs, encompassing the types of resources recovered and the environmental benefits associated with this approach.

5.1 *Concept of Resource Recovery*

The concept of resource recovery in the context of Microbial Fuel Cells involves the extraction and reuse of materials or energy from the treated water, transforming the wastewater treatment process into a sustainable and circular system (Kumar et al. 2019). MFCs utilize the metabolic activities of microorganisms to break down organic matter, generating electrical energy in the process. Simultaneously, various resources present in the wastewater are recovered and repurposed. This dual-purpose approach aligns with the principles of a circular economy, promoting the idea that waste from one process can become a valuable resource for another.

5.2 *Types of Resources Recovered*

Microbial Fuel Cells enable the recovery of diverse resources from wastewater, enhancing their sustainability and contributing to a more circular approach to water treatment. The primary resource recovered is electrical energy, generated during the microbial oxidation of organic matter. This electrical energy can be harnessed for powering various applications, reducing the dependence on conventional energy sources (Lovley 2006). Additionally, MFCs facilitate the recovery of nutrients, particularly nitrogen and phosphorus, in the form of bioavailable compounds. These nutrient-rich byproducts can be utilized as fertilizers, closing the loop in nutrient cycles, and reducing the need for synthetic alternatives (Sengupta et al. 2015). Resources can be recovered in the following four major states of matter: gas (biomethane and hydrogen), liquid (biofuel, reuse water, acids, alkalis, and other

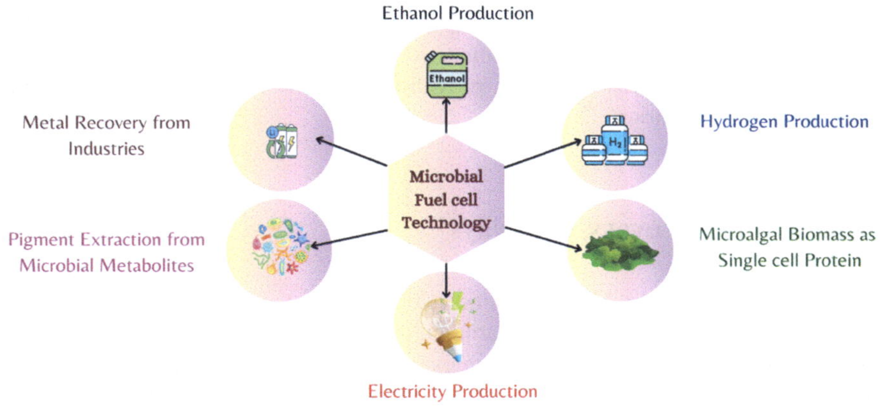

Fig. 2 Different types of resources recovered using the microbial fuel cell technology

chemical compounds) and solid (heavy metals and nutrients). MFC fed with potassium ferricyanide (MFC-FC) and MFC fed with phosphate buffer as catholyte (MFC-blank) were tested for their bioelectrochemical performance with MFC fed with artificial wastewater laden with silver (MFC-Ag). After 72 h of operation, high silver removal (83 ± 0.7%) and recovery (67.8 ± 1%) efficiency were obtained with MFC-Ag. For MFC-Ag, the highest power density (3006 mW/m^3) and current density (3410 mA/m^3) were discovered (Ali et al. 2019). Furthermore, the microbial processes within MFCs can result in the production of methane, offering another potential energy resource that can be utilized for heating or power generation (Virdis et al. 2009).

An innovative integrated drip hydroponics-microbial fuel cell system was designed with domestic wastewater and planted with lemon grass plantlet. It yielded power alongside the plant biomass. These are the effective way of recovering resources like small plants and herbs from MFC (Yadav et al. 2020). In another study, a novel MEC system was designed to produce hydrogen continuously under visible light, with an average hydrogen-production rate of 1.35 ± 0.15 mL h^{-1} and a maximum current density of 0.68 A m^{-2} (Wan et al. 2015) (Fig. 2).

5.3 *Environmental Benefits of Resource Recovery*

Resource recovery from Microbial Fuel Cells offers substantial environmental benefits, contributing to the overall sustainability of water treatment processes. The recovery of electrical energy from MFCs aligns with the global shift towards renewable and clean energy sources, reducing the carbon footprint associated with conventional power generation. The utilization of recovered nutrients as fertilizers promotes sustainable agricultural practices, minimizing the environmental impact of synthetic fertilizers and supporting nutrient cycling in ecosystems (Anex et al. 2007). Additionally, the potential recovery of methane as an energy resource further enhances the environmental benefits of MFCs, providing an additional avenue for reducing greenhouse gas emissions.

6 Resource Recovery Pathways

Resource recovery from Microbial Fuel Cells (MFCs) represents a transformative approach in sustainable water treatment, demonstrating versatility in extracting various valuable resources. This section examines the key resource recovery pathways associated with MFCs, including electricity generation, nutrient reclamation, metal, and chemical extraction, as well as biohydrogen and biogas production.

6.1 Electricity Generation

One of the primary and most studied resource recovery pathways from MFCs is electricity generation. The microbial oxidation of organic matter at the anode generates electrons, which can be harnessed as electrical power (Zhang et al. 2013). MFCs have shown promising potential for producing sustainable and renewable energy through this process. The electrical output from MFCs has been explored for powering small electronic devices, sensors, and even contributing to the grid. As advancements continue in MFC technology and system optimization, the scalability and efficiency of electricity generation from MFCs are expected to increase, promoting their integration into broader energy systems (Du et al. 2007).

6.2 Nutrient Reclamation

Microbial Fuel Cells offer a significant avenue for the reclamation of nutrients, particularly nitrogen and phosphorus, from wastewater (Kelly and He 2014). The microbial processes involved in MFCs lead to the formation of bioavailable compounds that can be utilized as fertilizers. By recovering and repurposing these nutrients, MFCs contribute to sustainable agriculture practices (Kour et al. 2020). The nutrient-rich byproducts from MFCs can be employed in crop cultivation, reducing the reliance on synthetic fertilizers, and minimizing the environmental impact associated with nutrient runoff into water bodies (Mahmud et al. 2020).

6.3 *Metal and Chemical Extraction*

MFCs have demonstrated the ability to assist in the extraction of metals and chemicals from contaminated water (Sophia and Sai 2016). The microbial-driven redox reactions at the anode can influence the speciation and solubility of certain metal ions, leading to their precipitation or adsorption onto electrode surfaces. This mechanism offers a potential avenue for the remediation of water polluted with heavy metals (Brown et al. 1999). Additionally, MFCs can be coupled with electrochemical processes for enhanced metal recovery, showcasing their potential in contributing to the circular economy by recycling valuable metals.

6.4 *Biohydrogen and Biogas Production*

In addition to electricity generation, MFCs exhibit potential to produce biohydrogen and biogas. Hydrogen gas can be produced at the cathode through the reduction of protons, offering a clean and renewable energy source (Sun et al. 2008). Similarly, the microbial degradation of organic matter in MFCs can result in the production of methane, constituting a form of biogas. Both biohydrogen and biogas hold promise as alternative energy carriers, and the optimization of MFC conditions can enhance their yield for practical applications in the future (Bharathiraja et al. 2016). The resource recovery pathways associated with Microbial Fuel Cells extend beyond electricity generation, encompassing nutrient reclamation, metal and chemical extraction, as well as biohydrogen and biogas production. These pathways collectively contribute to the sustainability of water treatment processes, offering solutions to energy production and resource management challenges.

7 Applications and Case Studies

Microbial Fuel Cells (MFCs) have emerged as versatile and sustainable tools for various water treatment applications, showcasing their potential in municipal wastewater treatment, industrial effluent remediation, agricultural runoff treatment, and successful resource recovery projects. The following subsection explores the current state of MFC applications through case studies and highlights their efficacy in addressing diverse water treatment challenges.

7.1 Municipal Wastewater Treatment

Municipal wastewater treatment is a critical aspect of urban infrastructure, and MFCs have demonstrated promise in this domain. Case studies reveal the effectiveness of MFCs in treating municipal wastewater by harnessing the microbial-driven degradation of organic matter (Kumar et al. 2019). The microbial processes not only contribute to pollutant removal but also generate electrical energy. Successful implementations have shown that MFCs can enhance overall treatment efficiency, providing a sustainable and cost-effective solution for urban wastewater management (Guo et al. 2020). The integration of MFCs in municipal wastewater treatment plants holds the potential to reduce energy consumption and contribute to a more environmentally friendly approach.

7.2 Industrial Effluent Remediation

Industries generate complex and diverse effluents containing pollutants that often challenge conventional treatment methods. MFCs have been successfully employed in the remediation of industrial effluents, showcasing their adaptability to different types of contaminants (Abbasi et al. 2016). Case studies highlight the ability of MFCs to address organic and inorganic pollutants present in industrial wastewaters. The microbial communities within MFCs facilitate the degradation of complex organic compounds, and the electrochemical reactions contribute to the removal of heavy metals (Chakraborty et al. 2020). The versatility of MFCs in handling industrial effluents underscores their potential for sustainable industrial wastewater treatment.

7.3 Agricultural Runoff Treatment

Agricultural runoff, containing nutrients and pesticides, poses a significant threat to water quality and ecosystems (Pavlidis and Tsihrintzis 2017). MFCs have been explored as a viable solution for treating agricultural runoff, with case studies demonstrating their efficacy in nutrient removal and pollutant degradation. The microbial processes within MFCs contribute to the recovery of nutrients, offering an opportunity for their reuse in agriculture (Shahid et al. 2021). Successful implementations in agricultural runoff treatment highlight the potential of MFCs in mitigating the environmental impact of agricultural activities and promoting sustainable water management practices (Meena et al. 2019).

7.4 Successful Resource Recovery Projects

Resource recovery projects utilizing Microbial Fuel Cells showcase the broader environmental and economic benefits of this technology. Case studies reveal successful resource recovery initiatives, including electricity generation, nutrient reclamation, and metal extraction. MFCs have been implemented in projects that harness electrical energy for powering remote sensors, providing a sustainable energy source for off-grid applications (Yang et al. 2012). Table 1 list the performance of MFC and its corresponding resources recovered after treatment. Nutrient recovery from MFCs has been utilized in agriculture, reducing the dependence on synthetic fertilizers (Xiao et al. 2016). Additionally, successful metal extraction projects demonstrate the potential of MFCs in contributing to circular economy practices by recycling valuable metals from contaminated water.

8 Environmental and Economic Assessment

Microbial Fuel Cells (MFCs) have gained recognition for their potential in water treatment, with environmental and economic assessments being integral components of their evaluation.

8.1 Environmental Impacts of MFC-Based Water Treatment

The environmental impacts of MFC-based water treatment are critical considerations for assessing the sustainability of this technology. MFCs offer a unique advantage by promoting biological degradation of pollutants, reducing the need for chemical treatments. This aspect contributes to a lower carbon footprint and minimizes the generation of sludge associated with traditional treatment methods (Kaushik and Singh 2020). Moreover, the recovery of resources, such as energy and nutrients, adds to the environmental benefits of MFCs. However, it is essential to consider potential environmental impacts associated with MFC materials, electrode production, and the overall life cycle of the system (Pandit et al. 2021). A comprehensive life cycle assessment is crucial for understanding the net environmental benefits of MFC-based water treatment.

8.2 Economic Viability and Cost–Benefit Analysis

The economic viability of Microbial Fuel Cells in water treatment is a key factor in determining their practical utility. Cost–benefit analysis considers factors such

Table 1 MFC Performance with different wastewater and its resource recovery

S.N.	Industry type	Waste pollutant	Microorganism used	Anode	Cathode	Power generated	Resource recovered	References
1	Agro industry	Food crops wheat straw, corn straw, rice straw vegetable waste	*Cellulomonas*	Carbon cloth	Carbon cloth	123 mW/m^2	Bioelectricity, cellulose, hemicellulose, energy	Roy et al. (2023) and Pandit et al. (2021)
2	Agro industry	Winery/ Brewery wastewater		Titanium	Platinum coated carbon cloth		Energy, COD, BOD removal electricity	Pandit et al. (2021)
3	Agro industry	Lignocellulosic wastes such as sugarcane laggase, wood chips, crop residues	*Lignolytic fungi, white rot fungi*			8.78 W/m^3	Lignin, cellulose, hemicellulose	Kumar et al. (2020)
4	Agro industry	Fertilizer wastewater		Microorganism	Ferrinitrate		Nitrogen, ammonia, water	Xiao et al. (2016)
5	Agro industry	Mustard tuber and molasses wastewater	*Brevibacillus borstelensis STR II*			246 mW/m^2	COD removal Bioelectricity	Pandit et al. (2021)
6	Agro industry	Palm oil mill effluent		Carbon cloth	Carbon cloth	45–304 mW/m^2	COD removal Bioelectricity	Roy et al. (2023) and Pandit et al. (2021)

(continued)

Table 1 (continued)

S.N.	Industry type	Waste pollutant	Microorganism used	Anode	Cathode	Power generated	Resource recovered	References
7	Agro industry	Cassava mill effluents	*Aspergillus niger, Aspergillus flavus, E. coli*			1.7 W/m^2	Starch, large effluents rich in COD	Pandit et al. (2021)
8	Agro industry	Plant and yard waste			Air cathode	0.43 W/m^3	Energy, COD removal	Pandit et al. (2021)
9	Oil industry-Palm oil	Oil, grease and fat	*P. aeruginos*	Polyacroni trile carbon		451.26 mW/m^2	Electricity 654.90 mA/m^2	Nor et al. (2015)
10	Oil industry-olive oil	Polypheno l, TSS, VS	*Geobacter spp*	Platinum	Open air	124.6 mW/m^2	Electricity	Sciarra et al. (2013)
11	Oil industry-tannery	Heavy metal, tannins, oils, grease	*K. pneumoniae, Bacillus firmus, Pseudomo nas, K. quasipneu moniae*	Carbon material		7371 mW/cm^3	Electricity, biogas	Chauhan et al. (2022)
12	Oil industry-palm oil	Oil, fat	*Clostridium, Bacillus*	Carbon material		55 mW/m^3	Electricity	Tee et al. (2016)
13	Oil industry-spend engine oil	Oil, grease, nitrogen, ammonia, phosphate	*Isolated cellulolytic bacteria and microalgae*	Graphite		16–34.88 mW/m^2	Electricity	Verma et al. (2023)

(continued)

Table 1 (continued)

S.N.	Industry type	Waste pollutant	Microorganism used	Anode	Cathode	Power generated	Resource recovered	References
14	Dairy wastewater		*Acutodesmus dimorphus*				Biodiesel and bioethanol 195 g of biodiesel and 78 g of bioethanol	Chokshi et al. (2016)
15	Dairy wastewater					192, 161 mW/m^2	Bio-energy generation COD removal of 91%	Elakkiya and Matheswaran (2013)
16	Farm wastewater		*Chlorophyta, Cyanophyta*				Biodiesel	Hena et al. (2015)
17	Dairy waste water		*Botryococcus braunii*				Bioethanol Reduced value of COD, BOD, TSS	Nurmayani et al. (2016)
18	Fresh cheese		*Chlorella pyrenoidosa*				230–270 mg L^{-1} day^{-1} biomass productivity and 58.84–77.41 mg^{-1} day^{-1} lipid productivity	Pandey et al. (2020)
19	Swine waste water	Nitrogen and carbon dioxide	*Exoelectrogenic Bacteria*	Carbon	Open air	37.5 W	Electricity	Ding et al. (2017)
20	Organic contaminated water	Organic load	*Anaerobic bacteria*	Cylindrical polyacryl pipes	Graphite	8 mW/m^2	Electricity	An et al. (2009)
21	Sulphur containing effluents	Sulfite and thiosufite	*Desulfovibrio, desulfuricans*	Activated carbon cloth + carbon fiber veil composite	Open air	2.68 mW	Electricity	Zhao et al. (2009)

(continued)

Table 1 (continued)

S.N.	Industry type	Waste pollutant	Microorganism used	Anode	Cathode	Power generated	Resource recovered	References
22	Wastewater	Synthetic chemicals	*Anaerobic bacteria*	Stainless steel	Graphite rod	0.008 mA/m^2	Electricity	Jadhav and Ghangrekar (2009)
23	Starch industry	Starch	*Clostridium*	Graphite	Ferrocyanide	1.3 mA/cm^2	Decrease in BOD	Niessen et al. (2004)
24	Sewage	Domestic waste	*Geobacter spp.*	Graphite	Air	26 mW/m^3	Removes 80% of organic load	Liu et al. (2004)
25	Dye manufacturing industry	Textile dye wastewater	*Bacterial community DC5*	Anaerobic anodes	Aerobic cathodes	197.94 mw/m^2	Electricity	Patel et al. (2021)
	Dye manufacturing industry	Textile dye wastewater	*Bacterial community DC5*	95% (v/v) textile dye wastewater inoculated with bacterial community DC5	Catholyte medium: 0.1 M phosphate buffer (pH 7.0)	127.21 mw/m^2	Electricity	Patel et al. (2021)
26	Textile industry	Aromatic intermediates of Congo red (di-azo dyes)	*DC5*	Anaerobic anodes	Aeration at the cathodes	35.94 2 mW/m^3	Electricity	Saket et al.(2022)

(continued)

Table 1 (continued)

S.N.	Industry type	Waste pollutant	Microorganism used	Anode	Cathode	Power generated	Resource recovered	References
27	Cronulla wastewater treatment plant	Wastewater with phosphate and ammonium ions	*Anaerobic microorganisms*	Graphite felt 3 cm in diameter and 0.6 cm in thickness	A carbon-fiber brush coated with a titanium bar		Electricity phosphate and ammonium is used as fertilizers	Ye et al. (2019)
28	Textile industry	Real dye textile wastewater (RTW)	*Microalgae Eg: Chlorella vulgaris*	Carbon fibers	Carbon fibers with microalgae immobilized on the cathode	123.2 ± 27.5 mW m^{-3}	Electricity Biodiesel and biomass production. Heavy metal recovery	Logroño et al. (2017)

as initial setup costs, operational expenses, and potential revenue from resource recovery (Logan and Regan 2006). While MFCs offer advantages in terms of reduced energy consumption and resource recovery, challenges may arise concerning material costs, maintenance, and system optimization (He 2013). Ongoing research and technological advancements aim to enhance the economic feasibility of MFCs, making them more competitive with conventional water treatment methods. An accurate economic assessment is essential for the widespread adoption of MFC technology and its integration into existing water treatment infrastructures (Li et al. 2013).

9 Future Perspectives and Innovations

Microbial Fuel Cells continue to evolve and future perspectives focus on emerging technologies, potential research directions and innovations that can further enhance their capabilities.

9.1 Emerging Technologies in Microbial Fuel Cells

The field of MFCs is witnessing the emergence of various technologies aimed at improving performance and expanding applications. Integration with advanced materials such as nanomaterials and conductive polymers holds promise for enhancing electrode efficiency and microbial interactions. Additionally, the incorporation of sensor technologies enables real-time monitoring and control of MFC processes. Nanotechnology, synthetic biology, and advances in electrode materials are among the emerging technologies that may shape the future of MFC-based water treatment.

9.2 Potential Research Directions

Future research directions in Microbial Fuel Cells encompass a wide range of areas. Understanding and optimizing microbial communities for specific pollutants, exploring novel electrode materials, and improving system scalability are ongoing research focuses. Enhancing resource recovery pathways, particularly the recovery of valuable metals and optimization of biohydrogen production, presents exciting avenues for exploration. Furthermore, interdisciplinary collaborations may unlock new insights, connecting MFC technology with fields such as materials science, microbiology, and environmental engineering.

In conclusion, the environmental and economic assessment of Microbial Fuel Cells is essential for determining their viability in water treatment applications. The environmental benefits, coupled with ongoing advancements in technology, position MFCs as promising tools for sustainable water treatment. As research progresses, exploring emerging technologies and identifying optimal research directions will contribute to the continued innovation and integration of Microbial Fuel Cells into mainstream water treatment practices. Overall, the future of MFCs holds great potential for addressing water pollution challenges while simultaneously recovering valuable resources in an environmentally and economically sustainable manner.

10 Conclusion

Water pollution is a significant challenge with far-reaching impacts on the environment, economy, and public health. However, MFCs offer a promising solution for water treatment by harnessing microbial activity to purify water and extract valuable resources from wastewater. The dual functionality of MFCs aligns with the principles of a circular economy, where waste is transformed into valuable inputs, contributing to sustainable resource management practices. By generating energy and nutrients, MFCs not only offset treatment costs but also support sustainable resource management. The integration of resource recovery in MFC-based water treatment processes shows potential for addressing economic and environmental burdens associated with water pollution, offering a pathway to more sustainable and economically viable solutions.

- Water pollution poses significant challenges with environmental, economic, and public health implications.
- MFCs offer a promising solution for water treatment by purifying water and extracting valuable resources from wastewater.
- MFCs align with the principles of a circular economy, transforming waste into valuable inputs and supporting sustainable resource management practices.

In conclusion, the continued research, innovation, and collaboration are crucial to fully unlock the potential of MFCs in resource recovery and water treatment. By leveraging advanced materials, sensor technologies, and interdisciplinary approaches, we can enhance the efficiency and versatility of MFC systems, paving the way for a more sustainable future.

Acknowledgements Author (AP) would like to give his sincere gratitude to NRF-SARChi South Africa for financial support as Post doctoral fellowship at Institute for Water and Wastewater Technology, Durban University of Technology Durban-4001, South Africa.

References:

Abbasi U, Jin W, Pervez A, Bhatti Z, Tariq M, Shaheen S, Iqbal A, Mahmood Q (2016) Anaerobic microbial fuel cell treating combined industrial wastewater: correlation of electricity generation with pollutants. Biores Technol 200:1–7. https://doi.org/10.1016/j.biortech.2015.09.088

Ali J, Wang L, Waseem H, Sharif HMA, Djellabi R, Zhang C, Pan G (2019) Bioelectrochemical recovery of silver from wastewater with sustainable power generation and its reuse for biofouling mitigation. J Clean Prod 235:1425–1437. https://doi.org/10.1016/j.jclepro.2019.07.065

An J, Kim D, Chun Y, Lee S, Ng H, Chang I (2009) Floating-type microbial fuel cell (FT-MFC) for treating organic-contaminated water. Environ Sci Technol 43(5):1642–1647. https://doi.org/10.1021/ES8024337

Anex R, Lynd L, Laser M, Heggenstaller A, Liebman M (2007) Potential for enhanced nutrient cycling through coupling of agricultural and bioenergy systems. Crop Sci 47:1327–1335. https://doi.org/10.2135/CROPSCI2006.06.0406

Bélafi-Bakó K, Vajda B, Bakonyi PN (2014). Removal of COD by two-chamber microbial fuel cells. https://doi.org/10.5772/58373

Ben W, Zhu B, Yuan X, Zhang Y, Yang M, Qiang Z (2018) Occurrence, removal and risk of organic micropollutants in wastewater treatment plants across China: comparison of wastewater treatment processes. Water Res 130:38–46. https://doi.org/10.1016/j.watres.2017.11.057

Bharathiraja B, Sudharsanaa T, Bharghavi A, Jayamuthunagai J, Praveenkumar R (2016) Biohydrogen and Biogas—an overview on feedstocks and enhancement process. Fuel 185:810–828. https://doi.org/10.1016/J.FUEL.2016.08.030

Bolong N, Ismail A, Salim M, Matsuura T (2009) A review of the effects of emerging contaminants in wastewater and options for their removal. Desalination 239:229–246. https://doi.org/10.1016/J.DESAL.2008.03.020

Brown G, Foster A, Ostergren J (1999) Mineral surfaces and bioavailability of heavy metals: a molecular-scale perspective. Proc Natl Acad Sci USA 96(7):3388–3395. https://doi.org/10.1073/PNAS.96.7.3388

Chakraborty I, Sathe S, Khuman C, Ghangrekar M (2020) Bioelectrochemically powered remediation of xenobiotic compounds and heavy metal toxicity using microbial fuel cell and microbial electrolysis cell. Mater Sci Energy Technol 3:104–115. https://doi.org/10.1016/j.mset.2019.09.011

Chauhan S, Sharma V, Varjani S, Sindhu R, Bhargava PC (2022) Mitigation of tannery effluent with simultaneous generation of bioenergy using dual chambered microbial fuel cell. Biores Technol 351:127084

Chokshi K, Pancha I, Ghosh A, Mishra S (2016) Microalgal biomass generation by phycoremediation of dairy industry wastewater: an integrated approach towards sustainable biofuel production. Biores Technol 221:455–460. https://doi.org/10.1016/j.biortech.2016.09.070

Coppola L, Trevisan M (2012) Biopurification system between research and technology transfer. J Biorem Biodegrad. https://doi.org/10.4172/2155-6199.1000E118

Das M (2013) Optimization and energy production by microbial fuel cell. Int J ChemTech Res 5:2193–2198

Ding W, Cheng S, Yu L, Huang H (2017) Effective swine wastewater treatment by combining microbial fuel cells with flocculation. Chemosphere 182:567–573. https://doi.org/10.1016/j.chemosphere.2017.05.006

Du Z, Li H, Gu T (2007) A state of the art review on microbial fuel cells: a promising technology for wastewater treatment and bioenergy. Biotechnol Adv 25(5):464–482. https://doi.org/10.1016/J.BIOTECHADV.2007.05.004

Ekubo AT, Abowei JFN (2011) Aspects of aquatic pollution in Nigeria. Res J Environ Earth Sci 3(6):673–693

Elakkiya E, Matheswaran M (2013) Comparison of anodic metabolisms in bioelectricity production during treatment of dairy wastewater in microbial fuel cell. Biores Technol 136:407–412. https://doi.org/10.1016/j.biortech.2013.02.113

Fadzli FS, Rashid M, Yaqoob AA, Ibrahim MNM (2021) Electricity generation and heavy metal remediation by utilizing yam (*Dioscorea alata*) waste in benthic microbial fuel cells (BMFCs). Biochem Eng J 172:108067

Fetzner S (2000) Enzymes involved in the aerobic bacterial degradation of N-heteroaromatic compounds: molybdenum hydroxylases and ring-opening 2,4-dioxygenases. Naturwissenschaften 87:59–69. https://doi.org/10.1007/s001140050011

Gude V (2016) Wastewater treatment in microbial fuel cells—an overview. J Clean Prod 122:287–307. https://doi.org/10.1016/J.JCLEPRO.2016.02.022

Guerra-Rodríguez S, Oulego P, Rodríguez E, Singh D, Rodríguez-Chueca J (2020) Towards the implementation of circular economy in the wastewater sector: challenges and opportunities. Water. https://doi.org/10.3390/w12051431

Guo Y, Wang J, Shinde S, Wang X, Li Y, Dai Y, Ren J, Zhang P, Liu X (2020) Simultaneous wastewater treatment and energy harvesting in microbial fuel cells: an update on the biocatalysts. RSC Adv 10:25874–25887. https://doi.org/10.1039/d0ra05234e

Hampel M, Blasco J, Segner H (2015) Molecular and cellular effects of contamination in aquatic ecosystems. Environ Sci Pollut Res 22:17261–17266. https://doi.org/10.1007/s11356-015-5565-5

He Z (2013) Microbial fuel cells: now let us talk about energy. Environ Sci Technol 47(1):332–333. https://doi.org/10.1021/es304937e

Heijne AT, Liu F, Weijden RVD, Weijma J, Buisman CJN, Hamelers HVM (2010) Copper recovery combined with electricity production in a microbial fuel cell. Environ Sci Technol 44:4376–4381. https://doi.org/10.1021/es100526g

Hena S, Fatimah S, Tabassum S (2015) Cultivation of algae consortium in a dairy farm wastewater for biodiesel production. Water Resour Ind 10:1–14. https://doi.org/10.1016/J.WRI.2015.02.002

Hu Z (2008) Electricity generation by a baffle-chamber membraneless microbial fuel cell. J Power Sources 179:27–33. https://doi.org/10.1016/J.JPOWSOUR.2007.12.094

Huang W, Chen J, Hu Y, Zhang L (2018) Enhancement of Congo red decolorization by membrane-free structure and bio-cathode in a microbial electrolysis cell. Electrochim Acta 260:196–203. https://doi.org/10.1016/j.electacta.2017.12.055

Jadhav GS, Ghangrekar MM (2009) Performance of microbial fuel cell subjected to variation in pH, temperature, external load and substrate concentration. Biores Technol 100:717–723

Jadhav D, Ghangrekar M, Duțeanu N (2018) Recent progress towards scaling up of MFCs, pp 443–457. https://doi.org/10.1007/978-3-319-66793-5_23

Kaushik A, Singh A (2020) Metal removal and recovery using bioelectrochemical technology: the major determinants and opportunities for synchronic wastewater treatment and energy production. J Environ Manag 270:110826. https://doi.org/10.1016/j.jenvman.2020.110826

Kelly P, He Z (2014) Nutrients removal and recovery in bioelectrochemical systems: a review. Biores Technol 153:351–360. https://doi.org/10.1016/j.biortech.2013.12.046

Kim B, Chang I, Gadd G (2007) Challenges in microbial fuel cell development and operation. Appl Microbiol Biotechnol 76:485–494. https://doi.org/10.1007/s00253-007-1027-4

Klöpffer W (1996) Environmental hazard assessment of chemicals and products. Part V. Anthropogenic chemicals in sewage sludge. Chemosphere 33:1067–1081. https://doi.org/10.1016/0045-6535(96)00247-0

Kour D, Rana K, Yadav A, Yadav N, Kumar M, Kumar V, Vyas P, Dhaliwal H, Saxena A (2020) Microbial biofertilizers: bioresources and eco-friendly technologies for agricultural and environmental sustainability. Biocatal Agric Biotechnol 23:101487. https://doi.org/10.1016/j.bcab.2019.101487

Kumar S, Kumar V, Malyan S, Sharma J, Mathimani T, Maskarenj M, Ghosh P, Pugazhendhi A (2019) Microbial fuel cells (MFCs) for bioelectrochemical treatment of different wastewater streams. Fuel. https://doi.org/10.1016/J.FUEL.2019.05.109

Li W, Yu H, He Z (2013) Towards sustainable wastewater treatment by using microbial fuel cells-centered technologies. Energy Environ Sci 7:911–924. https://doi.org/10.1039/C3EE43106A

Li M, Zhou S, Xu Y, Liu Z, Ma F, Zhi L, Zhou X (2018) Simultaneous Cr(VI) reduction and bioelectricity generation in a dual chamber microbial fuel cell. Chem Eng J 334:1621–1629. https://doi.org/10.1016/j.cej.2017.11.144

Liu H, Ramnarayanan R, Logan B (2004) Production of electricity during wastewater treatment using a single chamber microbial fuel cell. Environ Sci Technol 38(7):2281–2285. https://doi.org/10.1021/ES034923G

Logan B, Regan J (2006) Electricity-producing bacterial communities in microbial fuel cells. Trends Microbiol 14(12):512–518. https://doi.org/10.1016/J.TIM.2006.10.003

Logroño W, Pérez M, Urquizo G, Kadier A, Echeverría M, Recalde C, Rákhely G (2017) Single chamber microbial fuel cell (SCMFC) with a cathodic microalgal biofilm: a preliminary assessment of the generation of bioelectricity and biodegradation of real dye textile wastewater. Chemosphere 176:378–388. https://doi.org/10.1016/j.chemosphere.2017.02.099

Lovley D (2006) Microbial fuel cells: novel microbial physiologies and engineering approaches. Curr Opin Biotechnol 17(3):327–332. https://doi.org/10.1016/J.COPBIO.2006.04.006

Mahmud K, Makaju S, Ibrahim R, Missaoui A (2020) Current progress in nitrogen fixing plants and microbiome research. Plants. https://doi.org/10.3390/plants9010097

Marcus A, Torres C, Rittmann B (2007) Conduction-based modeling of the biofilm anode of a microbial fuel cell. Biotechnol Bioeng. https://doi.org/10.1002/BIT.21533

Meena R, Kannah R, Sindhu J, Ragavi J, Kumar G, Gunasekaran M, Jeyakumar R (2019) Trends and resource recovery in biological wastewater treatment system. Bioresour Technol Rep. https://doi.org/10.1016/J.BITEB.2019.100235

Mink J, Rojas J, Logan B, Hussain M (2012) Vertically grown multiwalled carbon nanotube anode and nickel silicide integrated high performance microsized (1.25 μL) microbial fuel cell. Nano Lett 12(2):791–795. https://doi.org/10.1021/nl203801h

Niessen J, Schröder U, Scholz F (2004) Exploiting complex carbohydrates for microbial electricity generation—a bacterial fuel cell operating on starch. Electrochem Commun 6:955–958

Nor M, Mubarak M, Elmi H, Ibrahim N, Wahab M, Ibrahim Z (2015) Bioelectricity generation in microbial fuel cell using natural microflora and isolated pure culture bacteria from anaerobic palm oil mill effluent sludge. Biores Technol 190:458–465. https://doi.org/10.1016/j.biortech.2015.02.103

Nurmayani S, Sugiarti Y, Putra R (2016) Utilization of industrial dairy waste as microalgae cultivation medium: a potential study for sustainable energy resources. IOP Conf Ser Mater Sci Eng. https://doi.org/10.1088/1757-899X/128/1/012033

Oh S, Logan B (2006) Proton exchange membrane and electrode surface areas as factors that affect power generation in microbial fuel cells. Appl Microbiol Biotechnol 70:162–169. https://doi.org/10.1007/s00253-005-0066-y

Pandey A, Srivastava S, Kumar S (2020) Development and cost-benefit analysis of a novel process for biofuel production from microalgae using pre-treated high-strength fresh cheese whey wastewater. Environ Sci Pollut Res 27:23963–23980. https://doi.org/10.1007/s11356-020-08535-4

Pandit S, Savla N, Sonawane J, Sani A, Gupta P, Mathuriya A, Rai A, Jadhav D, Jung S, Prasad R (2021) Agricultural waste and wastewater as feedstock for bioelectricity generation using microbial fuel cells: recent advances. Fermentation. https://doi.org/10.3390/fermentation7030169

Pasqualino J, Meneses M, Castells F (2011) Life cycle assessment of urban wastewater reclamation and reuse alternatives. J Ind Ecol. https://doi.org/10.1111/j.1530-9290.2010.00293.x

Passos V, Neto S, Andrade A, Reginatto V (2016) Energy generation in a microbial fuel cell using anaerobic sludge from a wastewater treatment plant. Sci Agric 73:424–428. https://doi.org/10.1590/0103-9016-2015-0194

Patel D, Bapodra SL, Madamwar D, Desai C (2021) Electroactive bacterial community augmentation enhances the performance of a pilot scale constructed wetland microbial fuel cell for treatment of textile dye wastewater. Biores Technol 332:125088. https://doi.org/10.1016/j.biortech.2021.125088

Pavlidis G, Tsihrintzis V (2017) Environmental benefits and control of pollution to surface water and groundwater by agroforestry systems: a review. Water Resour Manag 32:1–29. https://doi.org/10.1007/s11269-017-1805-4

Polzin F, Sanders M, Taübe F (2017) A diverse and resilient financial system for investments in the energy transition. Curr Opin Environ Sustain 28:24–32. https://doi.org/10.1016/J.COSUST.2017.07.004

Rabaey K, Rozendal R (2010) Microbial electrosynthesis—revisiting the electrical route for microbial production. Nat Rev Microbiol 8:706–716. https://doi.org/10.1038/nrmicro2422

Rabaey K, Verstraete W (2005) Microbial fuel cells: novel biotechnology for energy generation. Trends Biotechnol 23(6):291–298. https://doi.org/10.1016/J.TIBTECH.2005.04.008

Rathinavel L, Jothinathan D, Sivsankar V, Agastian P, Prabhakaran M (2018) Algal microbial fuel cells—nature's perpetual energy resource. https://doi.org/10.1007/978-3-319-92904-0_5

Roy H, Rahman TU, Tasnim N, Arju J, Rafid MM, Islam MR, Pervez MN, Cai Y, Naddeo V, Islam MS (2023) Microbial fuel cell construction features and application for sustainable wastewater treatment. Membranes 13(5):490. https://doi.org/10.3390/membranes13050490

Saket P, Mittal Y, Bala K, Joshi A, Kumar Yadav A (2022) Innovative constructed wetland coupled with microbial fuel cell for enhancing diazo dye degradation with simultaneous electricity generation. Biores Technol 345:126490. https://doi.org/10.1016/J.BIORTECH.2021.126490

Saratale G, Saratale R, Shahid M, Zhen G, Kumar G, Shin H, Choi Y, Kim S (2017) A comprehensive overview on electro-active biofilms, role of exo-electrogens and their microbial niches in microbial fuel cells (MFCs). Chemosphere 178:534–547. https://doi.org/10.1016/j.chemosphere.2017.03.066

Sciarria T, Tenca A, D'Epifanio A, Mecheri B, Merlino G, Barbato M, Borin S, Licoccia S, Garavaglia V, Adani F (2013) Using olive mill wastewater to improve performance in producing electricity from domestic wastewater by using single-chamber microbial fuel cell. Biores Technol 147:246–253. https://doi.org/10.1016/j.biortech.2013.08.033

Sengupta S, Nawaz T, Beaudry J (2015) Nitrogen and phosphorus recovery from wastewater. Curr Pollut Rep 1:155–166. https://doi.org/10.1007/s40726-015-0013-1

Shahid K, Ramasamy D, Haapasaari S, Sillanpää M, Pihlajamäki A (2021) Stainless steel and carbon brushes as high-performance anodes for energy production and nutrient recovery using the microbial nutrient recovery system. Energy 233:121213. https://doi.org/10.1016/J.ENERGY.2021.121213

Sirés I, Brillas E, Oturan M, Rodrigo M, Panizza M (2014) Electrochemical advanced oxidation processes: today and tomorrow. A review. Environ Sci Pollut Res 21:8336–8367. https://doi.org/10.1007/s11356-014-2783-1

Sophia A, Sai S (2016) Modified microbial fuel cell for Cr(VI) reduction and simultaneous bioelectricity production. J Environ Chem Eng 4:2402–2409. https://doi.org/10.1016/J.JECE.2016.04.025

Sun M, Sheng G, Zhang L, Xia C, Mu Z, Liu X, Wang H, Yu H, Qi R, Yu T, Yang M (2008) An MEC-MFC-coupled system for biohydrogen production from acetate. Environ Sci Technol 42(21):8095–8100. https://doi.org/10.1021/ES801513C

Suresh R, Rajendran S, Kumar PS, Dutta K, Vo DVN (2022) Current advances in microbial fuel cell technology toward removal of organic contaminants—a review. Chemosphere 287:132186. https://doi.org/10.1016/j.chemosphere.2021.132186

Tee P, Abdullah M, Tan I, Amin M, Nolasco-Hipólito C, Bujang K (2016) Performance evaluation of a hybrid system for efficient palm oil mill effluent treatment via an air-cathode, tubular upflow microbial fuel cell coupled with a granular activated carbon adsorption. Biores Technol 216:478–485. https://doi.org/10.1016/j.biortech.2016.05.112

Thung WE, Ong SA, Ho LN, Wong YS, Ridwan F, Oon YL, Lehl HK (2015) A highly efficient single chambered up-flow membrane-less microbial fuel cell for treatment of azo dye acid orange 7-containing wastewater. Biores Technol 197:284–288. https://doi.org/10.1016/j.biortech.2015.08.078

Verma M, Singh V, Mishra V (2023) Bioelectricity generation by using cellulosic waste and spent engine oil in a concentric photobioreactor-microbial fuel cell. J Environ Chem Eng 11(5):110566

Virdis B, Rabaey K, Yuan Z, Rozendal R, Keller J (2009) Electron fluxes in a microbial fuel cell performing carbon and nitrogen removal. Environ Sci Technol 43(13):5144–5149. https://doi.org/10.1021/ES8036302

Wan LL, Li XJ, Zang GL, Wang X, Zhang YY, Zhou QX (2015) A solar assisted microbial electrolysis cell for hydrogen production driven by a microbial fuel cell. RSC Adv 5(100):82276–82281. https://doi.org/10.1039/C5RA16919D

Wang X, Hu J, Chen Q, Zhang P, Wu L, Li J, Yang J (2019) Synergic degradation of 2,4,6-trichlorophenol in microbial fuel cells with intimately coupled photocatalytic–electrogenic anode. Water Res 156:125–135. https://doi.org/10.1016/j.watres.2019.03.001

Wehrer M, Rennert T, Mansfeldt T, Totsche K (2011) Contaminants at former manufactured gas plants: sources, properties, and processes. Crit Rev Environ Sci Technol 41:1883–1969. https://doi.org/10.1080/10643389.2010.481597

Wen Q, Kong F, Zheng H, Cao D, Ren Y, Yin J (2011) Electricity generation from synthetic penicillin wastewater in an air-cathode single chamber microbial fuel cell. Chem Eng J 168(2):572–576. https://doi.org/10.1016/j.cej.2011.01.025

Xiao Y, Zheng Y, Wu S, Yang Z, Zhao F (2016) Nitrogen recovery from wastewater using microbial fuel cells. Front Environ Sci Eng 10:185–191. https://doi.org/10.1007/s11783-014-0730-5

Xue A, Shen ZZ, Zhao B, Zhao HZ (2013) Arsenite removal from aqueous solution by a microbial fuel cell–zerovalent iron hybrid process. J Hazard Mater 261:621–627. https://doi.org/10.1016/j.jhazmat.2013.07.072

Yadav RK, Chiranjeevi P, Patil SA (2020) Integrated drip hydroponics-microbial fuel cell system for wastewater treatment and resource recovery. Bioresource Technology Reports 9:100392. https://doi.org/10.1016/j.biteb.2020.100392

Yang Y, Sun G, Guo J, Xu M (2011) Differential biofilms characteristics of *Shewanella decolorationis* microbial fuel cells under open and closed circuit conditions. Biores Technol 102(14):7093–7098. https://doi.org/10.1016/j.biortech.2011.04.073

Yang F, Zhang D, Shimotori T, Wang K, Huang Y (2012) Study of transformer-based power management system and its performance optimization for microbial fuel cells. J Power Sources 205:86–92. https://doi.org/10.1016/J.JPOWSOUR.2012.01.025

Ye Y, Ngo HH, Guo W, Liu Y, Chang SW, Nguyen DD, Ren J, Liu Y, Zhang X (2019) Feasibility study on a double chamber microbial fuel cell for nutrient recovery from municipal wastewater. Chem Eng J 358:236–242. https://doi.org/10.1016/j.cej.2018.09.215

Zhang B, Zhang J, Liu Y, Hao C, Tian C, Feng C, Lei Z, Huang W, Zhang Z (2013) Identification of removal principles and involved bacteria in microbial fuel cells for sulfide removal and electricity generation. Int J Hydrog Energy 38:14348–14355. https://doi.org/10.1016/J.IJHYDENE.2013.08.131

Zhang M, Wang Y, Liang P, Zhao X, Liang M, Zhou B (2019) Combined photoelectrocatalytic microbial fuel cell (PEC-MFC) degradation of refractory organic pollutants and in-situ electricity utilization. Chemosphere 214:669–678. https://doi.org/10.1016/j.chemosphere.2018.09.085

Zhao F, Rahunen N, Varcoe J, Roberts A, Avignone-Rossa C, Thumser A, Slade R (2009) Factors affecting the performance of microbial fuel cells for sulfur pollutants removal. Biosens Bioelectron 24(7):1931–1936. https://doi.org/10.1016/j.bios.2008.09.030

Application of Microalgae-MFC to Mitigate Water Pollution and Resource Recovery

Ricky Rajamanickam and Rangabhashiyam Selvasembian

Abstract Microbial fuel cells (MFCs) are simple bio-electrochemical devices that have the potential to treat wastewater and generate power through anaerobic respiration of microorganisms. Microalgae can grow in wastewater conditions utilizing CO_2, nutrients, and other pollutants from the medium, and it has good symbiotic interactions with bacteria; this synergetic interaction has gained its application in integrating with microalgae-based MFCs. Microalgae produce O_2 through photosynthesis, can reduce the need for external oxygen supply, and can take part in the cathodic process. The bacteria from the anodic compartment capture O_2 and provide bioelectricity. Microalgae-integrated MFC can treat wastewater and provide sustainable bioenergy conversion through biomass production. This approach has led to the application of MFCs with algae-assisted cathodes to treat wastewater and generate power. The harvested biomass from this integrated treatment system can produce value-added products through biorefinery approaches. This chapter aims to cover the advantages associated with micro-algaeintegrated MFC for the application of electricity generation, carbon capture, wastewater treatment, algal biomass production, operating conditions, and its impacts on power generation.

Keywords Microalgae · Bioenergy · Wastewater treatment · MFC

1 Introduction

The issue of water pollution and resource depletion is a significant concern the world is grappling with. These challenges have far-reaching consequences for human well-being, the natural environment, and the ability to achieve sustainable development (Selvasembian et al. 2023; Wang et al. 2021). Water contamination may originate from a wide variety of sources, including commercial and industrial discharges,

R. Rajamanickam · R. Selvasembian (✉)
Department of Environmental Science and Engineering, School of Engineering and Sciences, SRM University-AP, Amaravati, Andhra Pradesh 522240, India
e-mail: rambhashiyam@gmail.com

R. Selvasembian et al. (eds.), *Emerging Trends in Microbial Electrochemical Technologies for Sustainable Mitigation of Water Resources Contamination*,
https://doi.org/10.1007/978-3-031-74636-9_7

runoff from agricultural land, wastewater from metropolitan areas, and inappropriate disposal of waste. Various pollutants, including harmful chemicals and excessive nutrients, pollute surface water bodies, groundwater reserves, and marine ecosystems, posing a threat to aquatic biodiversity and the overall functioning of ecosystems (Morin-Crini et al. 2022; Nandan and Rangabhashiyam 2023). In addition, water pollution presents significant dangers to human populations, impacting the quality of drinking water, the security of food sources, and the overall public health by spreading waterborne illnesses and the buildup of harmful substances in the food chain (Akhtar et al. 2021). At the same time, the depletion of resources worsens the degradation of the environment and poses a threat to long-term sustainability. It is essential to recognize that natural resources, including freshwater, fossil fuels, minerals, and arable land, are limited and becoming scarcer due to overexploitation, population growth, and unsustainable consumption habits (Mondal and Palit 2021). The depletion of these resources hurts the resilience of ecosystems and the services they provide. It also worsens socio-economic disparities, especially in vulnerable communities that depend on natural resources. The depletion of resources also contributes to the worsening of environmental degradation, which in turn results in the destruction of ecosystems and a reduction in biodiversity (Azwardi et al. 2023). Consequently, this further exacerbates human activities' influence on the environment and contributes to the worsening of global concerns such as the loss of biodiversity and climate change.

MFCsand microalgae are cutting-edge biotechnological methods that have the potential to combat water pollution and resource depletion effectively. Microbial fuel cells are a bio-electrochemical device that harnesses the metabolic processes of microbes to generate power from organic materials (Obileke et al. 2021; Saranya and Rangabhashiyam 2023). Microorganisms, usually bacteria, oxidize organic compounds as fuel within an MFC, releasing electrons that generate electrical power through an external circuit. This electrochemical process provides a sustainable and renewable method of energy production, using organic waste streams, wastewater, or biomass as feedstock (Idris et al. 2022; Rajamanickam and Selvasembian 2024). However, microalgae are microscopic photosynthetic organisms capable of converting the energy from the sun into chemical energy via photosynthesis. There are a variety of aquatic settings in which these single-celled creatures can flourish, ranging from freshwater lakes and rivers to marine ecosystems (Dolganyuk et al. 2020). Microalgae are essential for maintaining global biogeochemical cycles, as they help with carbon sequestration, oxygen production, and nutrient cycling (Leong et al. 2021). In addition, microalgae display an impressive array of biochemical diversity, producing various bioactive compounds such as lipids, carbohydrates, proteins, and pigments. These compounds have numerous applications in food, feed, pharmaceuticals, and biofuels (Shaikh et al. 2022). Through the cultivation of microalgae in wastewater, the process not only helps with nutrient uptake but also plays a crucial role in sequestering carbon dioxide, a significant greenhouse gas linked to climate change (Ighalo et al. 2022). Through photosynthesis, microalgae can capture carbon dioxide from the environment and transform it into essential organic compounds like carbohydrates, lipids, and proteins (Dolganyuk et al. 2020). This process helps reduce carbon dioxide levels in the environment and generates biomass rich in energy-dense molecules, making it suitable for bioenergy production. One of the most promising uses for microalgae biomass is the extraction of lipids in biodiesel, carbohydrates in bioethanol, and proteins in animal feed additives (Siddiki et al. 2022). Anaerobic digestion may also be used to produce biogas (methane) from microalgae biomass, which can generate energy through electricity or heat (Vargas-Estrada et al. 2022).

The combination of microalgae and MFCs shows excellent potential in tackling issues related to water pollution, resource depletion, and energy security by utilizing synergistic bioprocesses (Bolognesi et al. 2021; Ribeiro et al. 2022). Integrated systems can achieve improved performance in water treatment, nutrient recovery, carbon capture, and sustainable energy generationby combining microorganisms and microalgae (Sharma et al. 2022). Microalgae cultivated in wastewater offers a sustainable and cost-effective solution for nutrient removal, carbon dioxide sequestration, and biomass production in enhanced wastewater treatment. Reduced eutrophication and improved water quality are two outcomes of microalgae's ability to use photosynthesis to absorb nitrogen, phosphorous, and other pollutants from wastewater (Hussain et al. 2021). In addition, microalgae biomass can be used as a substrate for bacterial metabolism in MFCs, which improves the degradation of organic matter and the generation of electricity while also facilitating the wastewater treatment process (Yadav et al. 2020). The biomass of microalgae is rich in carbohydrates, lipids, and proteins, making it an excellent source for nutrient extraction (Yaakob et al. 2021). These include agricultural fertilizers, aquaculture feed supplements, and biopolymer production. In addition, the organic matter found in microalgae biomass can be used as a feedstock for microbial metabolism in MFCs, which helps produce bioenergy and utilize resources (Don and Babel 2021).

Integrating microalgae with microbial fuel cells (MFCs) provides a comprehensive and integrated solution to tackle issues related to water pollution, resource depletion, and energy security. Through the strategic combination of microorganisms and microalgae, integrated systems have a wide range of applications, such as water treatment, nutrient recovery, carbon capture, and renewable energy generation. The process and use of microalgae-integrated MFCs for wastewater treatment and resource recovery via microalgal biomass generation are discussed in this chapter.

2 Microalgae-MFC Integration: Mechanism and Process

MFC are able to convert organic materials into electrical energy by using the metabolic activity of microorganisms, which are often bacteria (Garbini et al. 2023). Within an MFC, bacteria utilize organic compounds as a fuel source, releasing electrons that can be harnessed to generate electricity through an external circuit. This process relies on the redox reactions that take place at the anode (where oxidation occurs) and cathode (where reduction occurs) of the MFC (Prathiba et al. 2022). MFC comprises three chambers: an anode chamber, a cathode chamber, and a proton exchange membrane (PEM) that divides the two rooms. Both of these chambers are illustrated in the diagram Fig. 1a (Arun et al. 2020).

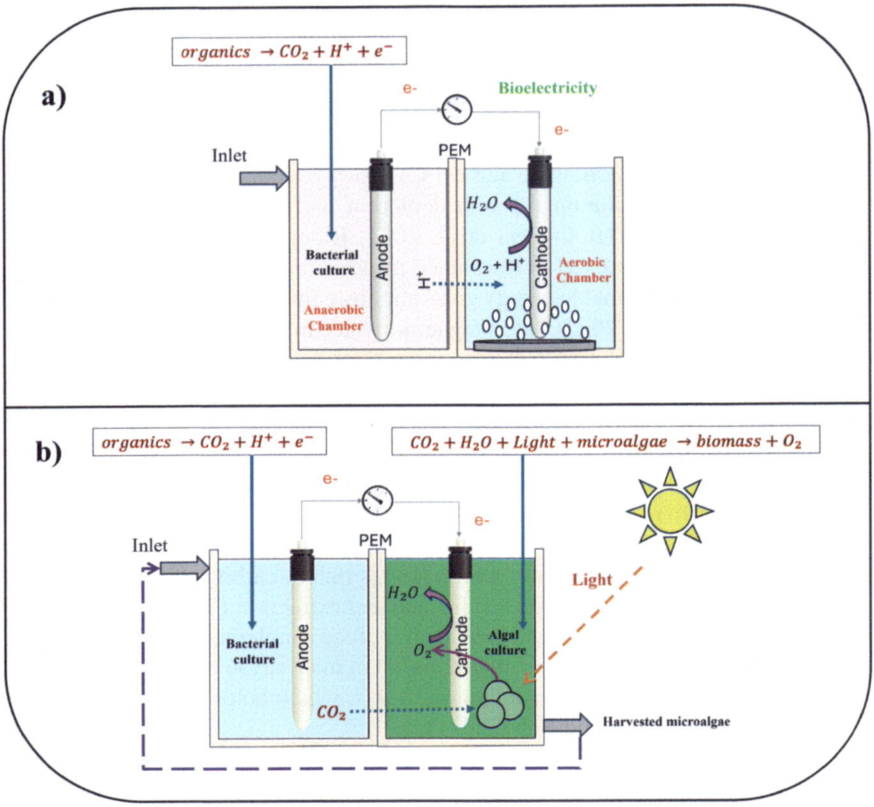

Fig. 1 Schematic illustration of **a** traditional MFC and **b** microalgae integrated MFC system

In the anode chamber of an MFC, there are electroactive bacteria that can oxidize organic matter and transfer electrons to an electrode surface, which is crucial for electricity generation (Cai et al. 2020). When bacteria break down organic compounds like acetate, glucose, or wastewater-derived substrates, they naturally produce electrons as a result of their metabolic processes (Kumar et al. 2022). The anode electrode captures the electrons, which starts a flow of electrons through the external circuit towards the cathode electrode. Meanwhile, protons (H^+) produced during bacterial metabolism move through the proton exchange membrane to the cathode chamber (Rossi and Logan 2021), where they react with electrons and oxygen to create water. Electrons flowing through the external circuit produce electrical current, which can be utilized for a range of purposes such as generating electricity, treating wastewater, and powering sensors.

In an MFC, carbon dioxide can be generated in the anodic chamber as a result of the oxidation of organic feeds. Meanwhile, in the cathode chamber, an electron acceptor like O_2 or ferricyanide is required to receive the surplus protons from the anodic chamber and electrons from the external circuit (Lee et al. 2015). The reaction equations are as follows.

At anode

$$organics \rightarrow CO_2 + H^+ + e^- \tag{1}$$

At cathode

$$O_2 + H^+ + e^- \rightarrow H_2O \tag{2}$$

In Eq. 1, carbon dioxide reacts with hydrogen ions and electrons from the external circuit at the cathode. At the cathode, oxygen reacts with hydrogen ions from the anode and electrons from the external circuit. The two reactions from Eqs. 1 and 2 mentioned above combine to form the overall reaction from the Eq. 3.

Overall reaction

$$Organics + O_2 \rightarrow CO_2 + H_2O + external\,power \tag{3}$$

MFCs have traditionally relied on bacterial metabolism for electricity generation. However, recent research has been investigating the potential synergy between microorganisms and microalgae in integrated systems. In these systems, microalgae act as primary producers, harnessing solar energy through photosynthesis and generating organic compounds that can be used as substrates for bacterial metabolism (Tanvir et al. 2021). The organic matter produced by microalgae, including carbohydrates, lipids, and proteins, can be used by bacteria in MFCs to boost their metabolic activity and generate electricity as shown in Fig. 1b. In addition, the oxygen released by microalgae during photosynthesis creates an oxygen-rich environment at the cathode electrode, which improves reduction reactions and boosts the performance of MFCs (Yadav et al. 2020). This harmonious connection between microorganisms and microalgae in integrated MFC systems shows potential for enhancing energy production efficiency and wastewater treatment effectiveness.

Within a microalgae photoreactor, the single-celled algal cells utilize sunlight and CO_2 from the medium and microbial biomassand O_2. Figure 1b depicts the microalgae integrated MFC process (Arun et al. 2020).

$$CO_2 + H_2O + Light + microalgae \rightarrow biomass + O_2 \tag{4}$$

Chlorella vulgaris generates oxygen in the medium, which can serve as an electron acceptor for the MFCs (Kakarla and Min 2014; Powell et al. 2009). Studies have revealed a positive correlation between the cell voltage and dissolved oxygen in MFC, and the generation of oxygen by the microalgae through photosynthesis favours the production of electrons (Enamala et al. 2020; Xiao and He 2014). Hence, microalgae may be deemed appropriate for usage in microbial fuel cells (MFCs), whether in a suspended or attached state inside the cathodic chamber. Numerous investigations have been conducted on the integration of microalgae in microbial fuel cells (MFCs), whereby living green algae are used in the cathode chamber, and deceased microalgae biomass serves as the substrate for the anodic biofilm, as seen in Fig. 1b (Greenman et al. 2019; Saba et al. 2017).

Integration mechanisms are essential for successfully combining microalgae cultivation with MFCs and the interactions between microalgae and bacteria within MFCs. Incorporating microalgae and electroactive bacteria in a single bioreactor or sequential reactors is common in integrated systems (Sharma et al. 2022). In these systems, wastewater acts as the growth medium for microalgae, supplying vital nutrients and carbon sources for photosynthesis. When microalgae undergo photosynthesis, they generate organic matter that can be harnessed by electroactive bacteria in MFCs to generate electricity.

3 Water Pollution Mitigation and Bioelectricity Generation

Microalgae are essential in the process of wastewater treatment as they naturally help in bioremediation. As they develop, microalgae absorb nutrients like nitrogen and phosphorus from the wastewater, efficiently eliminating these pollutants from the water column. Microalgae assimilate nitrogen through different metabolic pathways, such as nitrate reductase and glutamine synthetase, primarily in the form of ammonium and nitrate (Su 2021; Vega 2018). In a similar manner, microalgae absorb phosphorus, primarily in the form of phosphate, and use it to support their growth and reproduction. Through the utilization of microalgae's nutrient uptake capabilities, wastewater treatment plants can effectively reduce nutrient concentrations, leading to enhanced water quality and improved ecosystem health (Su 2021). Research has shown that a photo microbial cell made with an algae bacterial consortium can efficiently eliminate nitrogen (87%), phosphorus (69%), and organic carbon (99%) respectively (Ravikumar et al. 2021), and also they play a crucial role in boosting electricity generation and improving the efficiency of wastewater treatment (Sharma et al. 2022). This symbiotic partnership between microorganisms and microalgae in MFCs promotes a collaborative bioprocess that optimizes resource utilization and energy generation (Sharma et al. 2022).

Microalgal strains like from the *Chlorella* and *Scenedesmus* species are highly efficient and effective in the treatment of wide variety of pollutants and when used in MFC for biodegradation and bioenergy production shows promising results, with organic wastes being effectively degraded and electricity production being increased as a result as shown in Table 1. *Chlorella vulgaris* demonstrated significant nitrogen and phosphorus removal when utilized in an MFC. Furthermore, the stored inorganic carbon dioxide (CO_2) plays a crucial role in enhancing biomass yield and lipid content, resulting in a substantial increase ranging from 40 to 60% (Molazadeh et al. 2019).

In addition, the effective showcase of using microalgae as a prospective power generator via MFC was accomplished utilizing several categories of wastewater, such as household wastewater, contaminated lake water, dye-contaminated water,

Table 1 Bioelectricity generated from the wastewater in the algae integrated MFC system

S. no.	Wastewater type	Microalgae	Anode	Cathode	Cell voltage	References
1	Municipal	*Chlorella vulgaris*	Stainless steel mesh	Stainless steel mesh	126 mW/m^3	Bazdar et al. (2018)
2		*Scenedesmus obliquus*	Toray carbon paper	Toray carbon paper	951 mW/m^3	Kondaveeti et al. (2014)
3	Swine	*Chlorella vulgaris*	Carbon fibre pretreated with acetone	Carbon fibre modified with Pt	3720 mW/m^3	Zhang et al. (2019)
4	Anaerobic digested effluent	*Golenkinia* sp	Graphite	Sheet copper	6150 mW/m^3	Hou et al. (2016)
5	Pharmaceutical	*Tetraselmis indica*	Graphite	Graphite	560 mW/m^3	Amit et al. (2020)
6	Textile	*Chlorella vulgaris*	Carbon fibre brush	Carbon fibre brush	123 mW/m^3	Logroño et al. (2017)
7	Leachate	*Algal Consortium*	Carbon fibre brush	Carbon fibre brush	330 mV	Nguyen and Min (2020)
8	Food	*Scenedesmus* sp	Graphite plate	Graphite plate	28.5 mW/m^3	Naina Mohamed et al. (2020)v
9	Domestic	*Scenedesmus Quadricauda*	Carbon cloth woven with Ti	Carbon cloth woven with Ti	62 mW/m^3	Yang et al. (2018)
10	Anaerobic sludge	*Chlorella vulgaris*	Carbon felt	Carbon fibre cloth	1108 mW/m^3	Li et al. (2019)
11	Dairy	*Scenedesmus dimorphus*	Graphite felt stainless steel	Graphite felt stainless steel	285 mV	Mehrotra et al. (2021)

and synthetic wastewater. *Scenedesmus acutus* (catholyte) and polybenzimidazole (PBI) composite were used as a membrane separator for the treatment of wastewater. The system was run for a duration of 100 days, during which it achieved a maximum volumetric power output of around 400 mW/m^3. Notably, the MFC successfully eliminated both chemical oxygen demand (COD) and total organic carbon (TOC) (Ravikumar et al. 2021). Microalgae can also be used as a biofilm to minimize the biomass loss in the treatment of textile wastewater *Chlorella* sp biomass decolorized the textile wastewater with the 98% removal of COD from the medium and generated bioelectricity of 123 mW/m^3 (Logroño et al. 2017). Synthetic wastewater was used to evaluate Chlorella sp. biomass waste as the only carbon source in the anolyte and catholyte. The data showed that biomass content increased electricity production, COD elimination, and CO_2 fixation (Elshobary et al. 2021).

Microalgae have shown great potential in addressing heavy metal contamination in water bodies through their ability to accumulate and remove these metals. Microalgae have distinct biochemical and physiological traits that allow them to gather heavy metals from water sources via bioaccumulation mechanisms (Danouche et al. 2021). Microalgae can absorb heavy metals like cadmium, lead, mercury, and chromium through various mechanisms, both passive and active (Manikandan et al. 2022). Passive uptake happens when metal ions diffuse or adhere to the cell surface, and then get transported into the cell interior. In order to facilitate the uptake and intracellular sequestration of heavy metals, specific metal transporters and binding proteins are involved in the process of active uptake (Ayele et al. 2021). Once inside the cell, heavy metals are stored in vacuoles or attached to cellular components like proteins, polysaccharides, and cell walls. This helps to decrease their bioavailability and toxicity (Umamaheswari et al. 2023). Different species of microalgae have varying abilities when it comes to absorbing and storing metals. Some species are more tolerant to metals and are better at efficiently accumulating them compared to others.

In microalgae integrated MFC system, *Tetraselmis indica* in the treatment of pharmaceutical wastewater, during the culture period of 16 days, *Tetraselmis indica* was able to remove heavy metals such as aluminium (94%), iron (56%), calcium (79%), sodium (79%), magnesium (85%), zinc (65%), copper (88%), manganese (94%), and potassium (90%) and also the integrated system removed nitrate (81%), phosphate (94%), TOC (97%), and COD (90%) respectively (Amit et al. 2020). To summarise, utilizing microalgae for nutrient removal and heavy metal remediation provides sustainable solutions for tackling water pollution and environmental contamination. Through the utilization of microalgae's metabolic activities and the collaborative effects of bacteria in integrated systems, we can enhance nutrient removal and heavy metal remediation processes. This ultimately results in better water quality, improved ecosystem health, sustainable environment and it provides good strategy for resource recovery.

4 Resource Recovery and the Production of Value Added Bioproducts

Utilizing microalgae cultivation integrated with MFC offers a sustainable solution to tackle nutrient pollution and carbon emissions, while also generating valuable bioproducts (Das et al. 2023). Understanding the process of nutrient recycling is crucial for harnessing the potential of microalgae biomass. By recovering vital elements like nitrogen and phosphorus, we can find new ways to reuse them in agricultural fertilizers, aquaculture feed supplements, and other value added bioproducts (Ali et al. 2021). The biomass harvested from MFC systems contains substantial amounts of nitrogen and phosphorus, which are vital nutrients for the growth and development of plants (Bajracharya 2020). Microalgae biomass is processed using techniques like drying, mechanical extraction, and chemical treatment to extract proteins rich in nitrogen and phosphorus, in order to recover these valuable nutrients. These extracted nutrients can be concentrated, purified, and formulated into fertilizers or soil amendments for agricultural applications and the nutrient-rich microalgae biomass as a feedstock for anaerobic digestion or composting (Janeeshma et al. 2022). This process releases nitrogen and phosphorus in bioavailable forms that can be easily absorbed by plants. By implementing nutrient recycling, microalgae cultivation provides a sustainable approach to decrease dependence on synthetic fertilizers and limit nutrient runoff into water bodies.

Carbon capture and utilization (CCU) is the process of removing carbon dioxide (CO_2) from the atmosphere or industrial emissions and transforming it into valuable products or energy sources (Peres et al. 2022). Microalgae are incredibly effective at capturing CO_2 from the atmosphere or dissolved in water. They have a remarkable ability to sequester this greenhouse gas, making them an important tool in mitigating climate change. In the treatment of pharmaceutical wastewater in the algae integrated MFC, the system was able to reach 0.089 g CO_2/L per day respectively (Amit et al. 2020). This process not only helps to decrease carbon dioxide levels in the environment, but it also generates biomass that is abundant in energy-rich molecules, making it ideal for bioenergy production. Microalgae cultivation in wastewater treatment systems or open ponds provides a cost-effective and scalable method for carbon capture (Nguyen et al. 2023). It also offers a renewable feedstock for various applications like biofuels, pigments and bio-based chemicals as shown in Fig. 2 (Mehariya et al. 2021).

MFCs facilitate the transformation of organic matter into bioenergy via electrochemical processes, complementing carbon capture efforts. Various organic molecules, such as acetate, glucose, or substrates produced from wastewater, are oxidized by electroactive bacteria that are present in MFC. As a consequence of this process, electrons are released as a byproduct of the metabolic (Ramya and Senthil Kumar 2022). The anode electrode is responsible for capturing these electrons, which then initiates a flow of electrons across the external circuit, which ultimately results in the production of electrical current. Microalgae biomass contains organic matter that can be used as a renewable substrate for bacterial metabolism in MFCs (Ali et al. 2020). This helps to increase electricity generation and also aids in the completion of the wastewater treatment process. MFCs may also generate biogas (methane) from organic waste streams in anaerobic digestion systems. Biogas may be produced from biomass and surplus CO_2 that sustains algae. A study has shown that the MFC system can generate upto 62% of biomethane from the organic matter (Leng et al. 2021).

There are various stages involved in the MFC system utilization of microalgae for the production of value-added products. The first step is to remove the algal biomass

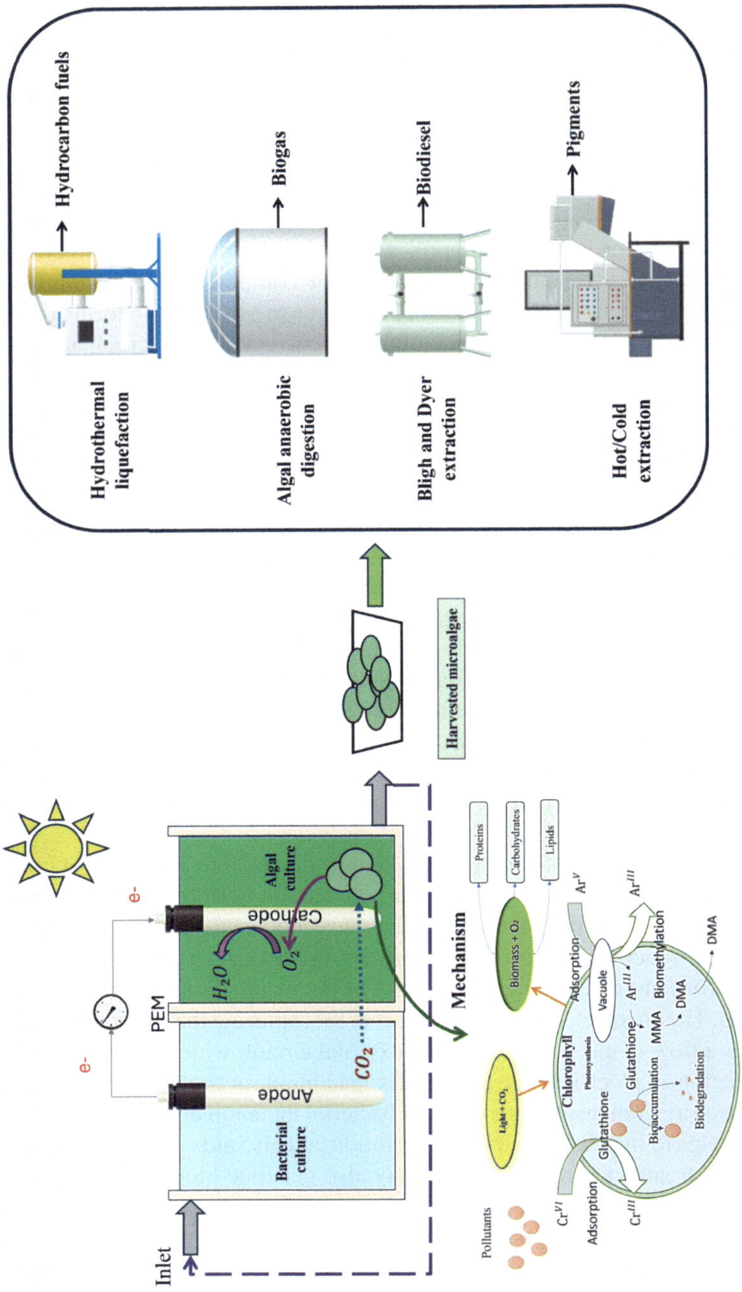

Fig. 2 Microalgae integrated MFC biorefinery system for the production of VAP

from the cathode compartment, which may be done by the formation of biofilm (Yadav et al. 2020). Using bio-flocculation, cell immobilization, biofilm formation, or membrane separation, it is possible to get algae biomass from the cathode section for a low cost (Sandaka et al. 2024). In the microalgae integrated MFC *Tetraselmis indica* generated 1 g/L of biomass after 16 days of operation with the doubling rate of 3.8 per day and produced lipid at the rate of 16 mg/L/day. In the transesterification process the FAME profile shown the presence of saturated and monounsaturated fatty acids and produced high-quality biodiesel with more than 50% palmitic and oleic acid in total FAME composition (Amit et al. 2020). Figure 2 depicts a bio-refinery system that utilizes microalgae to create electricity and value-added products (VAP) in MFCs. This system is responsible for the production of five pigments, namely chlorophyll, β-carotene, canthaxanthin, lutein, and zeaxanthin (Mehariya et al. 2021).

5 Conclusion with Future Perspective and Opportunities

Integrating microalgae with microbial fuel cells (MFCs) offers a hopeful solution to tackle the issues of water pollution and resource recovery. However, there are various factors that need to be taken into account in order to maximize system performance, promote environmental sustainability, and improve economic viability.

1. Optimizing the performance of integrated systems is a key challenge in microalgae-MFC integration. The goal is to achieve maximum efficiency in wastewater treatment, nutrient removal, and energy generation. Optimizing microalgae growth and metabolic activity requires careful adjustment of operational parameters, including hydraulic retention time, nutrient loading rates, light intensity, and electrode configuration.
2. In order to optimize nutrient uptake and electricity generation, it is essential to carefully choose and engineer appropriate strains of microalgae and electroactive bacteria. In addition, it is crucial to have advanced monitoring and control systems in place to optimize the performance of integrated systems in real-time. This allows for quick adaptation to environmental changes and variations in wastewater composition.
3. Scaling up microalgae-MFC integration from laboratory-scale prototypes to pilot-scale and full-scale applications for practical implementation in wastewater treatment plants and industrial facilities poses a significant technological challenge.
4. In addition, it is crucial to ensure that the combination of microalgae cultivation and MFCs can seamlessly fit into current infrastructure and wastewater treatment procedures, in order to keep retrofitting expenses and operational interruptions to a minimum.
5. In order to promote the widespread adoption and commercialization of integrated systems, it is important to consider factors such as land availability, energy requirements, and maintenance needs.

6. Consideration must be given to what happens to the pollutants and byproducts that are captured in integrated microalgae-MFC systems. Proper management of captured pollutants is crucial to prevent their release back into the environment or accumulation in biomass, as microalgae efficiently remove nutrients, heavy metals, and organic pollutants from wastewater. These byproducts, such as excess biomass, sludge, or wastewater effluent, need to be properly disposed of or treated to prevent any harm to the environment.
7. The economic feasibility of integrating microalgae-MFC relies on several factors, such as initial investment, ongoing expenses, energy efficiency, and income generated from resource recovery. Although there have been significant advancements in microalgae cultivation and MFC technology in recent years, the initial expenses for infrastructure, equipment, and maintenance could potentially hinder the widespread adoption and commercialization of these methods. In addition, it is crucial to carefully manage operational costs such as nutrient supplementation, energy consumption, and labour expenses to ensure that they are cost-effective and contribute to profitability.

There are various ways to generate revenue from integrated microalgae-MFC systems, such as resource recovery, energy production, and value-added bioproducts. Although there may be some upfront costs, these systems have the potential to bring in significant returns. Monetizing nutrient-rich microalgae biomass offers a range of potential applications, such as agricultural fertilizers, aquaculture feed supplements, pharmaceuticals, and biofuels. These diverse uses can generate additional revenue streams for system operators. In addition, the production of bioenergy from organic matter in MFCs, such as electricity, hydrogen, or methane, can be utilized on-site or sold to help reduce energy expenses and improve economic sustainability. This offers a flexible and environmentally friendly energy solution that can be used for heating, electricity, or as a transportation fuel. Through the utilization of organic matter in microalgae biomass, MFCs provide a practical solution to address greenhouse gas emissions, encourage the generation of renewable energy, and improve the extraction of resources from wastewater streams. However, with the help of technological advancements, process optimization, and economies of scale, capital and operational costs can be reduced. This makes integrated systems more economically competitive compared to conventional wastewater treatment methods.

Acknowledgements The authors acknowledge the financial support provided by the SRM University-AP under the Grant: SRMAP/URG/SEED/2023-2024/021

Author Contributions Ricky Rajamanickam: Writing, original draft preparation and visualization. Rangabhashiyam Selvasembian: Methodology, Supervision, Conceptualization, Writing—Review and Editing.

Declarations

Conflict of interest The authors declare they have no known conflict of interest.

References

Akhtar N, Syakir Ishak MI, Bhawani SA, Umar K (2021) Various natural and anthropogenic factors responsible for water quality degradation: a review. Water (Switzerland) 13:2660. https://doi.org/10.3390/w13192660

Ali J, Wang L, Waseem H, Song B, Djellabi R, Pan G (2020) Turning harmful algal biomass to electricity by microbial fuel cell: a sustainable approach for waste management. Environ Pollut 266:115373. https://doi.org/10.1016/j.envpol.2020.115373

Ali S, Paul Peter A, Chew KW, Munawaroh HSH, Show PL (2021) Resource recovery from industrial effluents through the cultivation of microalgae: a review. Bioresour Technol 337:125461. https://doi.org/10.1016/j.biortech.2021.125461

Amit, Nayak JK, Ghosh UK (2020) Microalgal remediation of anaerobic pretreated pharmaceutical wastewater for sustainable biodiesel production and electricity generation. J Water Process Eng. https://doi.org/10.1016/j.jwpe.2020.101192

Arun S, Sinharoy A, Pakshirajan K, Lens PNL (2020) Algae based microbial fuel cells for wastewater treatment and recovery of value-added products. Renew Sustain Energy Rev 132:110041. https://doi.org/10.1016/j.rser.2020.110041

Ayele A, Suresh A, Benor S (2021) Phycoremediation of heavy metals, factors involved and mechanisms related to functional groups in the algae cell surface—a review. Strateg Tools Pollut Mitig Ave Clean Environ. https://doi.org/10.1007/978-3-030-63575-6_13

Azwardi, Andaiyani S, Igamo AM, Wijaya WA (2023) The environmental impacts of natural resources depletion. In: Lecture notes in civil engineering. Springer, London, pp 705–714. https://doi.org/10.1007/978-981-99-1403-6_47

Bajracharya S (2020) Microbial fuel cell coupled with anaerobic treatment processes for wastewater treatment. In: Integrated microbial fuel cells for wastewater treatment. Elsevier, London, pp 295–311. https://doi.org/10.1016/B978-0-12-817493-7.00014-X

Bazdar E, Roshandel R, Yaghmaei S, Mardanpour MM (2018) The effect of different light intensities and light/dark regimes on the performance of photosynthetic microalgae microbial fuel cell. Bioresour Technol 261:350–360. https://doi.org/10.1016/j.biortech.2018.04.026

Bolognesi S, Cecconet D, Callegari A, Capodaglio AG (2021) Combined microalgal photobioreactor/microbial fuel cell system: performance analysis under different process conditions. Environ Res 192:110263. https://doi.org/10.1016/j.envres.2020.110263

Cai T, Meng L, Chen G, Xi Y, Jiang N, Song J, Zheng S, Liu Y, Zhen G, Huang M (2020) Application of advanced anodes in microbial fuel cells for power generation: a review. Chemosphere 248:125985. https://doi.org/10.1016/j.chemosphere.2020.125985

Danouche M, El Ghachtouli N, El Arroussi H (2021) Phycoremediation mechanisms of heavy metals using living green microalgae: physicochemical and molecular approaches for enhancing selectivity and removal capacity. Heliyon. https://doi.org/10.1016/j.heliyon.2021.e07609

Das S, Raj R, Das S, Ghangrekar MM (2023) Evaluating application of photosynthetic microbial fuel cell to exhibit efficient carbon sequestration with concomitant value-added product recovery from wastewater: a review. Environ Sci Pollut Res 30:98995–99012. https://doi.org/10.1007/s11356-022-21184-z

Dolganyuk V, Belova D, Babich O, Prosekov A, Ivanova S, Katserov D, Patyukov N, Sukhikh S (2020) Microalgae: a promising source of valuable bioproducts. Biomolecules 10:1–24. https://doi.org/10.3390/biom10081153

Don CDYA, Babel S (2021) Effects of organic loading on bioelectricity and micro-algal biomass production in microbial fuel cells using synthetic wastewater. J Water Process Eng 39:101699. https://doi.org/10.1016/j.jwpe.2020.101699

Elshobary ME, Zabed HM, Yun J, Zhang G, Qi X (2021) Recent insights into microalgae-assisted microbial fuel cells for generating sustainable bioelectricity. Int J Hydrog Energy 46:3135–3159. https://doi.org/10.1016/j.ijhydene.2020.06.251

Enamala MK, Dixit R, Tangellapally A, Singh M, Dinakarrao SMP, Chavali M, Pamanji SR, Ashokkumar V, Kadier A, Chandrasekhar K (2020) Photosynthetic microorganisms (Algae)

mediated bioelectricity generation in microbial fuel cell: concise review. Environ Technol Innov 19:100959. https://doi.org/10.1016/j.eti.2020.100959

Garbini GL, Barra Caracciolo A, Grenni P (2023) Electroactive bacteria in natural ecosystems and their applications in microbial fuel cells for bioremediation: a review. Microorganisms 11:1255. https://doi.org/10.3390/microorganisms11051255

Greenman J, Gajda I, Ieropoulos I (2019) Microbial fuel cells (MFC) and microalgae: photo microbial fuel cell (PMFC) as complete recycling machines. Sustain Energy Fuels 3:2546–2560. https://doi.org/10.1039/c9se00354a

Hou Q, Nie C, Pei H, Hu W, Jiang L, Yang Z (2016) The effect of algae species on the bioelectricity and biodiesel generation through open-air cathode microbial fuel cell with kitchen waste anaerobically digested effluent as substrate. Bioresour Technol 218:902–908. https://doi.org/10.1016/j.biortech.2016.07.035

Hussain F, Shah SZ, Ahmad H, Abubshait SA, Abubshait HA, Laref A, Manikandan A, Kusuma HS, Iqbal M (2021) Microalgae an ecofriendly and sustainable wastewater treatment option: Biomass application in biofuel and bio-fertilizer production. A review. Renew Sustain Energy Rev. https://doi.org/10.1016/j.rser.2020.110603

Idris MO, Kim HC, Yaqoob AA, Ibrahim MNM (2022) Exploring the effectiveness of microbial fuel cell for the degradation of organic pollutants coupled with bio-energy generation. Sustain Energy Technol Assess 52:102183. https://doi.org/10.1016/j.seta.2022.102183

Ighalo JO, Dulta K, Kurniawan SB, Omoarukhe FO, Ewuzie U, Eshiemogie SO, Ojo AU, Abdullah SRS (2022) Progress in microalgae application for CO_2 sequestration. Clean Chem Eng 3:100044. https://doi.org/10.1016/j.clce.2022.100044

Janeeshma E, Singh G, Puthur JT (2022) Microbial mediated waste management and bioenergy production. In: Utilization of waste biomass in energy, environment and catalysis. CRC Press, London, pp 177–198. https://doi.org/10.1201/9781003196358-9

Kakarla R, Min B (2014) Evaluation of microbial fuel cell operation using algae as an oxygen supplier: carbon paper cathode vs. carbon brush cathode. Bioprocess Biosyst Eng 37:2453–2461. https://doi.org/10.1007/s00449-014-1223-4

Kondaveeti S, Choi KS, Kakarla R, Min B (2014) Microalgae *Scenedesmus obliquus* as renewable biomass feedstock for electricity generation in microbial fuel cells (MFCs). Front Environ Sci Eng 8:784–791. https://doi.org/10.1007/s11783-013-0590-4

Kumar A, Chalotra T, Pathak AK, Raina N (2022) Role of microbes in wastewater treatment and energy generation potentials: a sustainable approach. Microb Biotechnol Role Ecol Sustain Res. https://doi.org/10.1002/9781119834489.ch15

Lee D-J, Chang J-S, Lai J-Y (2015) Microalgae–microbial fuel cell: a mini review. Bioresour Technol 198:891–895

Leng X, Usman M, Salama ES, Ni H, Zhou T, Jeon BH, Liu P, Zhang P, Li X (2021) Development of an innovative MFC-biosensor for real-time monitoring of anaerobic digestion for biogas production: controlled substrate feeding strategy. J Environ Chem Eng 9:106703. https://doi.org/10.1016/j.jece.2021.106703

Leong YK, Chew KW, Chen WH, Chang JS, Show PL (2021) Reuniting the biogeochemistry of algae for a low-carbon circular bioeconomy. Trends Plant Sci 26:729–740. https://doi.org/10.1016/j.tplants.2020.12.010

Li J, Hu J, Yang C, Pu W, Hou H, Xu J, Liu B, Yang J (2019) Enhanced detection of toxicity in wastewater using a 2D smooth anode based microbial fuel cell toxicity sensor. RSC Adv 9:8700–8706. https://doi.org/10.1039/c8ra10337b

Logroño W, Pérez M, Urquizo G, Kadier A, Echeverría M, Recalde C, Rákhely G (2017) Single chamber microbial fuel cell (SCMFC) with a cathodic microalgal biofilm: a preliminary assessment of the generation of bioelectricity and biodegradation of real dye textile wastewater. Chemosphere 176:378–388. https://doi.org/10.1016/j.chemosphere.2017.02.099

Manikandan A, Suresh Babu P, Shyamalagowri S, Kamaraj M, Muthukumaran P, Aravind J (2022) Emerging role of microalgae in heavy metal bioremediation. J Basic Microbiol 62:330–347. https://doi.org/10.1002/jobm.202100363

Mehariya S, Goswami RK, Verma P, Lavecchia R, Zuorro A (2021) Integrated approach for wastewater treatment and biofuel production in microalgae biorefineries. Energies. https://doi.org/10.3390/en14082282

Mehrotra S, Kiran Kumar V, Man Mohan K, Gajalakshmi S, Pathak B (2021) Bioelectrogenesis from ceramic membrane-based algal-microbial fuel cells treating dairy industry wastewater. Sustain Energy Technol Assess 48:101653. https://doi.org/10.1016/j.seta.2021.101653

Molazadeh M, Ahmadzadeh H, Pourianfar HR, Lyon S, Rampelotto PH (2019) The use of microalgae for coupling wastewater treatment with CO_2 biofixation. Front Bioeng Biotechnol. https://doi.org/10.3389/fbioe.2019.00042

Mondal S, Palit D (2021) Challenges in natural resource management for ecological sustainability. In: Natural resources conservation and advances for sustainability. Elsevier, London, pp 29–59. https://doi.org/10.1016/B978-0-12-822976-7.00004-1

Morin-Crini N, Lichtfouse E, Liu G, Balaram V, Ribeiro ARL, Lu Z, Stock F, Carmona E, Teixeira MR, Picos-Corrales LA, Moreno-Piraján JC, Giraldo L, Li C, Pandey A, Hocquet D, Torri G, Crini G (2022) Worldwide cases of water pollution by emerging contaminants: a review. Environ Chem Lett 20:2311–2338. https://doi.org/10.1007/s10311-022-01447-4

Naina Mohamed S, Ajit Hiraman P, Muthukumar K, Jayabalan T (2020) Bioelectricity production from kitchen wastewater using microbial fuel cell with photosynthetic algal cathode. Bioresour Technol 295:122226. https://doi.org/10.1016/j.biortech.2019.122226

Nandan A, Rangabhashiyam S (2023) Reuse of water treatment plant sludge for treatment of pollutants. In: Resource recovery in drinking water treatment. Elsevier, London, pp 187–203. https://doi.org/10.1016/B978-0-323-99344-9.00006-2

Nguyen HTH, Min B (2020) Leachate treatment and electricity generation using an algae-cathode microbial fuel cell with continuous flow through the chambers in series. Sci Total Environ 723:138054. https://doi.org/10.1016/j.scitotenv.2020.138054

Nguyen LN, Vu MT, Vu HP, Johir MAH, Labeeuw L, Ralph PJ, Mahlia TMI, Pandey A, Sirohi R, Nghiem LD (2023) Microalgae-based carbon capture and utilization: a critical review on current system developments and biomass utilization. Crit Rev Environ Sci Technol 53:216–238. https://doi.org/10.1080/10643389.2022.2047141

Obileke KC, Onyeaka H, Meyer EL, Nwokolo N (2021) Microbial fuel cells, a renewable energy technology for bio-electricity generation: a mini-review. Electrochem Commun 125:107003. https://doi.org/10.1016/j.elecom.2021.107003

Peres CB, Resende PMR, Nunes LJR, Morais LCD (2022) Advances in carbon capture and use (CCU) technologies: a comprehensive review and CO_2 mitigation potential analysis. Clean Technol 4:1193–1207. https://doi.org/10.3390/cleantechnol4040073

Powell EE, Mapiour ML, Evitts RW, Hill GA (2009) Growth kinetics of *Chlorella vulgaris* and its use as a cathodic half cell. Bioresour Technol 100:269–274. https://doi.org/10.1016/j.biortech.2008.05.032

Prathiba S, Kumar PS, Vo DVN (2022) Recent advancements in microbial fuel cells: a review on its electron transfer mechanisms, microbial community, types of substrates and design for bio-electrochemical treatment. Chemosphere 286:131856. https://doi.org/10.1016/j.chemosphere.2021.131856

Rajamanickam R, Selvasembian R (2024) A sustainable resource recovery approach through micro-algae integrated brine management in minimal liquid discharge system. Desalination 578:117424. https://doi.org/10.1016/j.desal.2024.117424

Ramya M, Senthil Kumar P (2022) A review on recent advancements in bioenergy production using microbial fuel cells. Chemosphere 288:132512. https://doi.org/10.1016/j.chemosphere.2021.132512

Ravikumar Y, Razack SA, Yun J, Zhang G, Zabed HM, Qi X (2021) Recent advances in microalgae-based distillery wastewater treatment. Environ Technol Innov 24:101839. https://doi.org/10.1016/j.eti.2021.101839

Ribeiro VR, Osório HDD, Ulrich AC, Rizzetti TM, Barrios AS, de Cassia de Souza Schneider R, Benitez LB (2022) The use of microalgae–microbial fuel cells in wastewater bioremediation

and bioelectricity generation. J Water Process Eng 48:102882. https://doi.org/10.1016/j.jwpe.2022.102882
Rossi R, Logan BE (2021) Using an anion exchange membrane for effective hydroxide ion transport enables high power densities in microbial fuel cells. Chem Eng J 422:130150. https://doi.org/10.1016/j.cej.2021.130150
Saba B, Christy AD, Yu Z, Co AC, Islam R, Tuovinen OH (2017) Characterization and performance of anodic mixed culture biofilms in submersed microbial fuel cells. Bioelectrochemistry 113:79–84. https://doi.org/10.1016/j.bioelechem.2016.10.003
Sandaka BP, Kumar J, Melo JS (2024) Biofuels from microalgae: growing conditions, cultivation strategies, and techno-commercial challenges. In: Microalgal biomass for bioenergy applications. Elsevier, London, pp 305–340
Saranya N, Rangabhashiyam S (2023) Microbial fuel cells as an energy-efficient alternative for pollution degradation. In: Resource recovery in industrial waste waters. Elsevier, London, pp 407–420. https://doi.org/10.1016/B978-0-323-95327-6.00014-2
Selvasembian R, Thokchom B, Singh P, Jawad AH, Gwenzi W (2023) Remediation of heavy metals: sustainable technologies and recent advances. Remediat Heavy Met Sustain Technol Recent Adv. https://doi.org/10.1002/9781119853589
Shaikh R, Rizvi A, Pandit S, Desai N, Patil R (2022) Microalgae: classification, bioactives, medicinal properties, industrial applications, and future prospectives. In: An integration of phycoremediation processes in wastewater treatment. Elsevier, London, pp 451–486
Sharma M, Salama ES, Zhang P, Zhang L, Xing X, Yue J, Song Z, Nan L, Yujun S, Li X (2022) Microalgae-assisted microbial fuel cells for electricity generation coupled with wastewater treatment: biotechnological perspective. J Water Process Eng 49:102966. https://doi.org/10.1016/j.jwpe.2022.102966
Siddiki SYA, Mofijur M, Kumar PS, Ahmed SF, Inayat A, Kusumo F, Badruddin IA, Khan TMY, Nghiem LD, Ong HC, Mahlia TMI (2022) Microalgae biomass as a sustainable source for biofuel, biochemical and biobased value-added products: an integrated biorefinery concept. Fuel 307:121782. https://doi.org/10.1016/j.fuel.2021.121782
Su Y (2021) Revisiting carbon, nitrogen, and phosphorus metabolisms in microalgae for wastewater treatment. Sci Total Environ 762:144590. https://doi.org/10.1016/j.scitotenv.2020.144590
Tanvir RU, Zhang J, Canter T, Chen D, Lu J, Hu Z (2021) Harnessing solar energy using phototrophic microorganisms: a sustainable pathway to bioenergy, biomaterials, and environmental solutions. Renew Sustain Energy Rev 146:111181. https://doi.org/10.1016/j.rser.2021.111181
Umamaheswari J, Rajamanickam R, Vilvanathan S, Shanthakumar S, Venkateswarlu K, Abinandan S, Megharaj M (2023) Removal of heavy metals and organic pollutants by marine microalgae. In: Encarnação T, Canelas Pais A (eds). Springer, Cham, pp 29–64. https://doi.org/10.1007/978-3-031-17226-7_3
Vargas-Estrada L, Longoria A, Arenas E, Moreira J, Okoye PU, Bustos-Terrones Y, Sebastian PJ (2022) A review on current trends in biogas production from microalgae biomass and microalgae waste by anaerobic digestion and co-digestion. Bioenergy Res 15:77–92. https://doi.org/10.1007/s12155-021-10276-2
Vega JM (2018) Nitrogen and sulfur metabolism in microalgae and plants: 50 years of research. Prog Bot 81:1–40. https://doi.org/10.1007/124_2018_26
Wang D, Hubacek K, Shan Y, Gerbens-Leenes W, Liu J (2021) A review of water stress and water footprint accounting. Water (Switzerland) 13:201. https://doi.org/10.3390/w13020201
Xiao L, He Z (2014) Applications and perspectives of phototrophic microorganisms for electricity generation from organic compounds in microbial fuel cells. Renew Sustain Energy Rev 37:550–559. https://doi.org/10.1016/j.rser.2014.05.066
Yaakob MA, Mohamed RMSR, Al-Gheethi A, Ravishankar GA, Ambati RR (2021) Influence of nitrogen and phosphorus on microalgal growth, biomass, lipid, and fatty acid production: an overview. Cells 10:1–19. https://doi.org/10.3390/cells10020393

Yadav G, Sharma I, Ghangrekar M, Sen R (2020) A live bio-cathode to enhance power output steered by bacteria-microalgae synergistic metabolism in microbial fuel cell. J Power Sources 449:227560. https://doi.org/10.1016/j.jpowsour.2019.227560
Yang Z, Pei H, Hou Q, Jiang L, Zhang L, Nie C (2018) Algal biofilm-assisted microbial fuel cell to enhance domestic wastewater treatment: nutrient, organics removal and bioenergy production. Chem Eng J 332:277–285. https://doi.org/10.1016/j.cej.2017.09.096
Zhang Y, Zhao Y, Zhou M (2019) A photosynthetic algal microbial fuel cell for treating swine wastewater. Environ Sci Pollut Res 26:6182–6190. https://doi.org/10.1007/s11356-018-3960-4

Enzymatic Fuel Cell for Treatment of Water Pollutants and Resources Recovery

Manisha Agrawal, Manisha Thakur, Meena Chakraborty, and Prachi Parmar

Abstract The ever increasing demand of clean water has stimulated intensive research activities in the field of treatment of waste water. Industrial development and urbanization become major factors for contamination of water sources. Different contaminant removal technologies such as adsorption, bio oxidation, aerobic treatment, chemical treatment, electro oxidation, electro coagulation, electro dialysis etc. requires large consumption of energy, higher operational cost and produces enormous amount of sludge. It is a need of hour to find out a sustainable solution for treatment of water pollutants. Enzymatic biofuel cells (EFCs) may provide an alternative to solve the treatment problems of water pollutants. Enzymatic biofuel cells are the category of fuel cell who converts chemical energy into other form of energy that is electrical, through electrochemical reactions with use of biochemical trajectory. Since enzymatic biofuel cells used renewable fuels and enzyme biocatalysts in benign operating conditions, they are recognized as sustainable energy sources. The conventional sludge activation technologies have difficulty degrading resistant contaminants such as oil, grease, chemicals, pesticides, medicines, and plastics in wastewater. Due to their widespread presence, toxicity, resistance to natural biodegradability, and other detrimental effects, these pollutants pose a major hazard to aquatic habitats and organisms. A highly effective, enzymatic fuel cell is proposed for the treatment of these resistant pollutants. As useful instruments of living things, enzymes provide a sustainable and biobased approach to removal of pollutants from contaminated water. Consequently, the use of enzymatic fuel cells has great promise for lowering pollution levels and creating a better atmosphere for both people and other animals. The functions and uses of enzymatic fuel cells in wastewater treatment and resource recovery are the main topics of this chapter. The impact of enzymes, their kinds, and

M. Agrawal (✉)
Rungta College of Engineering and Technology, Raipur, CG, India
e-mail: dr.manisha.9000@gmail.com

M. Thakur · M. Chakraborty
Government Naveen College Bori, Durg, CG, India

P. Parmar
Shri Shankaracharya Professional University, Durg, CG, India

R. Selvasembian et al. (eds.), *Emerging Trends in Microbial Electrochemical Technologies for Sustainable Mitigation of Water Resources Contamination*,
https://doi.org/10.1007/978-3-031-74636-9_8

their sources are covered, along with enzymatic techniques for remediating resistant pollutants, identifying and testing the eco-toxicity of enzymatic transformation products, and the use of enzymatic fuel cells in wastewater treatment systems.

Keywords Enzymatic fuel cells · Water pollutants · Recovery · Treatment

1 Preview of the Literature

Rapid industrialization and urbanization bring advancement and comfort in every aspect of our life and made things easier for survival. But on the other hand, it also creates a serious threat to our life in terms of environmental pollution. Water, air and food are the fundamental requirements of human beings but because of pollution it hardly available in purest form. Discharge of industrial wastewater and sewage water into water bodies worsen the water quality and make it unfit for use by us. Such kind of waste water contains large amount of toxic chemicals and biological substances in it (Tauseef et al. 2013; Lu et al. 2016). Because of the scarcity of freshwater resources, it is essential to treat wastewater properly and make it fit for use of society. Various technologies are available for treatment of wastewater such as adsorption, bio oxidation, aerobic treatment, chemical treatment, electro oxidation, electro coagulation, electro dialysis etc. these techniques required large consumption of energy, higher operational cost and produces enormous amount of sludge (Wei et al. 2003; Mantz and Panagiotopoulos 2015). Bio fuel cell is the category of fuel cell which converts chemical energy into electrical energy through electrochemical reactions by using biochemical trajectory (Bullen et al. 2006; Kim and Kim 2018). Development of biofuel cell has provided a wider research area for chemists and biologists (Willner 2009).

Fuel cells are good alternative for energy sources as they are the renewal energy sources and have environment friendly nature (Wang et al. 2019; Hameed et al. 2018). Water formation by the reaction of hydrogen and oxygen is the fundamental of fuel cell (Squadrito et al. 2016). The Fuel cell utilizes acidic or basic electrolytes and expensive metals as a catalyst. In contrast to fuel cells, bio fuel cells use moderate electrolytes, renewable biocatalysts and generate energy from bio materials or organic pollutants (Song et al. 2017; Wang et al. 2021). The biofuel cell's ability to transform gasoline from a sustainable source into electrical energy develops them as an important alternative source to meet the energy requirements. The process involved in bio fuel cell also reduces the waste treatment cost. So, the bio fuel cells are better than conventional fuel cell in terms of cost, storage and distribution of fuel material (Scott et al. 2012).

Biofuel cells are further classified into various categories based on materials used in them and their applications. The categories are bio-electrochemical fuel cell including microbial fuel cells, microbial electrolysis cell, microbial desalination cells and microbial solar cells, organelle fuel cell and enzymatic bio fuel cell (Logan et al. 2008; Cao et al. 2009; Strik et al. 2011; Pant et al. 2012). Among the above mentioned cells, the enzymatic fuel cell has a special category cell which uses an enzyme as catalyst for the oxidation of fuel. Enzymes are very good catalysts because of their activity, selectivity and specificity (Hernandez et al. 2011). Therefore, use of enzymatic fuel cell has many advantages in terms of energy conversion, renewable source of energy, flexibility of fuels and the ability to work at room temperature (Pant et al. 2012). Enzymatic fuel cell was invented in 1960s (Yahiro et al. 1964), but still, it is in its early stage of development because of its short life time, small power densities and inefficient oxidation of fuels in enzymatic fuel cell (Minteer et al. 2007). Enzymatic fuel cells utilize biological enzyme as electrode catalyst (Cooney et al. 2008a, b). Enzymes used in the cell without cell membrane, as a result degradation of enzyme occurs (Xiao et al. 2019). Biofuel cells are one of the best remedial methods for waste water treatment. They can treat the pollutants of waste water and help us to recover the water resources.

In enzymatic fuel Cell, fuels get oxidised through the enzymes, whereas in conventional fuel cell precious metals used as catalyst. Enzymatic fuel cells use complete cells to catalyze fuel; these fuel cells are also known as enzymatic bio-fuel cells since they use enzymes that are derived from living cells, albeit not within living cells. Because they are incredibly prevalent in nature, organic molecules like sugars and alcohols are specifically made for processing by enzymes. The majority of organic molecules produce carbon monoxide during operation, making them unsuitable for use as fuel in fuel cells using metal catalysts. Produced CO_2 will quickly "poisoned" the precious metals thus the cell become useless (Atanassov et al. 2007). Large-scale cultivation of sugars and other biofuels is possible to supply enzymatic biofuel cells with fuel. Due to ease in availability and economic benefit of bio fuels, it becomes an extraordinarily attractive option through the use of sustainable energy sources.

EFC when used in large scale to treat water in wastewater treatment plants, complete unit will be self-sufficient. As power get generated, instead of only power consumption. Since glucose is a necessary component of many living things' metabolic systems, it is the fuel most frequently employed in enzymatic biofuel cells (EFCs). As a result, glucose-based EFCs open the door to both in vivo and in vitro medical applications. Glucose EFCs and conventional hydrogen fuel cells work on the same principle. Here, an appropriate biocatalyst converts glucose into an oxidant that releases electrons when a voltage is supplied, allowing oxygen reduction to produce bioelectrical power. Anaerobic digestion is employed by EFCs to harvest energy for water treatment. Pathogens may be lessened by the procedure. To turn biogas into energy, though, requires temperatures above 30 °C and an additional step. By producing a helical flow inside the EFC, spacing spirals can be utilized to boost the production of electricity (Zhang et al. 2013).

Thus, chapter deals with the components and structure of the enzymatic fuel cell along with the advantages and weaknesses of current EFCs technologies used in various wastewater plants. The technical barriers in operation and the economic feasibility of EFCs for wastewater treatment are also discussed.

2 Introduction

Let's know the difference among Fuel Cell, bio-fuel cell and microbial fuel cell. A fuel cell is a kind of electrochemical cell that produces electrical energy by oxidizing fuel's chemical energy when oxidants are present. The process propagates with the redox reactions. If fuel cell may use biomaterial or living organisms a fuel, to produce electricity then such kind of arrangement will be an example of bio-fuel cell. If, go further ahead than, A fuel cell is referred to as a microbial fuel cell if it oxidizes the fuel using enzymes as a catalyst rather than metal. Thus, Microbial fuel cell is a variety of cell in which due to interaction between enzymes and fuels, energy get produced. Bio-enzymes work as catalyst under mild working conditions. Microbe based fuel cell has capability to be a source of sustainable energy.

Proteins make up the intricate structure of enzymes. The apoenzyme and cofactor, which are the enzyme's units for transporting electrons, are buried deep within its intricate structure. Therefore, it is challenging to achieve effective electrical connection between the enzyme biocatalyst and the electrode substrate. Small redox molecules act as mediators of electron transport between the electrode and the enzyme in mediated bio electrocatalysis. One important reaction in using the enzymes in biological fuel cells is mediated bio-electrocatalysis, which is helpful for researching the kinetics of the enzymes' electron transfer (Park et al. 2000).

2.1 Concept of Enzymatic Fuel Cell

Microbial fuel cells (MFCs) and enzymatic biofuel cells (EBFCs) are two major categories into which BFCs can be divided based on the biocatalysts used. While the latter kind is based on redox-active enzymes like glucose oxidase, laccase, etc., the former uses a whole microbial cell for the catalytic oxidation of fuel (Xiao et al. 2019) and hence requires continual maintenance of whole living cells for their survival (Rabaey and Verstraete 2005), wherein no such upkeep is required. To produce electrical energy, both kinds of BFCs use electrochemical oxidation of fuels like glucose, ethanol, methanol, etc. at the anode and oxygen reduction at the cathode. Due to their ease of use and plentiful natural supplies of redox enzymes, EBFCs have garnered a lot of attention recently (Mansoorian et al. 2013). Enzymatic biofuel cells, or EFCs for short, are bioelectronic devices that use oxidoreductase enzymes as electrocatalysts to reduce oxygen or peroxide and/or oxidize organic substrates in order to convert energy into electrical power. Because enzymes have a high degree of substrate selectivity, they can sometimes eliminate the requirement for membranes and precious metals, resulting in extremely small systems that are ideal for downsizing.

Applications for EFCs that need low power consumption and biocompatibility of the device include wearable or implantable biofuel cells, biosensors that run on their own power, and portable battery-free power solutions in general.

Effectively immobilizing enzymes on the electrodes is a significant problem in the design of enzymatic biofuel cells in order to achieve direct electron transfer without the need for mediators, leading to increased performance and better long-term stability. The number of connected enzymes per volume unit can be increased and the electron transport between the electrode and the active site of the enzyme can be facilitated by using conductive nanomaterials and various polymer types as electrodes.

2.2 *Construction and Components of EFCs*

What distinguishes implantable EFC devices is their incredibly simple architecture are shown in Fig. 1. Enzymes are used for glucose oxidation at the anodes and O_2 reduction at the cathodes. Unlike conventional electrochemical sensors, this design does not require a reference electrode or its calibration. The sugar-oxidizing and O_2-reducing oxidoreductases are thought to be ideal and alluring enzymes for the creation of implanted EFCs due to the quantity of carbohydrates and the availability of O_2 (Zhu et al. 2012; Mac Vittie et al. 2013; Onuki et al. 2008; Kotanen et al. 2012; Cadet et al. 2016; Barton et al. 2004). The enzymes are linked to the surface area of electrodeeither by direct electron transfer (DET) or by mediated electron transfer (MET) (Zhu et al. 2012; Leech et al. 2012; Ganzales-Solino et al. 2018). The modern method in EFC technology presently is the manufacture of bio-electrodes whichproceed through DET instead of MET, therefore avoids the problem of toxicity developed during the use of redox mediators (Neto and De Andrade 2013; Falk et al. 2012; Bollella et al. 2018; Minteer et al. 2007).

Glucose oxidase, which electro-oxidizes glucose either directly or indirectly (through a mediator), is the anodic enzyme that has received the most research. However, using this enzyme in implantable circumstances had some drawbacks, including the production of a harmful by product (H_2O_2) and the requirement for additional enzymatic device adjustments for its in vivo operation (Zhu et al. 2012; Leech et al. 2012). To keep away from such issues, Other glucose-oxidizing substances, such as NADH-subordinate glucose dehydrogenases, can be used because they do not produce hydrogen peroxide in an embedded environment. Blue multicopper oxidases like laccase, ceruloplasmin, bilirubin oxidase, and ascorbate oxidase, which catalyse the direct oxygen reduction to water, typically operate the bio cathodes of implanted EFCs (Zhu et al. 2012; Leech et al. 2012; Ganzales-Solino et al. 2018; Falk et al. 2014).

2.3 *Designing*

The Enzymatic fuel cell is a bio-electrical device that produces power by using microorganisms to break down organic materials. Like as other typical cell, The

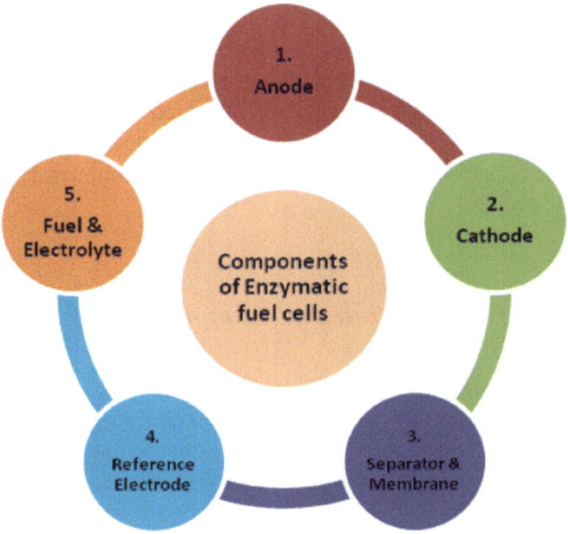

Fig. 1 Components of enzymatic fuel cells (EFCs)

proton exchange membrane, the anode, the cathode, and the external circuit make up the four components of the enzymatic fuel cell as well. The rules governing ordinary fuel cells also apply to enzymatic biofuel cells. It forces a parent molecule to cross an electrolyte barrier by removing its electrons using a catalyst. With the use of wire an electric current gets generated (Cooney et al. 2008a, b).

Enzymatic fuel cells typically consist of five components these are shown in Fig. 1.

1. *Anode*

   In anode region biocatalytic oxidation of a substrate takes place, which providedby a fuel. The power created is directed by the quantity of synergist destinations in direct relationship with the cathode and the substrate accessibility for the charge move response. Effective current generation necessitates an electrode with a high enzyme loading, which in turn necessitates a large surface area for enzyme immobilization and open porosity for mass transport. An optimal terminal material, for model, gives multifunctional porosity: mesopores for compound immobilization furthermore, macropores for mass vehicle of fuel. According to the current state of the art, biopolymers doped with conductive additives are typically used to create anode electrode materials (Cooney et al. 2008a, b; Konash et al. 2006).

   Cathode

   In EFCs, the cathode makes charge transfer reduction easier by consuming hydrogen protons and electrons. Hydrogen protons and electrons are shaped in oxidation responses in the anode, while protons are moved to the cathode through afluid based electrolyte. In an outer circuit, electrons are transferred simultaneously through an electric load.
2. *Separator and Membrane*

To limit power limitation caused by ion diffusion between the electrodes, the EFC anode and cathode should be separated as little as possible, similar to any other traditional electrochemical power source. Simultaneously, the two terminals should be electronically protected to keep away from interior short out. A straightforward thin separator, which must provide maximum ion transport and maximum electronic resistance, meets both requirements. Separators must also be able to withstand external operating conditions mechanically and thermally.

Particle specific films should be utilized in EFCs while the accompanying circumstances apply: The employment of diffusive mediators, the non-selectivity of charge transfer reactions, the incompatibility of the anode and cathode chemically, the anode's oxygen sensitivity, and the possibility of oxygen leakage from the cathode into the cell are all contributing considerations. In a membrane-free system, diffuse mediators, for example, function as an internal ionic short circuit. A membrane must be used to provide for fast proton transfer kinetics and, consequently, unhindered power output while preventing mediator diffusion to the opposite electrode.

Reference Electrode

EFC powers sources needn't concern with a reference electrode to work, yet the reference electrode is significant for the plan and improvement of EFCs as a way torecognize the presentation and lifetime restricting electrode. Saturated calomel electrodes (SCEs) or Ag/AgCl electrodes are typically used as reference electrodes for EFCs.

Fuel and Electrolyte. Aqueous solutions of the substrate and salts sufficient to form an electrolyte typically constitute the fuel. Fuel is buffered to maintain the necessary pH level since the stability and kinetics of the enzymes adsorbed to EFC electrodes are greatly impacted by pH.

2.3.1 Working

Theoretically, enzymatic fuel cells are interchangeable with other fuel cell types. The distinction stems from the type of catalysts used, which are enzymes (biocatalysts), as opposed to more traditional noble metal catalysts. When biocatalysts are used, different fuels and operating conditions are produced. The cells work on the same concept as traditional fuel cells: fuel is oxidized at the anode side, and the liberated electrons are driven via an external electrical circuit to produce electric current. At the cathode, where they mix with protons and an oxidant (usually oxygen) to produce a product (usually water), the electrons finally arrive. In Fig. 2, the basic elements and operating principle of an enzymatic fuel cell and a standard polymer electrolyte membrane (PEM) fuel cell are displayed in comparison.

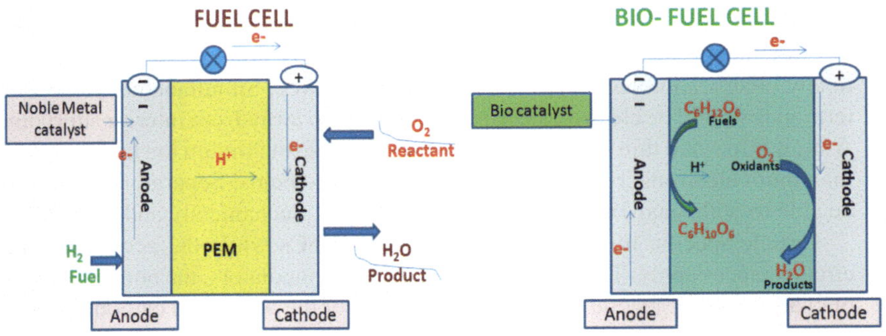

Fig. 2 Fuel cell and biofuel cell

2.3.2 Step Wise Working of Cell

While working of the enzymatic fuel cell stored chemical energy of the organic compound converted in to electrical energy in presence of the Enzyme catalyst, it can be briefed as below:

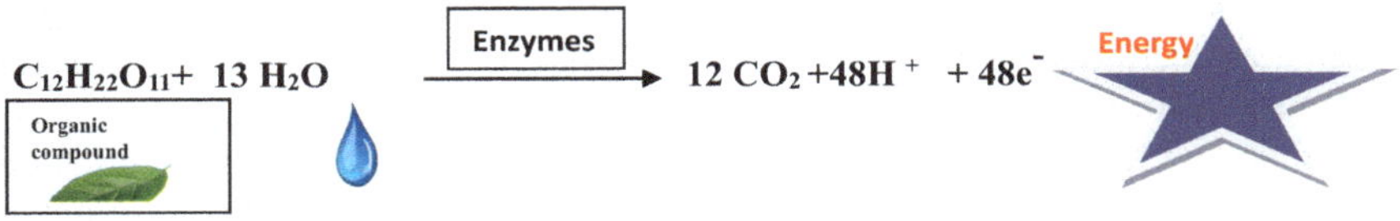

1. Reaction starts with an anode, where microbes metabolized. Organic compounds such as glucose, sugar which act as electron donor.
2. During metabolism, organic compound generates electrons and protons.
3. If Oxygen is not available these microbes produce Carbon Dioxide, Hydrogen Ions (hydrons) and electrons.
4. Generated electrons transferred to the surface of anode. It travels via an external electrical circuit to the cathode.
5. Whereas protons migrates though the electrolyte to the external circuit.
6. Electrons and protons get consumed in cathodic section.
7. Electrical power can measure through placing load between the two compartments.

An enzymatic fuel cell's voltage and current are what determine its power ($P = U \times I$). The potentials of a biofuel cell's electrodes control its voltage. Because of this, it is ideal to maintain the lowest possible oxidation potential and highest possible reduction potential. The rate of electron transfer and currents are frequently made faster or possible by the addition of mediators, but the fuel cell's capacity to create voltage is reduced. The potential of the enzymatic electrodes engaged in MET is mostly determined by the mediator's potential, which must be more positive than the enzyme potential in the case of oxidation or more negative in the case of reduction for electron transfer to occur. For biofuel cell applications, this makes using the DET technique more advantageous; nonetheless, there is frequently a trade-off between electron transfer rate and potential.

2.3.3 Types of Enzymatic Fuel Cells

Microbial fuel cells and enzymatic fuel cells are the two categories of biological fuel cells. We are only concerned about enzymatic fuel cells here. Proteins make up the intricate structure of enzymes. The apoenzyme and cofactor, which are the enzyme's units for transporting electrons, are buried deep within its intricate structure. Enzymatic fuel cells are of majorly classified on the basis of function of enzymes and methods of electron transfer. These two types are further classified as below:

1. *Types of Enzymatic Fuel Cells Based on function of Enzymes*

   Two categories can be distinguished between biofuel cells based on the roles played by the biocatalysts, or enzymes. These are mentioned in Fig. 3:

   1. *Product Type* Enzymes are not directly engaged in the production of energy in these fuel cells; instead, a conversional fuel cell produces the energy. Enzymes use a biocatalytic transformation or metabolic process to produce the fuel substrate for fuel cells (Nishizawa et al. 2008; Zhu et al. 2007). This type of biofuel cell is less common.
   2. *Direct Energy Production Type*

      In this type of enzymatic fuel cells, enzymes are directly engaged in the bioreactions for energy production. Enzymes take part in the process of electron transfers from the fuel to the anode. Similar to conventional fuel cells, enzymes work as catalysts for oxidant reduction on the cathode and produce electrons when they oxidize organic materials on the anode. The activity of

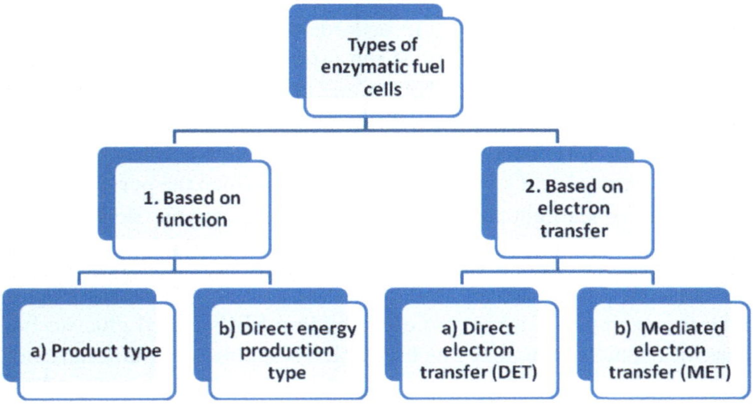

Fig. 3 Types of enzymatic fuel cells

the enzymes plays a major role on how well the fuel cell works. This kind of enzymatic fuel cell predominates. Establishing effective electron transport between enzymes and electrode supports is one of the biggest hurdles for this kind of fuel cell (Scott et al. 2012; Yu and Scott 2010).

2. *Types of Enzymatic Fuel Cells Based on Electron Transfer Methods*
   1. *Direct electron transfer (DET)* Enzymes with nicotinamide adenine dinucleotide (NADH/NAD+) or nicotinamide adenine dinucleotide phosphate (NADPH/NADP+) redox centers or enzymes with at least part of the redox center is placed at, or near, the periphery of the protein shell, are able to carry out direct electron transfer (DET) between the enzyme active center and the electrode surface.
   2. *Mediated Electron Transfer (MET)*

      Enzymes with a strongly bound redox center deeply bound in a protein or glycoprotein shell are not able to transfer electron directly between the active centres and electrodes due to the large distance, >21 A, between the enzyme active centres and the electrode surface (Heller 1990). Therefore, they transferred the electron from active centre to electrode by using electron transfer mediators. Mediators are the artificial electron donor or acceptor molecules which works on the basis of diffusional mechanism. Diffusional penetration of the oxidized orreduced relay into the protein can reduces the electron transfer distance between the enzyme active center and electrode (Katz et al. 2003).

      Enzymatic Fuel Cell

      In the case of quinohemeproteins, direct electron-transfer as well as mediated electrontransfer is possible between the enzyme and an electrode where the heme in the protein functions as a built-in mediator in an enzyme electrochemical reaction.

      Enzymes for the anodic reactions of biological fuel cells. Niacin is widely distributed in plant and animal tissues. The coenzyme forms of the vitamins are the pyridine nucleotides. In accordance with the recommendations of the Commission on Enzymes of the International Union of Biochemistry, coenzyme-I is referred to as NAD+ and coenzyme-II as NADP+. The pyridine nucleotides are coenzymes for the dehydrogenase group of enzymes that are responsible for catalysis of certain biological oxidation–reduction reactions. As examples of some redox reactions in which the pyridine nucleotides participate, one might consider the action of the enzyme alcohol dehydrogenase, which oxidizes ethanol to acetaldehyde. In ethanol-based biofuel cells, the aforesaid reaction is the basis for electron generation (Fig. 4).

Let us take the example of a subtype of traditional EFCs called glucose fuel cells (GFCs) has the capacity to oxidize glucose, which is supplied by numerous metabolic processes. The glucose oxidation process at the anode of glucose/O_2 EBFCs, the most common form of implantable EBFCs, was typically catalyzed by Gox (glucose oxidase), which was sensitive to O_2. However, O_2 is the only possible cathode-active

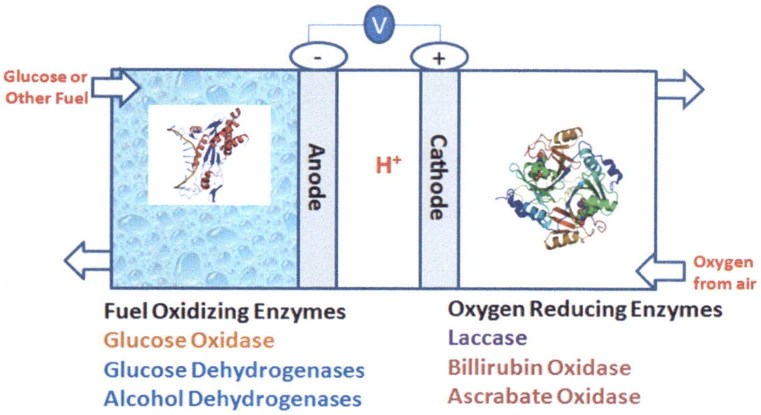

Fig. 4 Glucose enzymatic fuel cell

reactant. As a result, the anodic and cathodic reactions will interfere with one another, severely limiting the performance of EBFCs. Chemical reaction involved is shown in Fig. 2. EBFCs are primarily constructed as complex two-chamber structures with membranes to prevent O_2 interference. While laccase is the foundation of biocathodes, glucose oxidase (GOx) is the main component of bioanode designs (Figs. 5, 6).

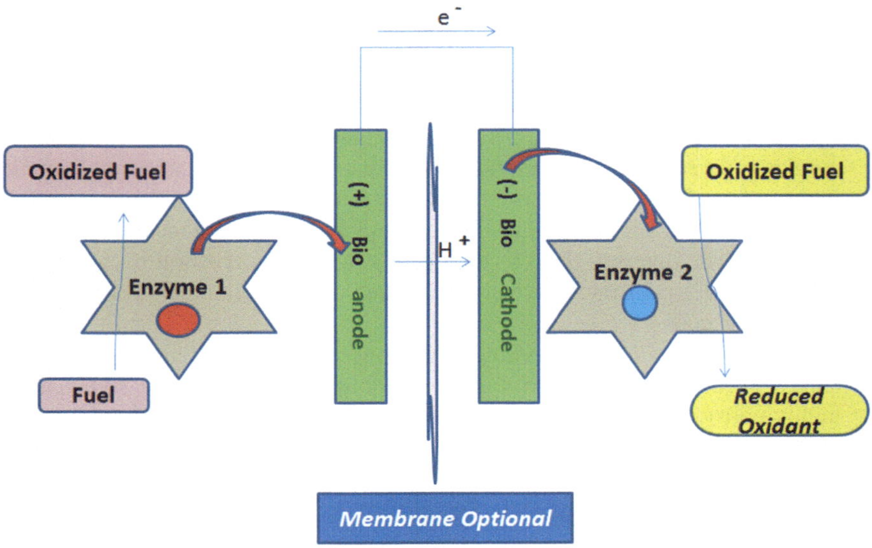

Fig. 5 Component structure of enzymatic fuel cell

$$\textbf{Anode: } C_6H_{12}O_6 \longrightarrow C_6H_{10}O_6 + 2e^- + 2H^+ \quad \textbf{Oxidation } E = -0.5\,V$$

$$\textbf{Cathode: } 0.5O_2 + 2e^- + 2H^+ \longrightarrow H_2O \quad \textbf{Reduction } E = 0.7\,V$$

$$\textbf{Overall Reaction: } C_6H_{10}O_6 + 0.5O_2 \longrightarrow C_6H_{10}O_6 + H_2O \Delta E = 1.2\,V$$

Fig. 6 Chemical reaction involved in glucose fuel cell

Enzymatic fuel cells use a wide range of organic molecules that are digested in living organisms or sourced from biomass to generate electricity from sustainable biological sources. EFCs differ greatly from traditional rare metal-catalyzed fuel cells (which are usually powered by hydrogen or methanol) in that they can be powered by a range of fuel sources. When it comes to the development of EFCs, a wide range of criteria need to be considered, including the fuels' availability, potential toxicity, energy density, and cost Wei et al. (2003).

2.3.4 Advantages and Disadvantages

1. *Lower in cost*—Enzymatic fuel cell have simple fuel cell design and lower cost of the main fuel cell components in comparison to conventional fuel cells. However the primary cause of the significant kinetic constraints observed in enzymatic fuel cells is the inefficiency of electron transport between the enzyme and the electrode surface.
2. *Utilization of Biocatalyst*—Biocatalysts are utilized in enzymatic fuel cells because they are affordable, and their extended use is anticipated to reduce production costs. While conventional noble metals are typically utilized as catalysts in fuel cells, there is a limited supply of these transition metals. *Inexpensive catalysts but High catalytic activity*—Use of inexpensive catalysts and anodic fuels in the cell. Urine and waste fluids are examples of complex organic waste, while simple organic compounds like glucose and acetate are included.
3. *High catalytic activity*—Due to high catalytic activity, reaction occurs at low overvoltage for substrate conversion, like ambient temperature and near-neutral pH and low cost.
4. *Soft working condition*—Working conditions of other cells have many challenges, such as requirement of high temperature, high cost and in some cases, highly corrosive media, but in the case of the enzymatic fuel cell because of the enzyme very mild conditions are applicable such as ambient operational temperature and pressure, employ neutral or circumneutral electrolytes.

2.3.5 Disadvantages

- Life time of cell has very short as compare to other conventional cell.

Table 1 Contaminants available in water

S. no.	Name of pollutants	Examples	References
1	Organic substrate	Herbicides, pesticides, dyes, chlorinated hydrocarbons, steroids, organ cyanides, Long chain organic molecules, Volatile organic compounds, aromatic molecules, polycyclic aromatic hydrocarbons (PAHs), drugs, and plastics,	Tsuchiya (2010)
2	Inorganic substrate	Arsenic, fluoride, iron, nitrate, Heavy metals, acids, bases	Kumar et al. (2023)
3	Industrial and mining containments	Electroplating, leather, paint, nuclear power, electric battery, paper, ceramic, mining and smelting	Lin et al. (2022)

- Power densities of the cell have low then others.
- Fuels used in the cell have generally faces the problem of inefficient oxidation

2.3.6 Application in Wastewater Treatment

The word water is within the need list of life guardian angel materials of living bodies. A list of contaminants in water bodies are briefed in Table 1 to note at a look.

Presence of pollutants as listed in Table 1 organic substrate, inorganic substrate and heavy metals are the major source of water pollution. This creates carcinogenic and lethal impact when they enter in the human body. Interaction with human cells in negative manner and disrupt their functioning. Due to their electropositive nature, heavy metals forms complexes with the ligand and become non-biodegradable. Through water they enter in food chain and affect our health adversely even though present in small concentration. Therefore, it is necessary to eradicate the metal contamination from wastewater before the water is released to environment. Precipitation, coagulation, solidification, oxidation, ion exchange process is a few normally used methods employed for extracting heavy metals from wastewater, but these methods generally suffer with demand of high energy supply, toxic sludge production, incomplete removal of pollutant, production of secondary contaminants and cost management etc. Therefore, enzyme-based fuel cell is one of the better options for removal of metal pollutants from wastewater.

2.3.7 Role of Enzymes for Removal of Organic Pollutants

Presence of polymers, any types of aromatic molecules, polycyclic aromatic hydrocarbons (PAHs), herbicides, insecticides, pesticides, drugs, dyes, and plastics, all comes under organic pollutants which are harmful for aquatic environment. Enzymes which are useful for removal of these organic pollutants are discussed below in Table 2.

Table 2 Enzyme useful for removal organic pollutants from wastewater

S. no.	Enzyme	Pollutant	References
1	*Hydrolases*	Organo-phosphorus compounds, nitrile compounds, polymers Herbicides and pesticides	Ufarte et al. (2015) and Eibes et al. (2015)
2	*Esterases*	Aromatic and aliphatic Polyesters, organophosphorus	Ufarte et al. (2015)
		Herbicides and pesticides. Polyurethanes and diethyl glycol adipate	
3	*Nitrilases*	Herbicides, polymers, and plastics	Nigam et al. (2017) and Martínková et al. (2017)
4	*Organophosphorus Hydrolase*	Organophosphorus	Mousavi et al. (2021)
		Compounds	
5	*Cytochrome p450 Monooxygenase*	Polychlorinated biphenyls, dioxins, polychlorinated dibenzo-p-dioxins	Mousavi et al. (2021)
		Polychlorinated dibenzofurans, Aliphatic hydrocarbons, and even Cr (VI)	
6	*Carboxylesterases*	Organophosphates, Carbamates, and various chlorinated organic compounds	Cummins et al. (2007)
7	*Phosphotriesterases*	Organophosphate	Romeh and Hendawi (2014)
8	*Peroxidase*	Phenol	Garg et al. (2020)
9	*Horseradish peroxidase and Pleurotus ostreatuslaccase*	Diclofenac and Sotalol	Stadlmair et al. (2017)
10	*Lipases*	Oil and grease	Meng et al. (2017)
11	*Oxidoreductases*	Various Dyes	Bilal et al. (2017)

Hydrolases enzymes like Esterases, nitrilases, aminohydrolases, lipase, cutinase, and organophosphorus hydrolase and *Nitrilases* enzymes like *P.* flfluorescens nitrilase, cyanide hydratase is useful for removal of organophosphorus compounds, nitrile compounds, polymers as well as herbicides and pesticides. Whereas Esterase enzymes are useful to break ester bonds in organophosphorus herbicides and pesticides, diethyl glycol adipate, polyurethanes, and aromatic and aliphatic polyesters (Eibes et al. 2015; Ufarte et al. 2015). Similarly, *Carboxylesterases, Phosphotriesterases, Peroxidase, lipases, Oxidoreductases* are useful for removal of organophosphate compounds, phenol, pharmaceutical waste, oil and grease and dyes (Table 1).

2.4 Role of Enzymes for Removal of Inorganic Pollutants

Microorganisms generate peptides which binds with metals and decreases their toxicity. Therefore, they produce some enzymes to resist against heavy metals, some of them are discussed below in Table 3.

Arsenite oxidase effectively transforms As^{3+} into the less harmful As^{5+} form, as demonstrated by Yamamura and Amachi (2014). Cupriavidus metallidurans utilizes p-type ATPase to eliminate Pb^{2+} ions, while concurrently generating inorganic phosphate to sequester Pb^{2+} within the periplasm. Staphylococcus epidermidis employs carbonate to biomineralize Pb^{2+}, with urease enzymes facilitating the formation of various carbonate crystalline Pb^{2+} species. Additionally, Pb^{2+} can undergo mineralization as oxalate and pyromorphite, as described by Rahman and Singh (2020).

Mercuric reductase is pivotal in reducing Hg, operating within the 1e meroperon, which comprises mercury-resistance genes activated by inducible levels of Hg. Collaborating with FAD and NADPH as electron donors, mercuric reductase converts Hg2 + into volatile Hg0, which disperses into the atmosphere. Kapahi and Sachdeva (2019) highlight dimethylmercury's volatility, suggesting biomethylation as a viable strategy for Hg bioremediation. Furthermore, meroperon-independent volatilization of mercury has been observed in Shewanella oneidensis, as documented by Xu

Table 3 Enzyme useful for removal inorganic pollutants from wastewater

S. no.	Enzyme	Pollutant	References
1	*Arsenite oxidase*	Arsenic	Yamamura and Amachi (2014)
2	*Cupriavidusmetallidurans, Staphylococcus epidermidis, Urease*	Lead	Naik and Dubey (2013)
3	*Mercuric reductase, organomercuriallyase*	Mercury	Mahbub et al. (2017) and Singh and Kumar (2020)
4	*Nitroreductase, iron reductase, flflavinreductases, and Quinonereductases*	Chromium	Thatoi et al. (2014)
5	*Catalase*	Cobalt, chromium, iron, nickel	Maphuhla et al. (2022)
6	*Phosphatase*	Cobalt, chromium, copper, iron, manganese, nickel	Maphuhla et al. (2022)
7	*Cupric reductase*	Copper	Naureen and Rehman (2016)
8	*Urease*	Nickel, cobalt, cadmium, copper and zinc	Naureen and Rehman (2016)

et al. (2015). Another enzyme aiding in mercury bioremediation is organomercuri-allyase, which cleaves carbon-mercury bonds in organo-Hg compounds, as reported by Mahbub et al. (2017) and Singh and Kumar (2020).

According to the study by Maphuhla et al. (2022) enzymes like Catalase and phosphataseshowed a negative correlation with metals like cobalt, chromium, iron, manganese, nickel. Therefore, they can be used to fix the metal concentration issue of waste water (Table 2).

Research by Naureen and Rehman (2016) indicates that degradation of heavy metals is quite difficult, but enzymes can biologically convert them into a non-harmful compound and reduces their toxic effect on environment. For instance, copper is reduced by cupric reductase enzymes into a nontoxic substance, while urease enzyme aids in the biomineralization of heavy metals such Ni, Co, Cd, Cu, and Zn (Table 2).

2.5 Other Applications of Fuel Cell

1. *Self-Powered Biosensors*

   Katz et al. (2001) were the first to introduce the concept of using enzymatic biofuel cells for self-powered biosensing applications. After that this trend continues. Researchers used a lactate oxidase-based biofuel cell, allowing real-time monitoring of lactate in sweat for on-body applications (Jeerapan 2021; Jeerapan et al. 2016).
2. *Biosensing Applications*

   A direct, quick, and easy way to test analytes in physiological fluids is through the use of biosensors. By connecting a biological response to a transducer that transforms the recognition event into a quantifiable signal, detection and monitoring are carried out (Rocchitta et al. 2016). To operate, the majority of biosensors need a power source, which is often provided by lithium batteries. However, depending on the energy needs of the device they power, such batteries are difficult to miniaturise, have a short lifespan of a few months to many years, and/or require regular charging (Barton et al. 2004). Enzymatic fuel cells (EFCs) have a lot of potential for autonomous biosensing in this situation. EFCs are electrochemical systems that use redox enzymes to convert chemical energy stored in biomolecules into electrical energy.
3. *Implantable Enzymatic Fuel Cells*

   Cinquin et al. (2010) implaned the first enzymatic fuel cell in retroperitoneal space of a rat which contains an anode modified with glucose oxidase, and a cathode modified with polyphenol oxidase and this cell generated a stable power inside the rat for 3 months without any problem (Cinquin et al. 2010). After that nanostructured materials and immobilization methods were used to increase the power output and stability of enzymatic fuel cells (Zaho et al. 2017). Different carbon materials like carbon nanotubes, graphene (Liu et al. 2010), pressed carbon nanotubes in buckypaper (Szczupak et al. 2012), carbon fibres (Sales et al. 2013) andgraphite modified with cellobiose dehydrogenase and bilirubin oxidase (Pankratov et al. 2016) were used as electrode material because of its high electrical conductivity and good biocompatibility (Zaho et al. 2017).

4. *Wearable Enzymatic Fuel Cells*

   Due to the difficulties in using implantable EFCs, research has been focused on wearable rather than implantable applications. As a result, physiological fluids other than blood, such as saliva, transdermal fluid, perspiration, tears, and urine, have been investigated. These liquids are easily accessible, and in vitro testing does not call for intrusive implantations or blood sampling.

   Du Toit et al. reported power generation using pig skin extracts derived by iontophoresis in (2014). The EFC's foundation was highly porous gold that had been functionalized with laccase and glucose oxidase (Du et al. 2016). In a work published in (2017) by Cho et al. glucose levels in perspiration were tracked. Two polymers, poly(3,4-ethylenedioxythiophene) and polystyrene sulfonate (PEDOT:PSS), were combined to modify the paper. A novel and flexible EFC that can be fully integrated with human skin was reported by Bandodkar et al. (2017) and (Bandodkar et al. 2017).

5. *Biomarkers with EFC-Based Sensors*—EFC-Based Sensors have enormous potential for self-powered, real-time, and in vivo monitoring of various interesting indicators. Only glucose has been the subject of investigation up to this point (Zhou et al. 2015). Additionally taken into consideration are biomarkers like lactate and cholesterol (Bandodkar et al. 2017; Sekretaryova et al. 2014). The usage of EFCs could also be expanded to detect medicines, other endogenous compounds like neurotransmitters, and metabolites other than glucose like lactate, creatinine, or uric acid by using the right enzymes (Gonzalez-Solino et al. 2018).

2.5.1 Resources Recovery

EFCs is a bio-electrochemical device that converts chemical energy contained in organic substrates into electrical energy by the activities of microbes. The use of organic material such as wastewater in EFC makes it an eco-friendly device that offers a dual benefit of bioelectricity generation and waste management.

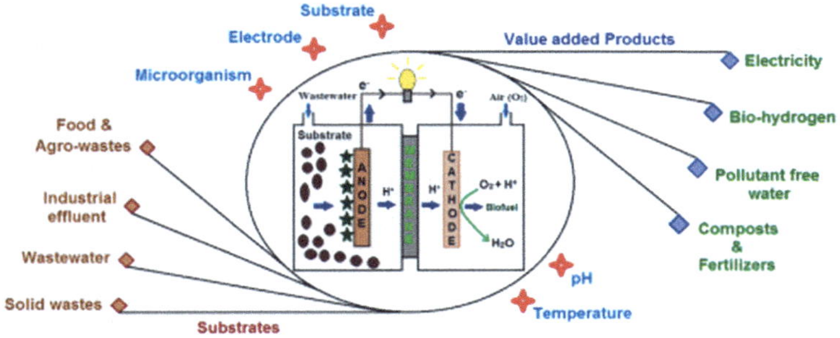

Source of figure—Chemosphere, Volume 290, March 2022, 133295

The huge number of pollutants from industrial activities represents a threat to the ecological and environmental balance. Chlorinated organic compounds, polycyclic aromatic hydrocarbons, pesticides, dioxins, industrial dyes, and other xenobiotics are among the most significant industrial pollutants. A great variety of these xenobiotics are substrates for different types of enzymes, consequently, susceptible to enzymatic transformation. The enzymes Isomerases, Liasas, Garters, Oxidoreductases, Transferases, and Hydrolases play crucial roles in catalyzing diverse reactions. Prior to extracting an enzyme from any source, it's essential to consider its intended application.

Enzymatic fuel cells (EFCs) are increasingly recognized as a promising technology for wastewater treatment. Researchers, such as Aelterman et al. have investigated their potential energy conversion efficiencies, particularly in the context of potato-processing wastewaters and hospital wastewater effluents. However, challenges persist, notably in achieving efficient removal of chemical oxygen demand (COD). The conventional aerobic treatment methods for low-strength wastewaters, like domestic sewage, not only entail significant initial investments but also pose ongoing operational and energy consumption burdens. For instance, the aeration of sewage demands approximately 0.5 kWh/m3, resulting in an annual energy expenditure of about 30 kWh per capita. Moreover, these processes generate substantial quantities of excess sludge, necessitating appropriate treatment and disposal methods. Nevertheless, research continues to explore various process configurations wherein microbial fuel cells (MFCs), a type of EFC, could prove beneficial in wastewater treatment (Wei et al. 2003).

EFC holds significant promise for energy recovery in wastewater treatment, potentially carving out a niche market as a standalone power source and in direct wastewater treatment. Additionally, they show promise in post-treating effluents from anaerobic digesters, even at ambient temperatures, as evidenced by this study. However, it's crucial to acknowledge the various challenges ahead, including the necessity to develop sustainable cathodes, address the fate of specific organics in sewage, and effectively remove residual nutrients.

Enzymatic fuel cell EFC is a novel technology that can convert organic matter in wastewater into electricity (Liu and Logan 2004; Cheng et al. 2006; Logan et al. 2008). Removal of metals, such as chromium, copper, vanadium and mercury has been studied using two-chamber MFCs, in which, heavy metals were removed in the anaerobic cathode chamber through cathode metal reduction while organics in the anodic chamber were used as carbon sources and electron donors (Wang et al. 2008; 2021; Huang et al. 2010; Heijne et al. 2010; Zhang and Dai 2012; Lefebvre et al. 2013). The solutions used as cathode liquid in these studies exclusively contained heavy metals without any organic compounds. However, many waste streams, such as those from animal farms, petrochemical factories, and the rayon industry, contain both heavy metals and organic substances (Malakahmad et al. 2011). While it is established that electricity can be generated from various organic substances in wastewater using Microbial Fuel Cells (MFCs), the influence of heavy metals on power output remains unclear. Additionally, it's uncertain whether and how heavy metals can be removed in the anodic chamber.

References

Atanassov P et al (2007) Enzymatic biofuel cells. Electrochem Soc Interface 16(2):28–31. https://doi.org/10.1149/2.F04072IF

Bandodkar AJ, You J-M, Kim N-H, Gu Y, Kumar R, Mohan AMV, Kurniawan J, Imani S, Nakagawa T, Parish B et al (2017) Soft, stretchable, high power density electronic skin-based biofuel cells for scavenging energy from human sweat. Energy Environ Sci 10:1581–1589. https://doi.org/10.1039/C7EE00865A

Barton SC, Gallaway J, Atanassov P (2004) Enzymatic biofuel cells for implantable and microscale devices. Chem Rev 104:4867–4886. https://doi.org/10.1021/cr020719k

Bilal M, Asgher M, Parra-Saldivar R, Hu H, Wang W, Zhang X, Iqbal HMN (2017) Immobilized ligninolytic enzymes: an innovative and environmental responsive technology to tackle dye-based industrial pollutants—a review. Sci Total Environ 576:646–659

Bollella P, Gorton L, Antiochia R (2018) Direct electron transfer of dehydrogenases for development of 3rd generation biosensors and enzymatic fuel cells. Sensors (Basel) 18:E1319

Bullen RA, Arnot TC, Lakeman JB, Walsh FC (2006) Biofuel cells and their development. Biosens Bioelectron 21:2015–2045

Cadet M, Gounel S, Stines-Chaumeil C et al (2016) An enzymatic glucose/O2 biofuel cell operating in human blood. Biosens Bioelectron 83:60–67

Cao X, Huang X, Liang P, Xiao K, Zhou Y, Zhang X, Logan BE (2009) A new method for water desalination using microbial desalination cells. Environ Sci Technol 43(18):7148–7152

Cheng S, Liu H, Logan BE (2006) Increased performance of single-chamber microbial fuel cells using an improved cathode structure. Electrochem Commun 8(3):489–494. https://doi.org/10.1016/j.elecom.2006.01.010

Cho E, Mohammadifar M, Choi S (2017) A self-powered sensor patch for glucose monitoring in sweat. In: Proceedings of the 2017 IEEE 30th international conference on micro electro mechanical systems (MEMS); Las Vegas, NV, USA, 22–26 January 2017, pp 366–369

Cinquin P, Gondran C, Giroud F, Mazabrard S, Pellissier A, Boucher F, Alcaraz J-P, Gorgy K, Lenouvel F, Mathé S et al (2010) A glucose biofuel cell implanted in rats. PLoS ONE 5:e10476. https://doi.org/10.1371/journal.pone.0010476

Cooney MJ, Lau C, Windmeisser M, Liaw BY, Klotzbach T, Minteer SD (2008a) Design of chitosan gel pore structure: towards enzyme catalyzed flow-through electrodes. J Mater Chem 18:667–674

Cooney MJ, Svoboda V, Lau C, Martin G, Minteer SD (2008b) Enzyme catalysed biofuel cells. Energy Environ Sci 1(3):320–337

Cummins I, Landrum M, Steel PG, Edwards R (2007) Structure activity studies with xenobiotic substrates using carboxylesterases isolated from *Arabidopsis thaliana*. Phytochemistry 68(6):811–818

Deepak Pant DP, Bogaert GV, Diels L, Vanbroekhoven K (2012) A comparative assessment of bioelectrochemical systems and enzymatic fuel cells. In: Microbial biotechnology: energy and environment. CABI, Wallingford, pp 39–57

Du Toit H, Rashidi R, Ferdani DW, Delgado-Charro MB, Sangan CM, Di Lorenzo M (2016) Generating power from transdermal extracts using a multi-electrode miniature enzymatic fuel cell. Biosens Bioelectron 78:411–417. https://doi.org/10.1016/j.bios.2015.11.074

Eibes G, Arca-Ramos A, Feijoo G, Lema JM, Moreira MT (2015) Enzymatic technologies for remediation of hydrophobic organic pollutants in soil. Appl Microbiol Biotechnol 99:8815–8829

Falk M, Blum Z, Shleev S (2012) Direct electron transfer based enzymatic fuel cells. Electrochim Acta 82:191–202

Falk M, Shleev S, Narváez Villarrubia CW et al (2014) Biological fuel cells for biomedical applications. In: Luckarift HR, Atanassov P, Johnson GR (eds) Enzymatic fuel cells: from fundamentals to applications. Wiley, Hoboken, pp 422–450

Ganzales-Solino C, Di Lorenzo M (2018) Enzymatic fuel cells: towards self-powered implantable and wearable diagnostics. Biosensors (Basel) 8(1):11

Garg S, Kumar P, Singh S, Yadav A, Dumee LF, Sharma RS, Mishra V (2020) *Prosopis juliflora* peroxidases for phenol remediation from industrial wastewater—an innovative practice for environmental sustainability. Environ Technol Innov 19:100865

Gonzalez-Solino C, Lorenzo MD (2018) Enzymatic fuel cells: towards self-powered implantable and wearable diagnostics. Biosensors (Basel) 8(1):11. https://doi.org/10.3390/bios8010011

Hameed A, Hussain SZ (2018) Enzymatic fuel cells for wastewater treatment and renewable energy: a review. Environ Technol Innov 9:225–240

Heijne AT, Strik DPBTB, Hamelers HVM, Buisman CJN (2010) Cathode potential and mass transfer determine performance of oxygen reducing biocathodes in microbial fuel cells. Environ Sci Technol 44(18):7151–7156. https://doi.org/10.1021/es100950t

Heller A (1990) Electrical wiring of redox enzymes. Acc Chem Res 23:128–134

Hernandez K, Fernandez-Lafuente R (2011) Control of protein immobilization: coupling immobilization and site-directed mutagenesis to improve biocatalyst or biosensor performance. Enzyme Microb Technol 48(2):107–122

Huang Y, He Z, Mansfeld F (2010) Performance of microbial fuel cells with and without Nafion solution as cathode binding agent. Bioelectrochemistry 79(2):261–264

Jeerapan I (2021) Sweat-powered sensor for sweaty socks, archived from the original on 2021-12-21, retrieved 2019-01-13

Jeerapan I et al (2016) Stretchable biofuel cells as wearable textile-based self-powered sensors. J Mater Chem A 4(47):18342–18353. https://doi.org/10.1039/C6TA08358G

Kapahi M, Sachdeva S (2019) Bioremediation options for heavy metal pollution. J Health Pollut 9(24):191203

Katz E, Bückmann AF, Willner I (2001) Self-powered enzyme-based biosensors. J Am Chem Soc 123(43):10752–10753. https://doi.org/10.1021/ja0167102

Katz E, Shipway AN, Willner I (2003) Handbook of fuel cells. In: Vielstich W, Gasteriger HA, Lamm A (eds) Fundamentals and survey of systems. Wiley, London

Kim JR, Kim HS (2018) Strategies for enhanced performance of enzymatic biofuel cells. Biotechnol Adv 36(6):1761–1772

Konash A, Cooney MJ, Liaw BY, Jameson DM (2006) Characterization of enzyme–polymer interaction using fluorescence. J Mater Chem 16:4107–4109

Kotanen CN, Moussy FG, Carrara S et al (2012) Implantable enzyme amperometric biosensors. Biosens Bioelectron 35(1):14–26

Kumar V, Sharma M, Sondhi S, Kaur K, Sharma D, Sharma S, Utreja D (2023) Removal of inorganic pollutants from wastewater: innovative technologies and toxicity assessment. Sustainability 15(23):16376

Leech D, Kavanagh P, Schuhmann W (2012) Enzymatic fuel cells: recent progress. Electrochim Acta 84:223–234

Lefebvre O, Tan Z, Shen Y, Ng HY (2013) Optimization of a microbial fuel cell for wastewater treatment using recycled scrap metals as a cost-effective cathode material. Bioresour Technol 127:158–164

Lin L, Yang H, Xu X (2022) Effects of water pollution on human health and disease heterogeneity: a review. Front Environ Sci 10:880246

Liu H, Logan BE (2004) Electricity generation using an air-cathode single chamber microbial fuel cell in the presence and absence of a proton exchange membrane. Environ Sci Technol 38(14):4040–4046

Liu C, Alwarappan S, Chen Z, Kong X, Li C-Z (2010) Membraneless enzymatic biofuel cells based on graphene nanosheets. Biosens Bioelectron 25:1829–1833. https://doi.org/10.1016/j.bios.2009.12.012

Logan BE, Call D, Cheng S, Hamelers HVM, Sleutels THJA, Jeremiasse AW, Rozendal RA (2008) Microbial electrolysis cells for high yield hydrogen gas production from organic matter. Environ Sci Technol 42:8630–8640

Lu X, Zhang X (2016) Enzymatic biofuel cells and their applications. Biosens Bioelectron 76:91–102

Mac Vittie K, Halamek J, Halamkova L et al (2013) From "cyborg" lobsters to a pacemaker powered by implantable biofuel cells. Energy Environ Sci 6:81–86
Mahbub KR, Bahar MM, Labbate M, Krishnan K, Andrews S, Naidu R, Megharaj M (2017) Bioremediation of mercury: not properly exploited in contaminated soils! Appl Microbiol Biotechnol 101:963–976
Malakahmad A, Basri NE, Zain SM (2011) Production of renewable energy by transformation of kitchen waste to biogas, case study of Malaysia. In: IEEE Symposium on Business, Engineering and Industrial Applications (ISBEIA); Langkawi, Malaysia, 25–28 September 2011. https://doi.org/10.1109/ISBEIA.2011.6088808
Mansoorian HJ, Mahvi AH, Jafari AJ, Amin MM, Rajabizadeh A, Khanjani N (2013) Bioelectricity generation using two chamber microbial fuel cell treating wastewater from food processing. Enzyme Microb Technol 52(6–7):352–357
Mantz YA, Panagiotopoulos G (2015) Enzymatic fuel cells: recent progress. Electrochim Acta 174:82–93
Maphuhla NG, Lewu FB, Adu A, Oyedeji OO (2022) Enzyme activities in reduction of heavy metal pollution from alice landfifill site in Eastern Cape, South Africa. Int J Environ Res Public Health 19:12054. https://doi.org/10.3390/ijerph191912054
Martínková L, Rucká L, Nešvera J, Pátek M (2017) Recent advances and challenges in the heterologous production of microbial nitrilases for biocatalytic applications. World J Microbiol Biotechnol 33:1–11
Meng Y, Luan F, Yuan H, Chen X, Li X (2017) Enhancing anaerobic digestion performance of crude lipid in food waste by enzymatic pretreatment. Bioresour Technol 224:48–55
Minteer SD, Liaw BY, Cooney MJ (2007a) Enzyme-based biofuel cells. Curr Opin Biotechnol 18:228–234
Mousavi SM, Hashemi SA, Iman Moezzi SM, Ravan N, Gholami A, Lai CW, Chiang WH, Omidifar N, Yousefi K, Behbudi G (2021) Recent advances in enzymes for the bioremediation of pollutants. Biochem Res Int 2021:1
Naik MM, Dubey SK (2013) Lead resistant bacteria: lead resistance mechanisms, their applications in lead bioremediation and biomonitoring. Ecotoxicol Environ Saf 98:1–7
Naureen A, Rehman A (2016) Arsenite oxidizing multiple metal resistant bacteria isolated from industrial effluent: their potential use in wastewater treatment. World J Microbiol Biotechnol 32(8):133
Neto SA, De Andrade AR (2013) New energy sources: the enzymatic biofuel cell. J Braz Chem Soc 24(12):1891–1912
Nigam VK, Arfi T, Kumar V, Shukla P (2017) Bioengineering of nitrilases towards its use as green catalyst: applications and perspectives. Indian J Microbiol 57:131–138
Nishizawa K, Konno T, Takai M, Ishihara K (2008) Bioconjugated phospholipid polymer biointerface for enzyme-linked immunosorbent assay. Biomacromol 9(1):403–407
Onuki Y, Bhardwaj U, Papadimitrakopoulos F et al (2008) A review of the biocompatibility of implantable devices: current challenges to overcome foreign body response. J Diabetes Sci Technol 2(6):1003–1015
Pankratov D, Ohlsson L, Gudmundsson P, Halak S, Ljunggren L, Blum Z, Shleev S (2016) Ex vivo electric power generation in human blood using an enzymatic fuel cell in a vein replica. RSC Adv 6:70215–70220. https://doi.org/10.1039/C6RA17122B
Park DH, Kim SK, Shin IH, Jeong Y (2000) Electricity production in biofuel cells using modified graphite electrode with neutral red. J Biotechnol Lett 22:1301–1304
Rabaey K, Verstraete W (2005) Microbial fuel cells: novel biotechnology for energy generation. Trends Biotechnol 23(6):291–298
Rahman Z, Singh VP (2020) Bioremediation of toxic heavy metals (THMs) contaminated sites: concepts, applications and challenges. Environ Sci Pollut Res 27:27563–27581
Rocchitta G, Spanu A, Babudieri S, Latte G, Madeddu G, Galleri G, Nuvoli S, Bagella P, Demartis MI, Fiore V, Manetti R, Serra PA (2016) Enzyme biosensors for biomedical applications: strategies for safeguarding analytical performances in biological fluids. Sensors 16(6):780

Romeh AA, Hendawi MY (2014) Bioremediation of certain organophosphorus pesticides by two biofertilizers, paenibacillus (Bacillus) polymyxa (Prazmowski) and Azospirillumlipoferum (Beijerinck). J Agric Sci Tech-IRAN 16(2):265–276

Sales FCPF, Iost RM, Martins MVA, Almeida MC, Crespilho FN (2013) An intravenous implantable glucose/dioxygen biofuel cell with modified flexible carbon fiber electrodes. Lab Chip 13:468–474. https://doi.org/10.1039/C2LC41007A

Scott K, Yu EH, Ghangrekar MM, Erable B, Duţeanu NM (2012) Biological and microbial fuel cells. In: Comprehensive renewable energy. Elsevier, New York, pp 277–300. ISBN 978-0-08-087872-0

Sekretaryova AN, Beni V, Eriksson M, Karyakin AA, Turner APF, Vagin MY (2014) Cholesterol self-powered biosensor. Anal Chem 86:9540–9547. https://doi.org/10.1021/ac501699p

Singh S, Kumar V (2020) Mercury detoxification by absorption, mercuric ion reductase, and exopolysaccharides: a comprehensive study. Environ Sci Pollut Res 27(22):27181–27201

Song RB, Wu Y, Lin ZQ, Xie J, Tan CH, Loo JSC, Zhang Q (2017) Back cover: living and conducting: coating individual bacterial cells with in situ formed polypyrrole (Angew. Chem. Int. Ed. 35/2017). Angew Chem Int Ed 56(35):10608–10608

Squadrito G, Cristiani P (2016) Microbial and enzymatic fuel cells. In: Compendium of hydrogen energy. Woodhead Publishing, London, pp 147–173

Stadlmair LF, Letzel T, Drewes JE, Graßmann J (2017) Mass spectrometry based in vitro assay investigations on the transformation of pharmaceutical compounds by oxidative enzymes. Chemosphere 174:466–477

Strik DPBTB, Timmers RA, Helder M, Steinbusch KJJ, Hamelers HVM, Buisman CJN (2011) Microbial solar cells: applying photosynthetic and electrochemically active organisms. Trends Biotechnol 29(1):41–49

Szczupak A, Halamek J, Halamkova L, Bocharova V, Alfonta L, Katz E (2012) Living battery—biofuel cells operating in vivo in clams. Energy Environ Sci 5:8891–8895. https://doi.org/10.1039/c2ee21626d

Tauseef SM, Abbasi T, Abbasi SA (2013) Energy recovery from wastewaters with high-rate anaerobic digesters. Renew Sustain Energy Rev 19:704–741

Thatoi H, Das S, Mishra J, Rath BP, Das N (2014) Bacterial chromate reductase, a potential enzyme for bioremediation of hexavalent chromium: a review. J Environ Manag 146:383–399

Tsuchiya Y (2010) Organical chemicals as contaminants of water bodies and drinking water. Water Qual Stand 150:1

Ufarté L, Laville É, Duquesne S, Potocki-Veronese G (2015) Metagenomics for the discovery of pollutant degrading enzymes. Biotechnol Adv 33(8):1845–1854

Wang Y, Basu S, Wang CY (2008) Modeling two-phase flow in PEM fuel cell channels. J Power Sources 179:603–617

Wang J, Zhao Y, Setzler BP, Rojas-Carbonell S, Ben Yehuda C, Amel A, Yan Y (2019) Poly(aryl piperidinium) membranes and ionomers for hydroxide exchange membrane fuel cells. Nat Energy 4(5):392–398

Wang L, Wu X, Su BQW, Song R, Zhang JR, Zhu JJ (2021) Enzymatic biofuel cell: opportunities and intrinsic challenges in futuristic applications. Adv Energy Sustain Res 2(8):2100031

Wei Y, Van Houten RT, Borger AR, Eikelboom DH, Fan Y (2003) Minimization of excess sludge production for biological wastewater treatment. Wat Res 37(18):4453–4467

Willner I (2009) Biofuel cells: harnessing biomass or body fluids for the generation of electrical power. Fuel Cells 9(1):5

Xiao X, Xia HQ, Wu R, Bai L, Yan L, Magner E, Liu A (2019) Tackling the challenges of enzymatic (bio) fuel cells. Chem Rev 119(16):9509–9558

Xu J, Bravo AG, Lagerkvist A, Bertilsson S, Sjöblom R, Kumpiene J (2015) Sources and remediation techniques for mercury contaminated soil. Environ Int 74:42–53

Yahiro AT, Lee SM, Kimble DO (1964) Bioelectrochemistry: I. Enzyme utilizing bio-fuel cell studies. Biochim Biophys Acta (BBA) Spec Sect Biophys Subj 88:375–383

Yamamura S, Amachi S (2014) Microbiology of inorganic arsenic: from metabolism to bioremediation. J Biosci Bioeng 118(1):1–9

Yu EH, Scott K (2010) Enzymatic biofuel cells—fabrication of enzyme electrodes. Energies 3(1):23–42

Zhang M, Dai L (2012) Carbon nanomaterials as metal-free catalysts in next generation fuel cells. Nano Energy 1(4):514–517

Zhang F, He Z, Ge Z (2013) Using microbial fuel cells to treat raw sludge and primary effluent for bioelectricity generation. Department of Civil Engineering and Mechanics; University of Wisconsin—Milwaukee

Zhao C-E, Gai P, Song R, Chen Y, Zhang J, Zhu J-J (2017) Nanostructured material-based biofuel cells: recent advances and future prospects. Chem Soc Rev 46:1545–1564. https://doi.org/10.1039/C6CS00044D

Zhou M (2015) Recent progress on the development of biofuel cells for self—powered electrochemical biosensing and logic biosensing: a review. Electroanalysis 27:1786–1810. https://doi.org/10.1002/elan.201500173

Zhu Z, Wang M, Gautam A et al (2007) Directed evolution of glucose oxidase from *Aspergillus niger* for ferrocenemethanol-mediated electron transfer. Biotechnol J 2(2):241–248

Zhu Z, Sun F, Zhang X et al (2012) Deep oxidation of glucose in enzymatic fuel cells through a synthetic enzymatic pathway containing a cascade of two thermostable dehydrogenases. Biosens Bioelectron 36(1):110–115

Application of CW-MFC to Mitigate Water Pollution and Resource Recovery

Abhinay Thakur, Ashish Kumar, Omar Dagdag, Elyor Berdimurodov, and Akshay Kumar

Abstract Constructed wetland–microbial fuel cell (CW-MFC) technology represents an innovative and sustainable approach to address two pressing environmental challenges simultaneously: water pollution mitigation and resource recovery. This book chapter offers a comprehensive overview of the utilizationof CW-MFC systems in the context of environmental remediation and wastewater treatment. The chapter begins by introducing the fundamental concepts of both microbial fuel cells and constructed wetlands, highlighting their capabilities in energy generation and pollutant removal. It then explores the synergistic integration of these two technologies within the CW-MFC framework, elucidating the intricate interactions between the biological, electrochemical, and chemical processes that underpin their functionality. Furthermore, the chapter delves into the mechanisms of pollutant removal within CW-MFCs, emphasizing the roles of microbial communities in breaking down organic matter, nutrient removal, and metal immobilization. It also explores the potential for resource recovery in the aspect of electricity generation and the production of value-added byproducts. Additionally, the chapter addresses challenges, future research directions, and the regulatory framework surrounding CW-MFCs, emphasizing the need for continued innovation and widespread adoption.

A. Thakur
Department of Chemistry, School of Chemical Engineering and Physical Sciences, Lovely Professional University, Phagwara, Punjab 144411, India

A. Kumar (✉)
Nalanda College of Engineering, Bihar Engineering University, Department of Science, Technology and Technical Education, Nalanda, Government of Bihar 803108, India
e-mail: drashishchemlpu@gmail.com

O. Dagdag
Department of Mechanical Engineering, Gachon University, Seongnam 13120, Republic of Korea

E. Berdimurodov
Faculty of Chemistry, National University of Uzbekistan, 100034 Tashkent, Uzbekistan

Akfa University, Milliy Bog' Street 264, 111221 Tashkent, Uzbekistan

A. Kumar
Department of Chemistry, Dyal Singh College, Karnal, Haryana 132001, India

R. Selvasembian et al. (eds.), *Emerging Trends in Microbial Electrochemical Technologies for Sustainable Mitigation of Water Resources Contamination*,
https://doi.org/10.1007/978-3-031-74636-9_9

Keywords CW-MFC technology · Wastewater treatment · Resource reclamation · Biological-electrochemical processes · Environmental sustainability

1 Introduction

Water pollution manifests itself through the introduction of a wide array of contaminants into natural water bodies, such asrivers, lakes, oceans, and groundwater. These contaminants encompass a spectrum of harmful substances, ranging from organic matter, nutrients, and heavy metals, to emerging pollutants (Jingyu et al. 2020; Wang et al. 2019a, b, c; Yadav et al. 2018). The resultsof such contamination are far-reaching and multifaceted. First and foremost, polluted water jeopardizes the quality and safety of our drinking water sources. As contaminants infiltrate aquifers and surface waters, they compromise the integrity of our freshwater supplies. Ingestion of contaminated water or contact with it can result in a variety of health issues, such as skin conditions, chronic illnesses, and gastrointestinal ailments. Beyond the direct human health impacts, water pollution wreaks havoc on aquatic ecosystems (Srivastava et al. 2018; Wang et al. 2022). Aquatic life, from fish to invertebrates, relies on clean water to thrive. Pollutants disrupt the delicate balance of these ecosystems, leading to fish kills, algal blooms, and the decline of biodiversity. The destruction of aquatic habitats has far-reaching consequences, not only for the species that inhabit them but also for the services these ecosystems provide to human societies, such as food, clean water, and climate regulation. Furthermore, polluted water bodies present a public health crisis (Li et al. 2020; Lu et al. 2015). Waterborne diseases, often linked to poor water quality, claim the lives of millions each year, particularly in developing regions with inadequate sanitation and limited access to clean water sources. Contaminated water exacerbates the burden of disease, placing significant strain on healthcare systems and economies. Agricultural runoff laden with fertilizers and pesticides, industrial discharges containing heavy metals and organic pollutants, and untreated municipal sewage are some of the primary contributors to water pollution. Here are some global examples illustrating the pervasive nature of water pollution and its far-reaching consequences:

- Flint, Michigan, USA: An important illustration of how water contamination directly affects human health is the Flint water crisis. Lead from corroded pipes leached into the drinking water when the city moved to the Flint River in 2014 in an attempt to save money. Lead levels in the bloodstreams of thousands of residents including childrenwere raised as a result of their exposure to lead-contaminated water. The catastrophe made clear how vulnerable marginalized populations are to water pollution and how crucial it is to preserve sources of clean drinking water (Zhang et al. 2021).
- Ganges River, India: One of the most venerated rivers in the world, the Ganges, is severely contaminated by industrial discharges, untreated sewage, and agricultural runoff. The contamination of this sacred river not only poses risks to human health

but also threatens the diverse aquatic life and ecosystems it supports. Efforts are ongoing to mitigate pollution and restore the river's ecological balance, reflecting the global challenge of managing water quality in densely populated regions.

- The Great Pacific Garbage Patch: The North Pacific Ocean's enormous buildup of marine garbage, mostly plastics, draws attention to the worldwide problem of non-point source pollution. Plastics decompose into tiny fragments that damage marine life, go into the food chain, and may even be harmful to human health. The Great Pacific Garbage Patch is an example of how pollution in one area of the world may affect ecosystems and communities in other areas of the world significantly.
- Yangtze River, China: The Yangtze River, China's longest and most economically significant waterway, faces severe pollution due to industrial discharges, urbanization, and agricultural runoff. Pollutants such as heavy metals and chemicals have led to water quality deterioration and harm to aquatic life (Yang et al. 2021a, b). The Chinese government has initiated ambitious cleanup efforts to address pollution and protect the river's ecosystem.
- The Citarum River, Indonesia: The Citarum has been severely polluted by domestic trash, agricultural, and textile industry, and is frequently listed as one of the worst polluted waterways in the world. The pollution in the river endangers the health of the surrounding populations and has an adverse effect on the livelihoods of those who rely on it for agricultural and water supplies. Efforts to restore the Citarum underscore the need for comprehensive pollution control in rapidly industrializing regions.
- These contaminants pose serious dangers to the environment and public health when they enter water bodies such as lakes, rivers, and groundwater (Xu et al. 2016). Traditionally, the remediation of wastewater and the mitigation of water pollution have been achieved through conventional methods, such as activated sludge systems and chemical treatments. While these methods have been successful to some extent, they often consume significant energy, require large infrastructure investments, and generate substantial quantities of waste sludge. Moreover, they may not effectively address emerging contaminants and nutrient-related water quality issues, such as eutrophication.58.54 $\pm$ 4.16 g·(d m^2)$^{-1}$, and nitrogen (N) at 13.61 $\pm$ 1.07 g·(d m^2)$^{-1}$
- For example, an investigation by Zhang et al. (2022) deliberatedon the impact of plant cultivation and the integration of MFCs upon the decontamination capabilities and remediationmechanisms of CWs. CW devoid plants (A1), CW including plants (A2), CW-MFC devoid plants (A3), and CW-MFC including plants (A4) were the four research setups that were set up. The results show that A4 had the largest daily decreases in chemical oxygen demand (COD) per unit areaat 58.54 $\pm$ 4.16 g·(d m^2)$^{-1}$, and nitrogen (N) at 13.61 $\pm$ 1.07 g·(d m^2)$^{-1}$. This configuration significantly enhanced pollutant removal efficiency, addressing the land footprint challenge associated with wetland processes. Additionally, A4 and A3 showed maximal power densities of 15.87 and 4.59 mW/m^3, correspondingly, along with median output voltages of 717.46 and 568.29 mV. The main technique

for removing phosphorous (P) in the integrated CW-MFC system was adsorption of chemical composition on the conductive substances. Figure 1a illustrates the P removal capabilityfor each system, denoted as A1, A2, A3, and A4, which yielded removal rates of 55.48%, 62.32%, 62.87%, and 67.86%, respectively. Comparatively, A2 and A3 exhibited similar phosphorus (P) removal rates, indicating the presence of planted vegetation. Comparable impacts of the system's closed-circuit design on P elimination led to both significant gains. In A2, the substrate was penetrated by plant roots, creating a naturally occurring barrier that performed two functions. It did two things: first, it took up P as a source of nutrients; secondly, it gave microorganisms a good place to attach themselves, which promoted microbial development and improved P absorption. The abundance of microbial communities on the electrode surface in A3 promoted microbial development and reproduction, which enhanced P uptake and use. In A4, where a combinatorial impact among plant growth and the production of electricity in the CW was improved, the most efficient removal of P was found. To improve P removal, this synergy brought together plant adsorption, microbiological sorption and digestion, and the electrode's outermost composition. A contrast of the P contents in the influent, anode region, and wastewater is shown in Fig. 1c. Interestingly, more P was taken from the wastewater downflow area than from the other three systems combined, with A1 demonstrating less P removal in this location.

Furthermore, the study underscores the significance of the downflow area in P removal, emphasizing the need to consider hydraulic dynamics within the system. This research contributes valuable insights into the intricate interplay of biological and electrochemical processes in enhancing wastewater treatment efficiency and nutrient removal, particularly in the context of swine wastewater treatment. Similarly, Ni et al. (2020) investigated the impact of different swine wastewater concentrations (800 mg/L, 1600 mg/L, and 2500 mg/L) on MFC performance and the response characteristics of the microbial community in the anode box. The results show that the amount of substrate and MFC output voltage, along with the rates at which COD is removed, are positively correlated. The performance of the battery and the variety of the microbial species in the anode compartment are both strongly impacted by changes in concentration. This knowledge can aid in the selective cultivation of adaptable dominant bacteria, enhancing MFC stability and decontamination efficiency. The dominant species in the anodic biofilm community structure include *Proteobacteria, Bacteroidetes, Firmicutes, Chloroflexi*, and *Spirochaetae*. These results have important implications for improving swine wastewater treatment and understanding the mechanisms by which varying concentrations impact microorganism production in MFCs.

The fusion of CW-MFC technologies represents a novel and sustainable approach to addressingthe dual challenges of water pollution mitigation and resource recovery. CWs, designed to mimic natural wetlands, have long been recognized for their ability to treat wastewater and remove contaminants through biological, chemical, and physical processes (Mu et al. 2020). They are cost-effective, environmentally friendly, and aesthetically pleasing, making them attractive options for wastewater treatment and

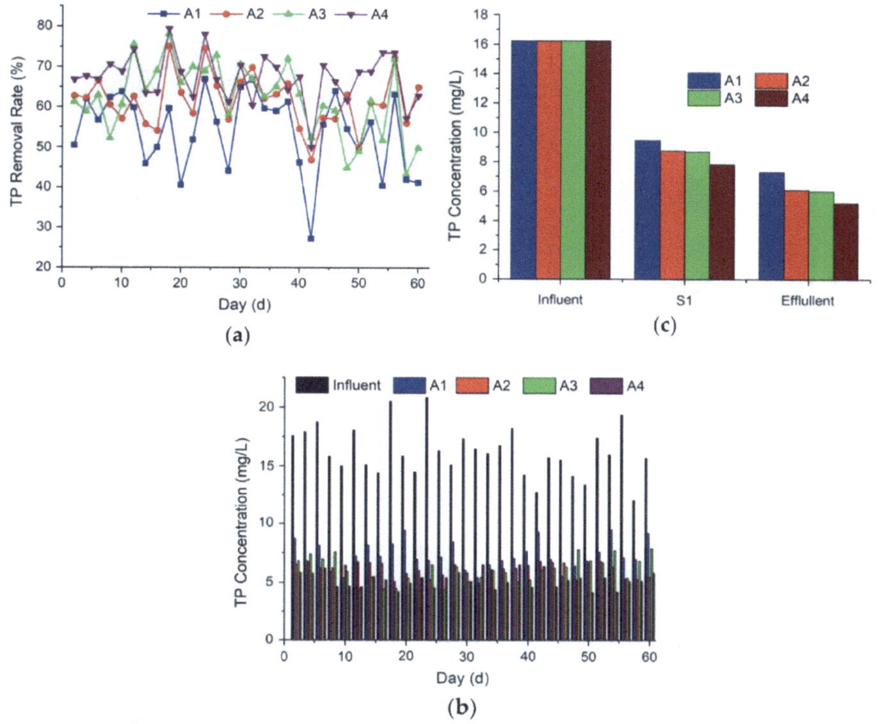

Figure 1 **a** Phosphorus (TP) eradication efficacies for A1, A2, A3, and A4; **b** Quantityof TP in the effluent and influent of A1, A2, A3, and A4; **c** TP quantityacrossthe water flow. Adapted from Ref. Zhang et al. (2022) under CCBY 4.0

environmental remediation. However, MFCs are state-of-the-art bioelectrochemical devices that use microorganisms' metabolisms for breaking down organic materials and produce energy. The ability of MFCs to transform the chemical energy of wastewater into electrical power has drawn interest (Fig. 2) (Roy et al. 2023).

The integration of these two technologies, CW and MFC, creates a symbiotic relationship that capitalizes on their respective strengths. The integration of CW and MFC technologies within the CW-MFC framework offers several advantages. As water passes through the wetland, CWs naturally purify wastewater by eliminating toxins; plants and microbes are essential to this purification mechanism. Meanwhile, as Fig. 3 illustrates, MFCs improve this treatment procedure by further decomposing organic material and producing power via the metabolic processes of microorganisms.

This symbiotic relationship results in improved water quality and the simultaneous recovery of valuable resources. The goal of this book chapter is to give a thorough description of how CW-MFC technology is applied to reduce water pollution and recover resources from wastewater. It explores the fundamental concepts underlying

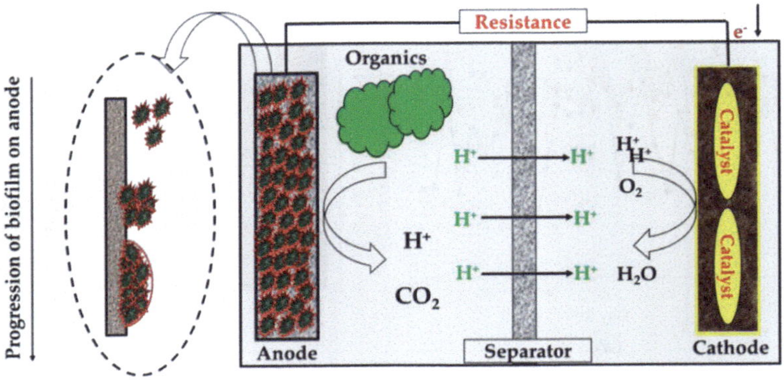

Fig. 2 Depiction of the operational mechanism of a dual-chamber microbial fuel cell. Adapted from Ref. Roy et al. (2023) under CCBY 4.0

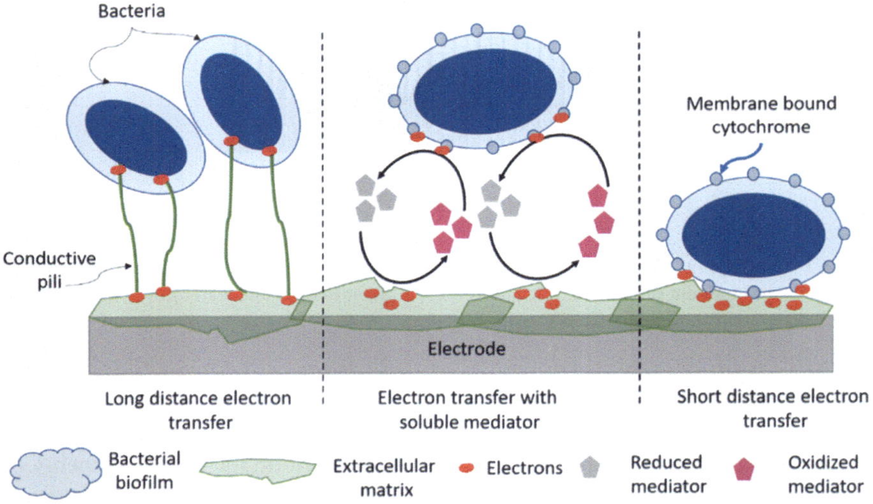

Fig. 3 Illustration of the mediators, biofilm, and electron transfer pathways pertinent to the MFC technique. Adapted from Ref. Afroz et al. (2021) under CCBY 4.0

constructed wetlands and microbial fuel cells, highlighting their capabilities in pollutant removal and energy generation. The chapter will delve into the intricate interactions between biological, electrochemical, and chemical processes that underpin the functionality of CW-MFCs. Furthermore, the chapter will investigate the mechanisms of pollutant removal within CW-MFCs, emphasizing the crucial roles played by microbial communities in breaking down organic matter, removing nutrients, and immobilizing heavy metals. Additionally, it will investigate the possibility of resource reclamation through the creation of value-added byproducts and electricity.

In addition to discussing the technological aspects, this chapter will address challenges associated with CW-MFC implementation, outline future research directions, and examine the regulatory framework surrounding these systems. It emphasizes the need for continued innovation and widespread adoption of CW-MFC technology as a sustainable and eco-friendly approach to address water pollution and resource recovery challenges.

2 Constructed Wetlands: Fundamentals and Applications

CWs are artificially manufacturedecosystems to replicate the natural processes of wetlands while providing controlled and efficient water treatment. They have gained prominence as sustainable solutions for addressing various environmental challenges, particularly in mitigating water pollution and improving water quality (Wang et al. 2017a, b, c). This section explores the fundamentals of constructed wetlands, their design principles, and their diverse applications in wastewater treatment and environmental management.

2.1 Overview of Constructed Wetlands

CWs stand as a testament to human ingenuity and ecological stewardship in the face of mounting water pollution challenges. These innovative systems harness the inherent capabilities of natural wetland ecosystems while optimizing them for enhanced water treatment and pollutant removal. This section provides an in-depth overview of constructed wetlands, delving into their fundamental principles, design considerations, and diverse applications in mitigating water pollution.

2.1.1 Fundamentals of Constructed Wetlands

At the heart of constructed wetlands lies the principle of biomimicry, where human-made systems seek to replicate and harness the natural processes observed in wetland environments. In essence, CWs are engineered to mimic the functions of natural wetlands while capitalizing on their ecological benefits. This biomimetic approach allows CWs to effectively treat and purify water, making them valuable tools in environmental restoration and wastewater treatment (Arvaniti and Fountoulakis 2021; Cheng et al. 2023).

- The Role of Natural Processes: Constructed wetlands owe their effectiveness to the intricate web of natural processes occurring within wetland ecosystems. These processes encompass a wide array of biological, physical, and chemical interactions that collectively lead to pollutant removal and water purification.

- Microbial Activity: Within CWs, microorganisms are essential for decomposing organic materials and converting pollutants. The communities of microbes that are found in the rhizosphere of wetland plants and the wetland substrate participate in various activities, including metal precipitation, denitrification, and aerobic and anaerobic decomposition.
- Plant–Microbe Synergy: The symbiotic relationship between wetland vegetation and microbial communities is a cornerstone of CW functionality. Wetland plants not only stabilize the substrate but also provide oxygen through their roots, creating microaerophilic conditions ideal for certain microbial activities. Additionally, plants take up nutrients, further enhancing water quality.
- Nutrient Cycling: Constructed wetlands facilitate nutrient cycling by capturing and transforming nutrients like nitrogen and phosphorus. This cycling occurs through mechanisms such as plant uptake, nitrification, and denitrification, resulting in reduced nutrient concentrations in the treated water.
- Contaminant Sequestration: Contaminants, including heavy metals, are adsorbed onto substrate particles or immobilized through various chemical reactions within CWs. By lowering the movement and bioavailability of pollutants, this sequestration lessens their environmental impact.
- Improvement of Water Quality: Through the intricate interplay of these natural processes, constructed wetlands substantially improve water quality (Ebrahimi et al. 2021; Wang et al. 2017a, b, c). Pollutants including organic matter, suspended solids, nutrients, pathogens, and heavy metals are effectively removed or transformed, resulting in treated water that is cleaner and safer for release into receiving waters or reuse.

2.1.2 Design Principles of Constructed Wetlands

Effective design is a cornerstone of constructed wetland performance. Several key design principles are considered to maximize the efficiency and efficacy of CWs:

Substrate Selection

The selectionof substrate materials is a vitalaspect of CW design. Typically composed of sand, gravel, and soil, the substrate provides a physical medium for wetland vegetation growth and microbial colonization. Furthermore, it exhibits a crucial role in adsorbing and immobilizing pollutants, contributing to water treatment.

Vegetation Selection

The selection of appropriate wetland vegetation is paramount. Plants serve multiple functions within CWs, stabilizing the substrate, enhancing oxygen exchange, and providing surfaces for microbes growth (Jingyu et al. 2020; Wang et al. 2020).

Different plant species exhibit varying degrees of tolerance to different pollutants and can also contribute to nutrient uptake, enhancing overall treatment performance.

Hydrological Considerations

The control of water flow through the wetland is a critical aspect of CW design. Factors such as flow rates, residence times, and hydraulic loading rates must be carefully considered to ensure that the wetland effectively treats incoming wastewater without becoming waterlogged or overwhelmed. Proper hydrological management is essential for pollutant removal and treatment efficiency.

Specific Wetland Types

Constructed wetlands come in various types, each with its unique design characteristics and treatment capabilities. The individual treatment aims, site characteristics, and resource availability all influence the selection of wetland type. Constructed wetlands with surface flow (SFCWs), subsurface flow (SSFCWs), and vertical flow (VFCWs) are common varieties that offer benefits specific to their respective applications.

Monitoring and Maintenance

An integral part of CW design is the incorporation of monitoring and maintenance strategies. The wetland will continue to function effectively if water quality measurements, vegetation wellness, and microbial activity are regularly monitored. Additionally, maintenance activities, such as vegetation management, substrate replenishment, and debris removal, are essential to prolong the wetland's lifespan and maintain treatment performance.

2.1.3 Types of Constructed Wetlands

CWs are a versatile family of treatment systems, offering various designs tailored to specific objectives and environmental conditions. The three primary types of constructed wetlands—SFCWs, SSFCWs, and VFCWs—each possess unique characteristics and treatment capabilities, allowing them to address a wide range of water pollution challenges.

Surface Flow Constructed Wetlands (SFCWs)

SFCWs, also known as free water surface (FWS) systems, are among the most common types of constructed wetlands. In SFCWs, water flows over the wetland's surface, creating a shallow pond-like environment (Wang et al. 2017a, b, c). This design allows for direct contact between the water and wetland vegetation, making it particularly effective for certain treatment objectives:

- Nutrient Removal: SFCWs are highly efficient at removing nutrients, especially nitrogen and phosphorus. Wetland plants in SFCWs take up these nutrients, promoting water quality improvement by reducing the risk of eutrophication in receiving waters.
- Polishing of Treated Effluents: SFCWs are often employed as tertiary treatment units in wastewater treatment plants to further improve water quality. They serve to polish already-treated effluents, ensuring that the water released into natural water bodies meets stringent quality standards.
- Habitat Creation: The shallow water environment of SFCWs provides habitat opportunities for various aquatic organisms and waterfowl, contributing to biodiversity and ecological restoration efforts (Xu et al. 2018; Zhong et al. 2020).

However, SFCWs may not be as effective for some contaminants that require extended contact times or specific redox conditions for removal. In such cases, alternative CW designs may be more suitable.

Subsurface Flow Constructed Wetlands (SSFCWs)

SSFCWs, also referred to as subsurface flow (SSF) systems, are designed with a subsurface layer where wastewater flows horizontally through gravel, sand, or other porous media. This design enhances contact between the water and the microbial communities residing within the substrate, making SSFCWs highly effective for several treatment objectives:

- Organic Matter Removal: SSFCWs excel at breaking down organic matter, including BOD and COD, through microbial degradation. This results in significant improvements in water quality by reducing organic pollutant levels (Ge et al. 2020).
- Nitrogen Removal: SSFCWs are particularly efficient at denitrification, the process of converting nitrate (NO_3^-) to harmless nitrogen gas (N_2). This is crucial for reducing nitrate pollution in groundwater and surface water bodies.
- Suspended Solids Removal: The substrate in SSFCWs acts as a natural filter, allowing suspended solids to settle out of the water column. This sedimentation process improves water clarity and removes particulate-bound contaminants.

While SSFCWs are highly effective for specific treatment objectives, they require careful design to maintain adequate oxygen levels for microbial processes. Oxygen supply can be achieved through proper hydraulic design and the presence of macrophytes or engineered aeration systems.

Vertical Flow Constructed Wetlands (VFCWs)

VFCWs represent a unique and efficient variant of constructed wetlands. They consist of stacked layers of substrate and wetland vegetation. Wastewater is introduced at the top of the system and percolates vertically through the layers, providing several advantages:

- Enhanced Treatment Efficiency: By lengthening the duration of contact across the wastewater and the ground surface, the vertical flow design improves the effectiveness of pollutant removal. This design is particularly effective for the removal of BOD, COD, nutrients, and some metals (Hartl et al. 2019).
- Aeration: The downward flow of wastewater naturally aerates the system, promoting aerobic microbial processes. This aeration enhances organic matter degradation and nutrient removal.
- Compact Footprint: VFCWs have a compact footprint, making them suitable for areas with limited space. This design feature allows for decentralized wastewater treatment in residential and small-scale applications.
- Reduced Odor Issues: The vertical flow design minimizes odor issues commonly associated with surface flow systems.

Despite their advantages, VFCWs may require periodic maintenance, such as substrate replacement or clogging prevention measures, to ensure long-term performance.

2.2 *Role of Constructed Wetlands in Water Pollution Mitigation*

CWs stand as powerful tools in the battle against water pollution, effectively removing a wide array of pollutantsfrom wastewater and stormwater. Their versatile pollutant removal mechanisms make them invaluable assets in environmental management (Wang and Tian 2021). In this section, we delve into the intricate mechanisms by which constructed wetlands contribute to water pollution mitigation.

2.2.1 Removal of Organic Matter

Constructed wetlands are champions in the eradicationof organic particulatesfrom wastewater, playing a pivotal role in improving water quality. Organic matter in wastewater is often quantified through indicators including COD and BOD, both of which are reliable measures of water pollution (Das et al. 2019; Srivastava et al. 2015). Within CWs, a symphony of microbial communities associated with wetland vegetation and the substrate actively engage in the degradation of organic matter. This degradation occurs through a range of biological processes, including aerobic and anaerobic decomposition (Bashir et al. 2019; Dhonchak and Agnihotri 2023; Thakur et al. 2021; Thakur and Kumar 2021, 2023a, b).

- Aerobic Decomposition: In the presence of oxygen, aerobic microorganisms break down organic matter into simpler compounds, converting it into biomass, carbon dioxide (CO_2), and water (H_2O). This process reduces BOD levels, reflecting a decrease in the oxygen-consuming potential of the water.
- Anaerobic Decomposition: In anaerobic zones of CWs, where oxygen is limited, anaerobic microorganisms thrive. They facilitate the breakdown of complex organic molecules, emergingcarbon dioxide (CO_2) and methane (CH_4) as byproducts. Anaerobic decomposition contributes to COD reduction and the removal of refractory organic compounds.

Through these microbial-driven processes, constructed wetlands significantly reduce BOD and COD levels in the treated water. This transformation of organic matter results in water that is cleaner and less burdensome on receiving waters. For example, in an investigation, González et al. (2020) examined three different setups to determine how cathode setup affected the efficiency of a membrane-less MFC: a wetland submerged in an activated carbon bed (MFCW), a cathode attached to an aeration tower alongside a water recirculation device (MFCT), and an activated carbon bed exposed to air (MFCE). The effectiveness of the MFC's biological matter elimination process, produced the voltage, power consumption, and resistance within are all evaluated. Based on the results of the experiment, the MFCT design performed the best, with an average removal effectiveness of 82%, as opposed to about 70% for MFCE and MFCW. However, the COD removal efficiency topped 60% in every instance (subsequent 40 days). The considerably better COD elimination effectiveness of MFCT was confirmed by statistical research. There was no discernible difference in the evaluated parameters between the various cathode topologies that were investigated.

Similarly, Wang et al. (2023) started a comprehensive investigation, using graphite rods and plates as electrodes, to examine the effects of hydraulic retention time (HRT) and electrode arrangement on the flow field characteristics of CW–MFCs. Additionally, the study optimized the electrode design using numerical simulations of computational fluid dynamics (CFD). The results highlighted a number of important insights: (1) The perceived HRT was found to be the most significant and influential element, accounting for more than 90% of the average HRT in CW-MFCs. Anode spacing was shown to have the greatest impact on the hydraulic performance of CW-MFCs. It was responsible for more than 50% of the water flow divergence, hydraulic efficiency (λ), and efficient volume proportion (e). (3) Anode size was significant for both e and λ, accounting for more than 20% of these factors. (4) In contrast, there was no statistically noteworthy difference in the hydraulic efficiency of CW-MFCs between cathode position and cathode length. Particularly, flow channel compression as well, distraction, blockage of water flow, and separation of boundary layers were the main processes by which MFC electrodes affected the hydraulic properties of the flow field within CW–MFCs. Within CW-MFCs, energy generation and pollution removal were efficiently enabled by optimizing the flow field by electrode configuration optimization. This work offers new research perspectives for bettering CW-MFCs' capacity to treat wastewater and generate electricity, in addition to offering insightful scientific information for boosting the hydraulic performance of CW-MFCs.

Similarly, Dai et al. (2021) examined the treatment of sulfamethoxazole in wastewater using both a CW and a CW-MFC system over 70 days. The research assesses the removal of sulfamethoxazole, changes in extracellular polymer content, evolution of microbial communities, and variations in the abundance of genes having resistance to antibiotics. Comparisons are made between the CW and CW-MFC regarding their removal stability, efficiencies, and performance in the inclusionof sulfamethoxazole. The study findings reveal significantly higher removal efficiencies for entireammonia nitrogen, nitrogen, and sulfamethoxazole in the CW-MFC, with increases of 6.87%, 21.07%, and 11.05%, respectively, compared to the CW. This enhanced performance, along with efficient power generation, demonstrates the superior stability of the CW-MFC in treating sulfamethoxazole-contaminated wastewater. It's interesting to note that the CW-MFC had less extracellular polymer content than the CW. The phylum Proteobacteria prevailed in the CW-MFC, according to microbial community research, whereas Pseudomonas was more prevalent in the CW, which may have contributed to the breakdown of organic materials but not the production of electricity. Additionally, the CW-MFC anode had 4.5 times more Methylotenera than the CW anode, indicating a possible role for Methylotenera in denitrification and a possible factor in the high rate of nitrogen removal seen in the CW-MFC. A noteworthy finding of the study was that the CW had a considerably higher copy number of sulfamethoxazole resistance genes than the CW-MFC, suggesting a higher probability of antibiotic resistance gene propagation in the CW. This indicates that, in comparison to other approaches, the CW-MFC provides a viable and sustainable strategy for eliminating sulfamethoxazole from wastewater, with potential advantages to the surroundings and the economy.

Odally et al. (2019) deliberatedon the influence of different indigenous South African wetland plants on bio-electricity production. Three plant species, namely *Wachendorfia thyrsiflora, Cyperus prolifer*, and *Phragmites australis*, were examined for their bio-electricity production potential evaluated in terms of voltage fluctuation and maximal power density. The results indicated that *Cyperus prolifer* demonstrated the highest bio-electricity production, with a maximum power density of 229 ± 52 mW/m^3 and a voltage of 510 mV. Additionally, the removal efficiency of COD was notably higher for *Cyperus prolifer* ($97 \pm 1\%$) compared to *P. australis* ($94 \pm 1\%$), *W. thyrsiflora* ($94 \pm 1\%$), and the unplanted control framework ($90 \pm 2\%$). The study also evaluated phosphorus removal, measured as orthophosphate removal efficiency. *Cyperus prolifer* exhibited superior orthophosphate removal efficacy ($98 \pm 0\%$) in comparison to the control ($72 \pm 7\%$), *P. australis* ($81 \pm 4\%$) and *W. thyrsiflora* ($58 \pm 6\%$). The results of the study demonstrated the significance of biomass and a high root-specific space for improved power generation in CW-MFCs. Rapid acclimatization to wastewater was identified as a critical factor in achieving efficient power generation. Based on the study's results, Cyperus prolifer emerged as the most suitable wetland plant for CW-MFC applications, offering advantages in bio-electricity generation and the removal of organic matter, ammonia, and phosphate. Consequently, the selection of Cyperus prolifer is recommended for the design of CW-MFC systems aimed at simultaneous wastewater treatment and bio-electricity production. Additionally, Yadav et al. (2012) focused on evaluating the performance of a CW-MFC system for both electricity generation and wastewater treatment, with a specific emphasis on dye removal. The anaerobic and aerobic chambers of the microbial fuel cell facilitate the oxidation and reduction processes. Anaerobic and aerobic zones are present in artificial wetlands as well, where oxidation and reduction activities take place. After these parallels between the two techniques were identified, a unique CW-MFC system was created. The study looks at COD and dye elimination in conjunction with the production of energy in this CW-MFC arrangement. Batch studies were carried out with synthetic wastewater that had different amounts of the dye methylene blue in it. The results demonstrate the effectiveness of this hybrid system in achieving significant dye removal rates. For wastewater samples with initial dye concentrations of 2000, 1500, 1000, and 500 mg/L, dye removal efficiencies of 76.2%, 80.87%, 69.29%, and 93.15%, respectively, were achieved after 96 h of treatment. Moreover, the CW-MFC system exhibited the capability to eliminate 75% of COD from wastewater with a primarydye dosageof 1500 mg/L. After treating wastewater with a baseline dye level of 1000 mg/L, the system generated the most electricity with the highest power density of 15.73 mW/m^2 and a maximal current density of 69.75 mA/m^2.

2.2.2 Nutrient Removal

One of the standout features of constructed wetlands is their remarkable efficiency in removing nutrients, particularly nitrogen and phosphorus, from water (Liu et al. 2013). Excess nutrients in aquatic environments can trigger eutrophication, a phenomenon characterized by harmful algal blooms, oxygen depletion, and ecological disruptions (Dhonchak et al. 2023; Sharma et al. 2023a, b; Sharma et al. 2023a, b; Thakur et al. 2023a, b; Thakur and Kumar 2023c; Verma et al. 2023). CWs address this issue through various nutrient removal mechanisms:

- Plant Uptake: Wetland plants actively take up and assimilate nutrients, primarily nitrogen and phosphorus. This process, known as plant uptake, contributes significantly to reducing nutrient concentrations in the water.
- Denitrification: Constructed wetlands, particularly those with anaerobic zones, support denitrification—a process where nitrate (NO_3^-) is converted into harmless nitrogen gas (N_2). This microbial-driven reaction helps mitigate nitrate pollution in groundwater and surface waters.
- Adsorption onto Substrate: The substrate within CWs, often composed of sand, gravel, and soil, can adsorb and retain nutrients. Nutrients become attached to substrate particles, reducing their mobility and bioavailability.

Through the careful design and operation of constructed wetlands, nutrient discharges into receiving waters are significantly curtailed, preventing eutrophication and its associated environmental issues (Kaur et al. 2022, 2023; Nidhi et al. 2023; Nidhi and Kaur 2022; Pathania et al. 2023; Rana et al. 2023; Thakur et al. 2019, 2020, 2023). In an investigation, Leng et al. (2023) presented the productionof aneffectiveCW-coupled MFC system designed for wastewater treatment and electricity generation. The primary focus was on the removal of total phosphorus from simulated domestic sewage. By analyzing differences in substrates, hydraulic retention durations, and populations of microbes, the study assessed the best phosphorus removal effectiveness and generation of electricity. The study also investigated the processes underlying the elimination of phosphorus. The two CW-MFC systems obtained remarkable efficiency of removal of 80.3% and 92.4%, accordingly, employing garnet and magnesia as substrates. While the magnesia system functioned via ion exchange processes, the elimination of phosphorus in the garnet-based medium was mainly achieved via an intricate adsorption procedure. Compared to the Magnesia system, the garnet system displayed greater maximal and stabilization voltages. At the genus level, the dominant genera were identified based on a relative abundance threshold of > 0.1%. In the D10 layer, the primary dominant genera included *Rhodobacter* (2.4%), *Nitrospira* (2.8%), *Sphingomonas* (7.2%), *Devosia* (1.3%), *Tabrizicola* (2.1%) and *Nevskia* (1.1%) (Fig. 4a). In the D30 layer, the predominant genera consisted of *Rhodobacter* (1.2%), *Nitrospira* (3.4%), *Sphingomonas* (2.1%), H16 (1.6%) and *Haliangium* (1.7%). Nitrospira, Haliangium, H16, Opitutus, and Pseudomonas all showed notable increases in relative prevalence (Fig. 4a) of 0.99%, 0.64%, 6.38%, and 0.36%, respectively. In comparison, the relative prevalence of the number of taxa reduced significantly, with Sphingomonas, Rhodobacter, Tabrizicola, Porphyrobacter and Devosiaseeing declines of 5.10%, 1.20%, 1.21%, 0.34% and Devosia, correspondingly (Fig. 4b).

Similarly, Han et al. (2021) investigated the efficacy of three highly efficient combination systems for the eradicationof contaminantsfrom domestic sewage while simultaneously generating clean energy. The CW-MFC, the Hydrolytic Filler Acidification paired with the CWMFC (HFA-CWMFC), and the Hydrolytic Sludge Acidification coupled with the CWMFC (HSA-CWMFC) are the systems that are being examined. The aggregate performance of all three systems showed that they could efficiently treat household sewage and generate sustainable electricity. In particular, 0.36 W/m^2, 0.40 W/m^2, and 0.43 W/m^2 have been determined as the power production rates for CWMFC, HSA-CWMFC, and HFA-CWMFC, correspondingly. These results demonstrate that treating the treatment of wastewater and the production of renewable energy at the same time is feasible. The study also shows that HFA-CWMFC was able to achieve the best rates of elimination for a number of pollutants, alongside removal proportions of 91.01%, 90.63%, 95.35%, 86.11%, and 91.01%, accordingly, for TOC, COD, Total Phosphorus (TP), and Ammonia Nitrogen (NH_3–N). Interestingly, HFA-CWMFC yielded more organic acids than HSA-CWMFC, which was advantageous for the reactions that followed. Firmicutes were shown to be important contributions to the higher pollution removal rates and

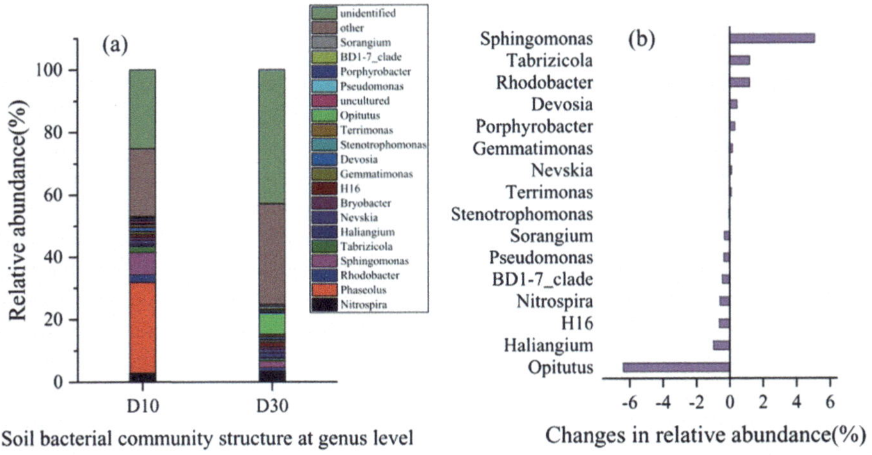

Fig. 4 An example of how soil microorganisms in wetland sediments change at the genus level. The bacterial communal architecture at the genus layer is shown in subfigure (**a**), and variations in relative abundance are shown in subfigure (**b**). Adapted from Ref. Leng et al. (2023) under CCBY 4.0

greater power generation capacity seen in HFA-CWMFC, according to Qualcomm sequencing study.

2.2.3 Heavy Metal Immobilization

Heavy metals, including cadmium (Cd), lead (Pb), and copper (Cu), pose severe risks to aquatic ecosystems and human health due to their toxicity. Constructed wetlands offer a unique solution by immobilizing these metals, rendering them less harmful (Xu et al. 2019; Zhou et al. 2018). This immobilization occurs through several processes:

- Adsorption onto Substrate: The substrate within CWs, with its high surface area and electrochemical properties, acts as a sorbent for heavy metals. Metals become adsorbed onto substrate particles, reducing their concentration in the water.
- Precipitation and Complexation: In certain conditions, heavy metals can precipitate as insoluble compounds or form complexes with other substances in the wetland, making them less bioavailable and less prone to entering the food chain.

By immobilizing heavy metals within the constructed wetland environment, the risk of these contaminants impacting aquatic life and human health is greatly diminished. In an experiment, Shi et al. (2023) introduced the *Leersia hexandra*-constructed wetland–microbial fuel cell (CW-MFC) system for the remediation of Cr(VI)-contaminated water as shown in Fig. 5.

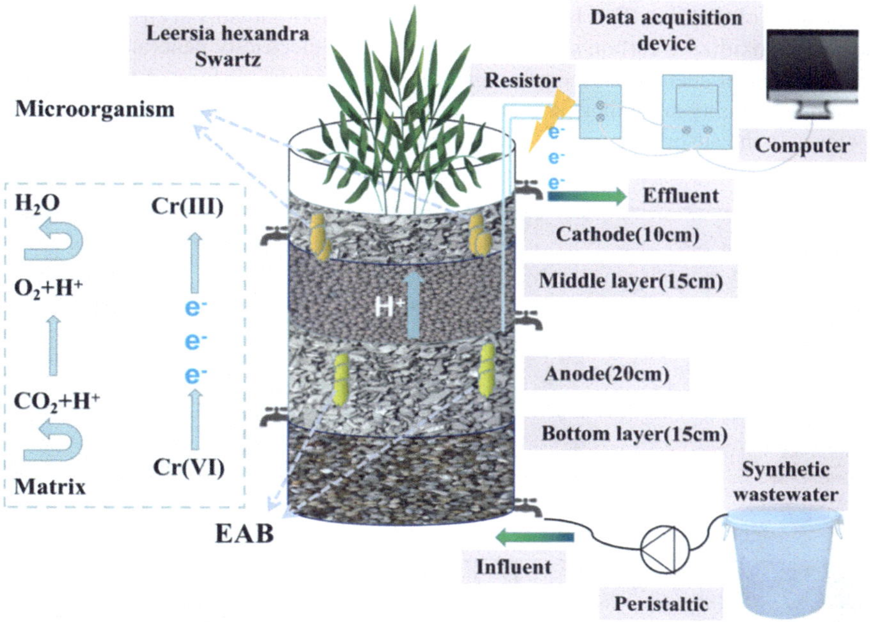

Fig. 5 A diagram illustrating the Leersia hexandra Swartz designed CW-MFC. Adapted from Ref. Shi et al. (2023) under CCBY 4.0

It also investigates how different dissolved oxygen (DO) levels affect this system's capacity to produce electricity, remove pollutants, and reduce Cr(VI). The findings show that when the amount of DO remains at 4.5 mg/L, cathode aeration considerably increases voltage output and contaminant removal effectiveness in the L. hexandra CW-MFC. The equipment reaches a high voltage of 520 mV, an elimination rate of 93.73% for COD, and an elimination rate of 97.77% for Cr(VI) under these circumstances. Moreover, increased levels of DO facilitate the transformation of Cr(VI) to Cr(III) by encouraging the uptake of heavy metal Cr by the substrate itself and L. hexandra. Thus, in L. hexandra and the substrate, chromium is primarily present in a less poisonous and less transportable form. This finding emphasizes how elevated DO concentrations drive redox processes in plant tissues and the system, which in turn convert Cr(VI) to Cr(III). Geobacter, a crucial microbe that improves the system's efficiency and Cr(VI) reduction, is also identified in the study. The surface architecture of the original activated carbon and the activated carbon at the cathode and anode after the film application are shown in Fig. 6. The original surface of the activated carbon was somewhat rough and had many layers of irregular holes. This roughness greatly increased the possibility for heavy metal adsorption and provided an environment that was favorable for bacterial development. On the other hand, the activated carbon surface showed the formation of a dense biofilm following the application of the film. This biofilm is crucial as microorganisms act as biocatalysts capable of oxidizing carbon sources, thus promoting electricity generation.

Moreover, it seemed as though metal ions covered the surface of the activated carbon, producing smoother pore architectures. This shift might be explained by a decrease in the adsorbent surface's non-uniformity, which would suggest that metals

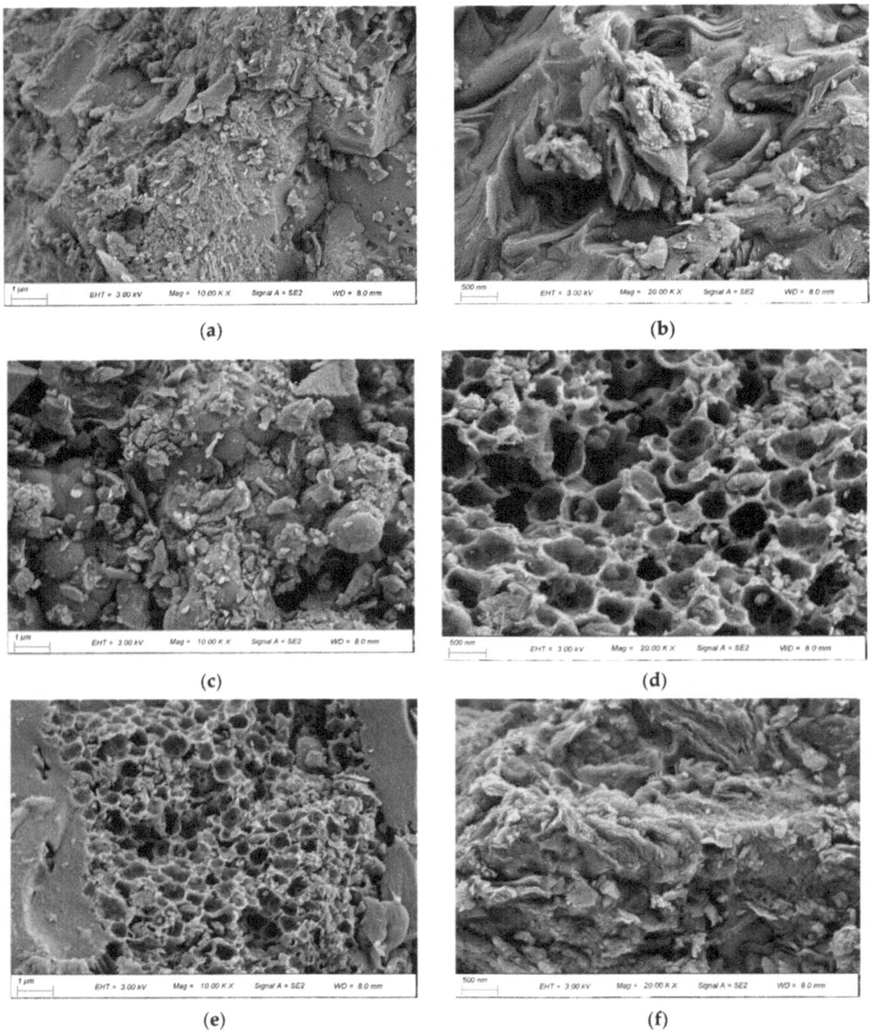

Fig. 6 Microstructure and surface topography of activated carbon, showcasing the following: **a**, **b** Pureactivated carbon; **c**, **d** Biofilm formed on the anode of the *Leersia hexandra* CW-MFC; **e**, **f** Biofilm developed on the cathode of the *Leersia hexandra* CW-MFC. Adapted from Ref. Shi et al. (2023) under CCBY 4.0

were adhering to functional groups inside the pores. The stability of the L. hexandra as a result of these changes CW-MFC system was kept up to date during the trial. These results provide important information for the use of CW-MFC systems in the treatment of wastewater that contains Cr in the real world. Furthermore, Mu et al. (2021) examinedthe influence of different filler materials in up-flow CW-MFC systems, such asbio-ceramic (CW-MFC1), zeolite (CW-MFC2), calcite (CW-MFC3), and volcanic rock (CW-MFC4), on Cr(VI) removal and electricity generation. The findings demonstrate that during the prolonged operating period, there was very little Cr(VI) absorption by the fillers. All systems achieved around93% chemical oxygen demand (COD) removal, indicating efficient organic matter removal. The systems differed in their rates of Cr(VI) elimination: CW-MFC4 (98.0%) > CW-MFC2 (94.6%) > CW-MFC3 (88.5%) > CW-MFC1 (71.6%). The anode layer was where the majority of Cr(VI) ions were transformed into Cr(III) deposition and immobilized, with the bottom layer following suit. The microorganisms in the anode layer and bottom layer were shown to be in an antagonistic interaction. Because the filler layer microorganisms consumed too much organic matter, which decreased the activity of anodic microbes, bio-ceramic filler (CW-MFC1) showed the lowest rate of Cr(VI) elimination. In addition, calcite inhibited microbial development, leading to decreased elimination of Cr(VI) in CW-MFC3, since it had a smaller effective area of surface and a less permeable structure than zeolite and volcanic rock. The following was the order of output voltage and highest power density: CW-MFC3 > CW-MFC4 > CW-MFC2 > CW-MFC1. This was mainly ascribed to greater organic matter intake (greater rate of COD elimination) in the anode layers, which caused exoelectrogenic bacteria to produce more energy. Volcanic rock therefore proved to be the best filler material, increasing the removal of Cr(VI) (99.0%) and producing electricity at a satisfactory voltage (0.595 ± 0.019 V) and power density (0.462 W/m^3).

2.2.4 Sediment and Suspended Solids Removal

Constructed wetlands serve as effective sedimentation basins, allowing suspended solids to settle out of the water column. This sedimentation process enhances water clarity and removes particulate-bound pollutants, including organic matter and some heavy metals. The settled solids can be periodically removed from the wetland, preventing their release back into the environment and further elevatingwater quality (Bashir et al. 2020a, b; Bashir et al. 2020a, b; Kumar and Thakur 2020; Parveen et al. 2020; Thakur et al. 2022a, b, c, d, e, f; Thakur et al. 2022a, b, c, d, e, f). In summary, constructed wetlands function as natural purifiers, effectively removing a broad array of pollutants from wastewater and stormwater. Their ability to address organic matter, nutrients, heavy metals, pathogens, and suspended solids makes them indispensable tools in the quest to mitigate water pollution and protect aquatic ecosystems and public health.

2.3 *Applications of Constructed Wetlands in Wastewater Treatment*

CWs have emerged as versatile and sustainable solutions in the field of wastewater treatment, finding applications in various sectors. Their adaptability, cost-effectiveness, and environmental benefits make them appealing choices for addressing diverse treatment objectives (Wang et al. 2019a, b, c). In this section, we deliberatethe wide-ranging applications of constructed wetlands in wastewater treatment.

2.3.1 Municipal Wastewater Treatment

Constructed wetlands are increasingly gaining recognition and adoption for the treatment of municipal wastewater. They can function as standalone treatment systems or be integrated into existing wastewater treatment plants (WWTPs) as tertiary or secondary treatment units. In municipal applications, CWs excel in several key aspects:

- Organic Load Reduction: One of the primary roles of CWs in municipal wastewater treatment is the reduction of organic loads. Microbial communities within the wetland environment actively break down organic matter, leading to a significant decrease in BOD and COD levels. This transformation results in cleaner and safer effluent.
- Nutrient Removal: Municipal wastewater often contains elevated levels of nutrients, primarily nitrogen and phosphorus. CWs efficiently address these nutrient concerns through processes such as plant uptake, denitrification, and adsorption onto substrate materials. This nutrient removal helps prevent eutrophication in receiving waters (Elhenawy et al. 2022; Gadegaonkar et al. 2020).
- Suspended Solids Reduction: Constructed wetlands function as sedimentation basins, allowing suspended solids to settle out of the water column. This sedimentation process leads to improved water clarity and the removal of particulate-bound pollutants.
- Low Maintenance: CWs are known for their low maintenance requirements, making them particularly suitable for small to medium-sized communities with limited resources. They offer an eco-friendly and cost-effective solution for municipal wastewater treatment.

2.3.2 Industrial Wastewater Treatment

Constructed wetlands have found a niche in industrial wastewater treatment, catering to the specific needs of various industries. They effectively treat industrial effluents, including those generated by food processing, textile manufacturing, mining operations, and more. In industrial applications, CWs offer the following advantages:

- Contaminant Specificity: CWs can be customized to address contaminants unique to each industry. This adaptability ensures efficient removal of industrial pollutants, contributing to regulatory compliance and reducing the environmental footprint of industrial operations (Nastro et al. 2021; Takavakoglou et al. 2022).
- Cost-Efficiency: Compared to conventional treatment methods, CWs often present a cost-efficient alternative for industries. They provide effective treatment while minimizing operational costs (Chen et al. 2022).
- Environmental Compliance: Industries are under increasing pressure to meet stringent environmental regulations. Constructed wetlands offer a sustainable and environmentally friendly solution for industries seeking to reduce their impact on water resources.

2.3.3 Stormwater Management

Constructed wetlands play a pivotal role in stormwater management, particularly in urban and industrial settings. Urban runoff and industrial stormwater can carry anarrayof pollutants, such asoil, heavy metals, and sediments (Erazo and Agudelo-Escobar 2023; Yadav et al. 2022). CWs act as natural filters, capturing and treating these contaminants before they are discharged into receiving waters. In the context of stormwater management, CWs provide the following benefits:

- Pollutant Capture: CWs capture and retain pollutants from stormwater runoff, preventing their entry into natural water bodies. This helps maintain water quality and protect aquatic ecosystems.
- Flood Mitigation: CWs assist in mitigating flooding by temporarily storing and slowly releasing stormwater. This reduces the risk of flash floods and erosion in urban areas.
- Ecosystem Services: Beyond pollution mitigation, constructed wetlands offer a range of ecosystem services in urban environments. They enhance biodiversity, provide recreational opportunities, and contribute to the aesthetic appeal of urban landscapes.

2.3.4 Agricultural Runoff Treatment

Agricultural runoff laden with nutrients and pesticides poses a significant challenge to water quality. Constructed wetlands strategically placed within agricultural landscapes can intercept and treat this runoff, preventing the transport of pollutants into downstream waters (Kesarwani et al. 2023; Liu et al. 2014). In agricultural runoff treatment, CWs offer the following advantages:

- Nutrient Reduction: Agricultural CWs effectively reduce nutrient loads in runoff, addressing issues of nutrient pollution and eutrophication in water bodies downstream.
- Sediment Control: These wetlands promote sediment settling, preventing soil erosion and reducing the transport of sediment-bound pollutants.

- Water Quality Enhancement: The treatment of agricultural runoff through CWs leads to improved water quality, benefiting both aquatic ecosystems and downstream water users.

2.3.5 Habitat Restoration and Conservation

Constructed wetlands serve as multifunctional assets, not limited to pollution mitigation (Roy et al. 2023). They are also powerful tools for habitat restoration and conservation. These wetlands provide vital ecosystems for wildlife, support diverse plant and animal communities, and offer several benefits:

- Biodiversity Conservation: Constructed wetlands contribute to biodiversity conservation by providing habitats for various species, including birds, amphibians, and aquatic invertebrates.
- Recreational Opportunities: Many constructed wetlands become valuable recreational spaces, allowing communities to connect with nature and engage in outdoor activities.
- Landscape Aesthetics: These wetlands enhance landscape aesthetics, making urban and industrial areas more visually appealing while improving overall environmental quality.

3 Microbial Fuel Cells: Principles and Operation

MFCs are cutting-edge biotechnological devices that use organic materials as fuel by using microorganisms' metabolic processes (Liu et al. 2019; Paucar and Sato 2021). They represent a promising technology with applications ranging from wastewater treatment to renewable energy production.

3.1 Introduction to Microbial Fuel Cells

MFCs are bioelectrochemical devices that use microorganisms' metabolic processes to directly transform organic chemicals into electrical energy. Their functioning is predicated on the utilization of electron transfer mechanisms that transpire during the procedure of microbial respiration (González et al. 2021; Yang et al. 2021a, b). The core components of an MFC typically include:

- External Circuit: Electrons can go from the anode to the cathode because of an external circuit connecting the two. Electrical current is produced by this electron flux and can be used in a variety of ways.

- Anode: Microorganisms oxidize organic material at the anode, generating electrons in the procedure. After that, these electrons are moved to an outside circuit, where they produce an electric current.
- Cathode: When electrons interact with protons and oxygen to generate water or other oxidized species, they are dissipated at the cathode electrode.
- Proton Exchange Membrane (PEM): In some MFC systems, a PEM keeps protons (H^+) from migrating between the anode and cathode chambers while limiting the combining of the contents. Sustaining the electrochemical gradient required for electron transport depends on this separation.

The operation of an MFC relies on the electrochemical processthat rendersat the anode and cathode. At the anode, microorganisms (typically bacteria or archaea) oxidize organic compounds, releasing electrons and protons (Doherty et al. 2015). Protons move via the PEM and into the cathode area, whereas these electrons are transported to the cathode via the external circuit. Protons, electrons, and an electron acceptor (often oxygen) come together at the cathode to finish the electrochemical reaction.

3.2 Energy Generation in MFCs

The energy generation process in MFCs is driven by the microbial metabolism of organic matter. This metabolism involves several key steps:

- Substrate Oxidation: Microorganisms at the anode oxidize organic substrates, such as glucose or organic wastewater contaminants, through enzymatic reactions. This process discharges protons andelectrons.
- Electron Transfer: An electron flow is started when the liberated electrons are moved to the anode electrode. This flow generates an electrical current in the external circuit, which can be used for a number of purposes, including battery charging and the powering of electronics.
- Proton Migration: Simultaneously, protons (H^+) are released during substrate oxidation. These protons migrate through the PEM to the cathode compartment (Wang et al. 2019a, b, c).
- Cathodic Reactions: At the cathode, electrons combine with oxygen (or another electron acceptor) and protons to form water or other oxidized products. This cathodic reaction consumes the electrons and protons, completing the overall electrochemical circuit.

The net result of these processes is the conversion of chemical energy stored in organic matter into electrical energy (Kaya et al. 2023; Thakur et al. 2022a, b, c, d; Thakur and Kumar 2022). MFCs hold significant promise for sustainable energy production, particularly in wastewater treatment plants where they can simultaneously treat wastewater while generating electricity.

3.3 Microbial Communities in MFCs

Microbial communities exhibita pivotal role in the operation and efficacyof MFCs. These communities consist of a diverse array of microorganisms, including bacteria, archaea, and even some fungi. There are several elements that affect the structure and activities of these microbial communities, such as the kind of organic substrate, the surrounding environment, and the design of the system (Samudro et al. 2021; Zhang et al. 2022). Key aspects of microbial communities in MFCs include:

- Electrogenic Microorganisms: Certain microorganisms, known as electrogenic bacteria, have the unique ability to transfer electrons to solid electrodes. *Geobacter* and *Shewanella* species are examples of electrogenic bacteria commonly found in MFCs.
- Microbial Diversity: MFCs often host diverse microbial communities, with different species contributing to various stages of organic matter degradation and electron transfer. This diversity can enhance the overall performance and stability of MFCs.
- Syntrophic Relationships: Microbial communities in MFCs engage in syntrophic interactions, where different species cooperate to carry out complex metabolic processes. Syntrophic relationships are vital for the degradation of complex organic substrates and the efficient transfer of electrons to the anode.
- Environmental Factors: The composition and activity of microbial communities in MFCs are influenced by many variables, including pH, temperature, accessibility to nutrients, and electrode material. Improving these environmental parameters can improve MFC efficiency.
- Biofilm Formation: Microorganisms in MFCs often form biofilms on the substateof the anode electrode. These biofilms exhibita structured and stable environment for microbial growth and electron transfer, improving the efficiency of MFCs.

4 Synergistic Integration of CW and MFC

The synergistic integration of CW and MFC represents an innovative technique which integrates the strengths of both technologies to address water pollution mitigation and resource recovery (Xu et al. 2020, 2021). In this section, we will explore the concept of CW-MFC technology, the mechanisms and benefits of integration, and important design considerations.

4.1 Concept of CW-MFC Technology

The innovative hybrid system known as CW-MFC technology combines the ideas of microbial fuel cells and built wetlands to produce a multipurpose foundation for the remediation of wastewater and the production of sustainable electricity. Fundamentally, CW-MFC technology uses natural procedures that take place in artificial wetlands to remove pollutants and cycle nutrients. At the same time, it uses the metabolic activity of the microorganisms in MFCs to produce energy from the organic matter found in wastewater. There are various benefits to this connection:

- Simultaneous Water Treatment and Energy Generation: CW-MFCs enable the dual-purpose use of wastewater. The artificial wetland component removes pollutants and treats water, but the organic matter in the wastewater is employed as fuel in the MFC to produce power.
- Enhanced Pollutant Removal: The combination of CWs and MFCs enhances pollutant removal efficiency. Constructed wetlands excel in removing organic matter, nutrients, and certain contaminants, while MFCs provide additional treatment and capture of organic compounds, improving overall water quality.
- Resource Recovery: CW-MFC technology allows for the recovery of valuable resources from wastewater, including electricity and the potential for the production of byproducts such as biogas or biofuels. This resource recovery contributes to sustainability and cost-effectiveness.

4.2 Mechanisms and Benefits of Integration

The integration of CWs and MFCs offers several mechanisms and benefits that enhance their combined performance:

- Synergistic Pollutant Removal: CWs provide a natural habitat for diverse microbial communities that contribute to organic matter degradation, nutrient removal, and pollutant immobilization. When integrated with MFCs, the additional degradation of organic compounds by electrogenic microorganisms further enhances pollutant removal efficiency (Villaseñor et al. 2013; Yang et al. 2020).
- Enhanced Nutrient Cycling: The capacity of constructed wetlands to cycle and eliminate pollutants like phosphorus and nitrogen is well recognized. MFCs can contribute to this process by promoting denitrification, which converts nitrate into nitrogen gas, reducing nutrient discharges into natural waters.
- Bioenergy Production: MFCs use the organic matter in wastewater as a source of energy. The electricity produced by CWs with MFC integration can be used for a number of purposes, including pumping and sensor power as well as adding to the grid's electrical supply.
- Reduced Environmental Footprint: CW-MFCs can have a smaller environmental footprint compared to standalone MFCs since they benefit from the existing

infrastructure of constructed wetlands. This integration promotes sustainable wastewater treatment practices.

4.3 Design Considerations

Designing a successful CW-MFC system requires careful consideration of various factors:

- Selection of CW Type: Different kindsof constructed wetlands (e.g., subsurface flow, surface flow, vertical flow) may be more suitable for specific treatment objectives. The choice should align with the pollutant removal goals and available space.
- MFC configuration: Consider the design and configuration of the MFC component, including the number of cells, electrode materials, and reactor geometry. These factors can impact electricity generation efficiency.
- Electrode Placement: For the best possible electron transmission among the anode and the microbial community, the electrodes must be positioned inside the created wetland. Materials for electrodes should be resistant to corrosion.
- Hydraulic Considerations: Maintain control over water flow rates, residence times, and hydraulic loading rates within the CW-MFC system to ensure effective treatment and prevent waterlogging.
- Microbial Communities: Foster the growth of suitable microbial communities in both the constructed wetland and MFC components. This may involve selecting appropriate wetland vegetation and promoting the development of electrogenic bacteria in the MFC (Wang et al. 2019a, b, c).
- Monitoring and Maintenance: Implement a monitoring and maintenance plan to ensure the continued performance of the CW-MFC system. Regular maintenance may include cleaning, substrate replacement, and electrode maintenance.

5 Pollutant Removal Mechanisms in CW-MFCs

CW-MFCs offer innovative and effective mechanisms for the removal of pollutants from wastewater, combining the natural processes of constructed wetlands with the electrochemical capabilities of microbial fuel cells (Paucar and Sato 2022). In this section, we will delve into the mechanisms behind pollutant removal in CW-MFCs, focusing on microbial processes, nutrient removal, and metal immobilization, and providing insights from case studies.

5.1 Microbial Processes in Pollutant Removal

Microbial processes are at the heart of pollutant removal in CW-MFCs. Within the integrated system, various microorganisms interact to break down organic matter and transform it into less harmful substances. Key microbial processes include:

- Organic Matter Degradation: Bacteria and archaea are two examples of microorganisms that are essential to the breakdown of organic chemicals found in wastewater. They emit protons and electrons when they eat organic materials, which serve as a carbon source. These particles are directed toward the microbial fuel cell's anode, where they produce energy.
- Nutrient Transformation: Microbial communities within the constructed wetland component facilitate the transformation of nutrients. Nitrogen, for example, undergoes denitrification, where nitrate (NO_3^-) is reduced to nitrogen gas (N_2) through a series of microbial reactions. This process removes excess nitrogen from the wastewater, mitigating eutrophication risks (Araneda et al. 2018).
- Pathogen Reduction: Microorganisms in CW-MFCs can contribute to pathogen reduction through various mechanisms, including predation and biological degradation. The presence of microorganisms in the system helps reduce pathogen concentrations in the treated effluent, enhancing water safety.
- Biofilm Formation: Microorganisms in CW-MFCs often form biofilms over the substrate of the anode electrode. These biofilms exhibita structured environment that enhances microbial activity and electron transfer, improving the overall treatment performance.

5.2 Nutrient Removal and Metal Immobilization

CW-MFCs are effective at removing nutrients and immobilizing heavy metals through various processes:

- Nutrient Removal: Constructed wetlands are known for their ability to remove nutrients, particularly nitrogen and phosphorus. Within CW-MFCs, this nutrient removal is enhanced by the electrogenic microorganisms in the MFC component (Han et al. 2021; Tamta et al. 2020). Denitrification, in particular, is promoted in anaerobic zones, reducing nitrate levels and preventing nitrogen pollution of receiving waters. Phosphorus removal occurs through adsorption onto substrate materials and plant uptake, further improving water quality.
- Heavy Metal Immobilization: Heavy metals, including lead, cadmium, and copper, pose environmental risks due to their toxicity. In CW-MFCs, these metals can be immobilized through various mechanisms. Substrate materials within the constructed wetland can adsorb heavy metals, reducing their mobility and bioavailability. Additionally, complexation reactions and precipitation can occur within the wetland environment, further reducing the risk of heavy metal contamination.

6 Resource Recovery Potential

CW-MFCs offer significant resource recovery potential, making them an sustainable and attractive alternative for wastewater treatment and environmental remediation (Mu et al. 2021; Teoh et al. 2020; Yadav et al. 2012). In this section, we will explore the resource recovery aspects of CW-MFCs, including electricity generation, the emergenceof value-added byproducts, and the associated economic and environmental implications.

6.1 Electricity Generation in CW-MFCs

Power generation is one of the main ways that CW-MFCs recover resources. Microbial fuel cells (MFCs) are incorporated into the system to facilitate the transformation of chemical energy found in organic matter in wastewater into electrical energy. This electricity generation has several key features:

- Renewable Energy Source: CW-MFCs provide a source of renewable energy. Consistent generation of electricity is made possible by the organic matter in wastewater, which provides the microbial communities in the MFCs with a steady and sustainable fuel source.
- Energy Independence: The electricity generated in CW-MFCs can be harnessed for on-site use, making wastewater treatment plants and other applications more energy-independent. This can lead to reduced operational costs and reliance on external energy sources (Gupta et al. 2021).
- Energy Recovery from Wastewater: CW-MFCs effectively recover energy from wastewater, transforming it from a liability into a valuable resource. This energy recovery aligns with the principles of a circular economy, where resources are efficiently utilized and recycled.
- Reduction of Greenhouse Gas Emissions: By utilizing the electricity generated from CW-MFCs, the need for conventional energy sources, often associated with greenhouse gas emissions, can be reduced. This contributes to a reduction in the carbon footprint of wastewater treatment processes.

6.2 Value-Added Byproducts

In addition to electricity generation, CW-MFCs have the potential to produce value-added byproducts, further enhancing their resource recovery capabilities. Some key byproducts and their benefits include:

- Biogas Production: In certain CW-MFC configurations, the microbial communities can produce biogas, primarily methane (CH_4). Biogas can be captured and used as an additional source of renewable energy or converted into heat for process applications, providing economic and environmental benefits (Dai et al. 2021; Oodally et al. 2019).

- Organic Matter Reduction: As CW-MFCs efficiently degrade organic matter in wastewater, the treated effluent often contains lower levels of organic compounds. This can simplify downstream treatment processes, reduce the volume of sludge generated, and lower disposal costs.
- Nutrient-Rich Sludge: The sludge produced in CW-MFCs may contain concentrated nutrients, particularly nitrogen and phosphorus. This nutrient-rich sludge could be utilized as a fertilizer or soil conditioner, contributing to agricultural sustainability and reducing the need for synthetic fertilizers.

6.3 Economic and Environmental Implications

The resource recovery potential of CW-MFCs carries several economic and environmental implications:

- Economic Sustainability: CW-MFCs can enhance the economic sustainability of wastewater treatment operations by reducing energy costs, creating revenue through electricity generation, and potentially generating income from the sale of value-added byproducts.
- Environmental Benefits: By recovering energy and producing value-added byproducts, CW-MFCs contribute to environmental sustainability (Tamta et al. 2023). They reduce the environmental footprint of wastewater treatment processes, lower greenhouse gas emissions, and promote responsible resource management.
- Circular Economy Principles: By effectively using and recycling resources from wastewater, such as energy and nutrients, CW-MFCs are in line with the concepts of a circular economy. This approach reduces waste and maximizes resource utilization.
- Energy Independence: The electricity generation capability of CW-MFCs can reduce the dependence on external energy sources, enhancing energy security for wastewater treatment facilities and other applications (Leng et al. 2023; Rojas-Flores et al. 2023; Rojas-Villacorta et al. 2023).
- Reduced Discharge of Pollutants: The enhanced pollutant removal capabilities of CW-MFCs contribute to improved water quality and reduce the discharge of contaminants into natural water bodies. This supports ecosystem health and public well-being.

7 Challenges and Future Directions

CW-MFCs offer innovative and sustainable solutions for water pollution mitigation and resource recovery. However, their practical implementation is accompanied by several challenges (Mao et al. 2023; Ramírez-Vargas et al. 2018). Firstly, scaling up these systems for municipal wastewater treatment plants poses a significant engineering challenge. Laboratory-scale success needs to be translated into consistent, efficient, and economically viable full-scale systems. Long-term stability is another pressing concern, as biofilm formation, electrode fouling, and shifts in microbial communities can impact system performance over time. Researchers are actively exploring strategies to ensure the reliability of CW-MFCs in the long run. Electrode materials, critical for microbial fuel cells, need to be cost-effective, durable, and corrosion-resistant, driving ongoing material research. Nutrient recovery, especially for nitrogen and phosphorus, presents an opportunity to enhance sustainability but requires optimization for practical reuse. Additionally, assessing the economic viability of CW-MFCs in comparison to conventional wastewater treatment methods is essential for their widespread adoption (Wang et al. 2023). Cold climate adaptation is necessary to ensure year-round functionality, particularly in regions with low temperatures. Overcoming these challenges is crucial for unlocking the full potential of CW-MFC technology.

7.1 Future Research Directions and Research Gaps

Looking ahead, several research directions and gaps offer opportunities for advancing CW-MFC technology:

- Microbial Communities: Understanding the dynamics of microbial communities within CW-MFCs is pivotal. Researchers aim to optimize these communities for specific pollutant removal and electricity generation goals, enhancing overall system performance and stability (Yaqoob et al. 2020).
- Hybrid Systems: An interesting direction to pursue is investigating new setups and hybrid systems that combine CW-MFCs with other therapeutic modalities. Such integration can maximize treatment efficiency, resource recovery, and versatility.
- Wastewater Characterization: Tailoring CW-MFC systems to different types of wastewater, such as industrial effluents or agricultural runoff, requires a deeper understanding of the unique challenges posed by each source. Future research will focus on characterizing diverse wastewater streams and optimizing CW-MFC configurations accordingly.
- Bioenergy Applications: Investigating the adaptability of CW-MFC technology to decentralized and off-grid applications, such as remote communities or disaster-stricken areas, can extend its reach and impact, providing clean water and energy solutions in diverse settings.

- Advanced Materials: The development of advanced materials for electrodes, membranes, and substrates is a priority. These materials should exhibit superior electrochemical properties, durability, and resistance to fouling, contributing to enhanced CW-MFC performance (Li et al. 2021).
- System Modeling: Advanced mathematical models that accurately predict CW-MFC performance under varying operating conditions are essential for optimization. Such models guide system design, operation, and control, resulting in more efficient and reliable CW-MFCs.

7.2 Regulatory Framework and Guidelines

As CW-MFC technology matures, the establishment of regulatory frameworks and guidelines becomes increasingly important:

- Environmental Regulations: Clear regulatory standards are needed to ensure CW-MFC systems comply with environmental regulations, encompassing effluent quality, pathogen removal, and discharge limits to safeguard water bodies and human health (Afroz et al. 2021; Shi et al. 2023).
- Safety Guidelines: The development of guidelines for safe CW-MFC operation and maintenance is paramount, particularly in the context of electricity generation and microbial management. Ensuring the safety of operators and the environment is crucial.
- Monitoring and Reporting: Implementing monitoring and reporting requirements is necessary to assess CW-MFC system performance and ensure long-term efficacy. Regular monitoring of water quality parameters, electricity generation, microbial activity, and system integrity provides valuable data for optimization and regulatory compliance.
- Incentives and Policies: Policymakers can incentivize CW-MFC adoption through supportive policies, grants, or tax incentives, facilitating the transition to sustainable wastewater treatment and resource recovery practices (Jiménez González et al. 2020; Ni et al. 2020). These incentives can accelerate the integration of CW-MFCs into existing wastewater infrastructure and promote responsible water resource management.

8 Conclusion

The importance of CW-MFC technology lies in its sustainability, resource recovery potential, contribution to ecosystem health, and innovation. Sustainability is embodied through the utilization of natural processes for water treatment and the generation of clean energy, aligning with global sustainability goals and circular economy principles. Resource recovery transforms wastewater into an asset through electricity generation and value-added byproduct production, enhancing economic and environmental sustainability. CW-MFCs support ecosystem health by efficiently removing pollutants and mitigating risks associated with nutrient pollution, heavy metals, and pathogens. Moreover, CW-MFC technology challenges conventional wastewater treatment methods and opens new avenues for research and technological advancement. As we conclude our exploration of CW-MFC technology, there is a call for sustainable practices in water management. This includes encouraging the adoption and integration of CW-MFC technology, collaboration among policymakers, engineers, and researchers, continued research to address challenges and optimize performance, development of clear regulatory frameworks, promotion of education and awareness, and advocacy for holistic and sustainable water management practices beyond CW-MFCs. These practices encompass efficient water use, pollution prevention, habitat restoration, and responsible stewardship of water resources.

References

Afroz AS, Romano D, Inglese F, Stefanini C (2021) Towards bio-hybrid energy harvesting in the real-world: pushing the boundaries of technologies and strategies using bio-electrochemical and bio-mechanical processes. Appl Sci (Switzerland) 11(5):1–44. https://doi.org/10.3390/app11052220

Araneda I, Tapia NF, Allende KL, Vargas IT (2018) Constructed wetland–microbial fuel cells for sustainable greywater treatment. Water (Switzerland) 10(7):1–9. https://doi.org/10.3390/w10070940

Arvaniti I, Fountoulakis MS (2021) Use of a graphite-cement composite as electrode material in up-flow constructed wetland–microbial fuel cell for greywater treatment and bioelectricity generation. J Environ Chem Eng 9(3):105158. https://doi.org/10.1016/j.jece.2021.105158

Bashir S, Thakur A, Lgaz H, Chung I-M, Kumar A (2019) Computational and experimental studies on Phenylephrine as anti-corrosion substance of mild steel in acidic medium. J Mol Liq 293:111539. https://doi.org/10.1016/j.molliq.2019.111539

Bashir S, Lgaz H, Chung IM, Kumar A (2020a) Effective green corrosion inhibition of aluminium using analgin in acidic medium: an experimental and theoretical study. Chem Eng Commun. https://doi.org/10.1080/00986445.2020.1752680

Bashir S, Thakur A, Lgaz H, Chung IM, Kumar A (2020b) Corrosion inhibition efficiency of bronopol on aluminium in 0.5 M HCl solution: insights from experimental and quantum chemical studies. Surf Interfaces 20(April):100542. https://doi.org/10.1016/j.surfin.2020.100542

Chen T, Zou C, Pan J, Wang M, Qiao L, Wang F, Zhao Q, Cheng H, Ding C, Yuan Y (2022) Mapping research on microbial fuel cells in wastewater treatment: a co-citation analysis. Processes. https://doi.org/10.3390/pr10010179

Cheng Z, Xu D, Zhang Q, Tao Z, Hong R, Chen Y, Tang X, Zeng S, Wang S (2023) Enhanced nickel removal and synchronous bioelectricity generation based on substrate types in microbial fuel cell coupled with constructed wetland: performance and microbial response. Environ Sci Pollut Res 30(8):19725–19736. https://doi.org/10.1007/s11356-022-23458-y

Dai M, Zhang Y, Wu Y, Sun R, Zong W, Kong Q (2021) Mechanism involved in the treatment of sulfamethoxazole in wastewater using a constructed wetland microbial fuel cell system. J Environ Chem Eng 9(5):106193. https://doi.org/10.1016/j.jece.2021.106193

Das B, Thakur S, Chaithanya MS, Biswas P (2019) Batch investigation of constructed wetland microbial fuel cell with reverse osmosis (RO) concentrate and wastewater mix as substrate. Biomass Bioenergy 122(August 2018):231–237. https://doi.org/10.1016/j.biombioe.2019.01.017

Dhonchak C, Agnihotri N (2023) Computational insights in the spectrophotometrically 4H-chromen-4-one Complex using DFT method. Biointerface Res Appl Chem 13(4):357

Dhonchak C, Agnihotri N, Kumar A (2023) Spectrophotometric investigation and computational studies pyrazolyl]-4*H*-chromen-4-one complex. J Anal Chem 78(7):856–865. https://doi.org/10.1134/S1061934823070055

Doherty L, Zhao Y, Zhao X, Hu Y, Hao X, Xu L, Liu R (2015) A review of a recently emerged technology: constructed wetland—microbial fuel cells. Water Res 85:38–45. https://doi.org/10.1016/j.watres.2015.08.016

Ebrahimi A, Sivakumar M, McLauchlan C, Ansari A, Vishwanathan AS (2021) A critical review of the symbiotic relationship between constructed wetland and microbial fuel cell for enhancing pollutant removal and energy generation. J Environ Chem Eng 9(1):105011. https://doi.org/10.1016/j.jece.2020.105011

Elhenawy S, Khraisheh M, Almomani F, Al-Ghouti M, Hassan MK (2022) From waste to watts: updates on key applications of microbial fuel cells in wastewater treatment and energy production. Sustainability (Switzerland) 14(2):1–23. https://doi.org/10.3390/su14020955

Erazo S, Agudelo-Escobar LM (2023) Determination of electrogenic potential and removal of organic matter from industrial coffee wastewater using a native community in a non-conventional microbial fuel cell. Processes. https://doi.org/10.3390/pr11020373

Gadegaonkar SS, Philippon T, Rogińska JM, Mander Ü, Maddison M, Etienne M, Barrière F, Kasak K, Lust R, Espenberg M (2020) Effect of cathode material and its size on the abundance of nitrogen removal functional genes in microcosms of integrated bioelectrochemical-wetland systems. Soil Systems 4(3):1–15. https://doi.org/10.3390/soilsystems4030047

Ge X, Cao X, Song X, Wang Y, Si Z, Zhao Y, Wang W, Tesfahunegn AA (2020) Bioenergy generation and simultaneous nitrate and phosphorus removal in a pyrite-based constructed wetland–microbial fuel cell. Bioresour Technol 296(October 2019):12350. https://doi.org/10.1016/j.biortech.2019.122350

González T, Puigagut J, Vidal G (2021) Organic matter removal and nitrogen transformation by a constructed wetland–microbial fuel cell system with simultaneous bioelectricity generation. Sci Total Environ. https://doi.org/10.1016/j.scitotenv.2020.142075

Gupta S, Nayak A, Roy C, Yadav AK (2021) An algal assisted constructed wetland–microbial fuel cell integrated with sand filter for efficient wastewater treatment and electricity production. Chemosphere 263:128132. https://doi.org/10.1016/j.chemosphere.2020.128132

Han J, Yang Z, Wang H, Zhong H, Xu D, Yu S, Gao L (2021) Decomposition of pollutants from domestic sewage with the combination systems of hydrolytic acidification coupling with constructed wetland microbial fuel cell. J Clean Prod 319(August):128650. https://doi.org/10.1016/j.jclepro.2021.128650

Hartl M, Bedoya-Ríos DF, Fernández-Gatell M, Rousseau DPL, Du Laing G, Garfí M, Puigagut J (2019) Contaminants removal and bacterial activity enhancement along the flow path of constructed wetland microbial fuel cells. Sci Total Environ 652:1195–1208. https://doi.org/10.1016/j.scitotenv.2018.10.234

Jiménez González ML, Benítez CH, Juarez ZA, Pérez EZ, Coutiño VÁR, Robles I, Godínez LA, Rodríguez-Valadez FJ (2020) Study of the effect of activated carbon cathode configuration on the performance of a membrane-less microbial fuel cell. Catalysts. https://doi.org/10.3390/catal10060619

Jingyu H, Miwornunyuie N, Ewusi-Mensah D, Koomson DA (2020) Assessing the factors influencing the performance of constructed wetland–microbial fuel cell integration. Water Sci Technol 81(4):631–643. https://doi.org/10.2166/wst.2020.135

Kaur H, Thakur RC, Pathania VS, Sharma S (2022) Effect of choline-based ionic liquid (*Cholinium ethanoate*) on volumetric and acoustic properties of aliphatic amino acids (glycine and l-alanine). J Mol Liq 346:118247. https://doi.org/10.1016/j.molliq.2021.118247

Kaur H, Thakur RC, Singh N, Kumar A, Gill R, Thakur A, Sharma S (2023) Influence of 1-butyl-3-propylimidazolium bromide on viscometric and conductometric behavior of proteinogenic

amino acids (glycine and l-alanine). AIP Conf Proc 2800(1):20092. https://doi.org/10.1063/5.0163118
Kaya S, Thakur A, Kumar A (2023) The role of in silico/DFT investigations in analyzing dye molecules for enhanced solar cell efficiency and reduced toxicity. J Mol Graph Model 124(June):108536. https://doi.org/10.1016/j.jmgm.2023.108536
Kesarwani S, Panwar D, Mal J, Pradhan N, Rani R (2023) Constructed wetland coupled microbial fuel cell: a clean technology for sustainable treatment of wastewater and bioelectricity generation. Fermentation. https://doi.org/10.3390/fermentation9010006
Kumar A, Thakur A (2020) Encapsulated nanoparticles in organic polymers for corrosion inhibition. In: Corrosion protection at the nanoscale. Elsevier, London. https://doi.org/10.1016/b978-0-12-819359-4.00018-0
Leng C, Yuan Y, Zhang Z, Shi Q, Li F, Wang H (2023) Decomposition of phosphorus pollution and microorganism analysis using novel CW-MFCs under different influence factors. Molecules. https://doi.org/10.3390/molecules28052124
Li H, Xu H, Song HL, Lu Y, Yang XL (2020) Antibiotic resistance genes, bacterial communities, and functions in constructed wetland–microbial fuel cells: responses to the co-stresses of antibiotics and zinc. Environ Pollut. https://doi.org/10.1016/j.envpol.2020.115084
Li K, Qi J, Zhang F, Miwornunyuie N, Amaniampong PS, Koomson DA, Chen L, Yan Y, Dong Y, Setordjie VE, Samwini AMN (2021) The role of wetland plants on wastewater treatment and electricity generation in constructed wetland coupled with microbial fuel cell. Appl Sci (Switzerland). https://doi.org/10.3390/app11167454
Liu S, Song H, Li X, Yang F (2013) Power generation enhancement by utilizing plant photosynthate in microbial fuel cell coupled constructed wetland system. Int J Photoenergy. https://doi.org/10.1155/2013/172010
Liu S, Song H, Wei S, Yang F, Li X (2014) Bio-cathode materials evaluation and configuration optimization for power output of vertical subsurface flow constructed wetland—microbial fuel cell systems. Biores Technol 166:575–583. https://doi.org/10.1016/j.biortech.2014.05.104
Liu F, Sun L, Wan J, Tang A, Deng M, Wu R (2019) Organic matter and ammonia removal by a novel integrated process of constructed wetland and microbial fuel cells. RSC Adv 9(10):5384–5393. https://doi.org/10.1039/c8ra10625h
Lu L, Xing D, Ren ZJ (2015) Microbial community structure accompanied with electricity production in a constructed wetland plant microbial fuel cell. Biores Technol 195(June):115–121. https://doi.org/10.1016/j.biortech.2015.05.098
Mao Y, Zhao Y, Cotterill S (2023) Examining current and future applications of electrocoagulation in wastewater treatment. Water (Switzerland). https://doi.org/10.3390/w15081455
Mu C, Wang L, Wang L (2020) Performance of lab-scale microbial fuel cell coupled with unplanted constructed wetland for hexavalent chromium removal and electricity production. Environ Sci Pollut Res 27(20):25140–25148. https://doi.org/10.1007/s11356-020-08982-z
Mu C, Wang L, Wang L (2021) Removal of Cr(VI) and electricity production by constructed wetland combined with microbial fuel cell (CW-MFC): influence of filler media. J Clean Prod 320(March):128860. https://doi.org/10.1016/j.jclepro.2021.128860
Nastro RA, Flagiello F, Silvestri N, Gambino E, Falcucci G, Chandrasekhar K (2021) Use of biochar-based cathodes and increase in the electron flow by pseudomonas aeruginosa to improve waste treatment in microbial fuel cells. Processes 9(11):1–14. https://doi.org/10.3390/pr9111941
Ni H, Kaixuan W, Shuaishuai L, Xingxing W, Lu Z, Zhang J (2020) Effects of concentration variations on the performance and microbial community in microbial. Energies 13(9):2231
Nidhi, Kaur H (2022) Advancement in field of Energy storage applications by using ionic liquids. J Phys Conf Ser. https://doi.org/10.1088/1742-6596/2267/1/012043
Nidhi, Kaur H, Thakur RC, Kumar A (2023) Effect of temperature on conductance studies of 1,4-dimethyl-4*H*-1,2,4-triazolium iodide in binary aqueous solutions of benzylamine and benzamide. AIP Conf Proc 2800(1):20229. https://doi.org/10.1063/5.0163221

Oodally A, Gulamhussein M, Randall DG (2019) Investigating the performance of constructed wetland microbial fuel cells using three indigenous South African wetland plants. J Water Process Eng. https://doi.org/10.1016/j.jwpe.2019.100930

Parveen G, Bashir S, Thakur A, Saha SK, Banerjee P, Kumar A (2020) Experimental and computational studies of imidazolium based ionic liquid 1-methyl-3-propylimidazolium iodide on mild steel corrosion in acidic solution experimental and computational studies of imidazolium based ionic liquid 1-methyl-3-propylimidazolium. Mater Res Express 7(1):016510. https://doi.org/10.1088/2053-1591/ab5c6a

Pathania V, Sharma S, Vermani BK, Kaur H, Thakur RC (2023) Investigating the solvation behavior of some perchlorates in non-aqueous medium and predicting their ion-solvation thermodynamics using Born formula. J Mol Liq 371:121067. https://doi.org/10.1016/j.molliq.2022.121067

Paucar NE, Sato C (2021) Microbial fuel cell for energy production, nutrient removal and recovery from wastewater: a review. Processes. https://doi.org/10.3390/pr9081318

Paucar NE, Sato C (2022) An overview of microbial fuel cells within constructed wetland for simultaneous nutrient removal and power generation. Energies. https://doi.org/10.3390/en15186841

Ramírez-Vargas CA, Prado A, Arias CA, Carvalho PN, Esteve-Núñez A, Brix H (2018) Microbial electrochemical technologies for wastewater treatment: principles and evolution from microbial fuel cells to bioelectrochemical-based constructed wetlands. Water (Switzerland) 10(9):1–29. https://doi.org/10.3390/w10091128

Rana S, Thakur RC, Kaur H, Sharma A, Sharma R, Thakur V, Singh H, Fatma I (2023) Investigating the solvation behavior of some lithium salts in binary aqueous mixtures of 1-ethyl-3-methylimidazolium tetrafluoroborate ([EMIM][BF4]) at equidistant temperatures (T = 298.15, 303.15, 308.15, 313.15, 318.15) K. J Chem Eng Data 68(6):1291–1304. https://doi.org/10.1021/acs.jced.3c00037

Rojas-Flores S, Ramirez-Asis E, Delgado-Caramutti J, Nazario-Naveda R, Gallozzo-Cardenas M, Diaz F, Delfin-Narcizo D (2023) An analysis of global trends from 1990 to 2022 of microbial fuel cells: a bibliometric analysis. Sustainability (Switzerland). https://doi.org/10.3390/su15043651

Rojas-Villacorta W, Rojas-Flores S, Benites SM, Nazario-Naveda R, Romero CV, Gallozzo-Cardenas M, Delfín-Narciso D, Díaz F, Murga-Torres E (2023) Preliminary study of bioelectricity generation using lettuce waste as substrate by microbial fuel cells. Sustainability 15(13):10339. https://doi.org/10.3390/su151310339

Roy H, Rahman TU, Tasnim N, Arju J, Rafid MM, Islam MR, Pervez MN, Cai Y, Naddeo V, Islam MS (2023) Microbial fuel cell construction features and application for sustainable wastewater treatment. Membranes 13(5):1–29. https://doi.org/10.3390/membranes13050490

Samudro G, Imai T, Hung YT (2021) Enhancement of power generation and organic removal in double anode chamber designed dual-chamber microbial fuel cell (Dac-dcmfc). Water (Switzerland). https://doi.org/10.3390/w13212941

Sharma D, Thakur A, Sharma-Kumar M, Jakhar K, Kumar A, Sharma Kumar A, Om H (2023a) Synthesis, electrochemical, morphological, computational and corrosion inhibition studies of 3-(5-naphthalen-2-yl-[1,3,4]oxadiazol-2-yl)-pyridine against mild steel in 1 M HCl. Asian J Chem 35(5):1079–1088

Sharma D, Thakur A, Sharma MK, Sharma R, Kumar S, Sihmar A, Dahiya H, Jhaa G, Kumar A, Sharma AK, Om H (2023b) Effective corrosion inhibition of mild steel using novel 1,3,4-oxadiazole-pyridine hybrids: synthesis, electrochemical, morphological, and computational insights. Environ Res 234(July):116555. https://doi.org/10.1016/j.envres.2023.116555

Shi Y, Tang G, You S, Jiang P (2023) Effect of external aeration on Cr(VI) reduction in the leersia hexandra swartz constructed wetland–microbial fuel cell system. Appl Sci (Switzerland). https://doi.org/10.3390/app13053309

Srivastava P, Yadav AK, Mishra BK (2015) The effects of microbial fuel cell integration into constructed wetland on the performance of constructed wetland. Biores Technol 195:223–230. https://doi.org/10.1016/j.biortech.2015.05.072

Srivastava P, Yadav AK, Garaniya V, Abbassi R (2018) Constructed wetland coupled microbial fuel cell technology: development and potential applications. In: Biomass, biofuels, biochemicals: microbial electrochemical technology: sustainable platform for fuels, chemicals and remediation. Elsevier, London. https://doi.org/10.1016/B978-0-444-64052-9.00042-X

Takavakoglou V, Pana E, Skalkos D (2022) Constructed wetlands as nature-based solutions in the post-COVID agri-food supply chain: challenges and opportunities. Sustainability (Switzerland). https://doi.org/10.3390/su14063145

Tamta P, Rani N, Yadav AK (2020) Enhanced wastewater treatment and electricity generation using stacked constructed wetland–microbial fuel cells. Environ Chem Lett 18(3):871–879. https://doi.org/10.1007/s10311-020-00966-2

Tamta P, Rani N, Mittal Y, Yadav AK (2023) Evaluating the potential of multi-anodes in constructed wetlands coupled with microbial fuel cells for treating wastewater and bioelectricity generation under high organic loads. Energies. https://doi.org/10.3390/en16020784

Teoh TP, Ong SA, Ho LN, Wong YS, Oon YL, Oon YS, Tan SM, Thung WE (2020) Up-flow constructed wetland–microbial fuel cell: influence of floating plant, aeration and circuit connection on wastewater treatment performance and bioelectricity generation. J Water Process Eng. https://doi.org/10.1016/j.jwpe.2020.101371

Thakur A, Kumar A (2021) Sustainable inhibitors for corrosion mitigation in aggressive corrosive media: a comprehensive study. J Bio Tribo Corros 7(2):1–48. https://doi.org/10.1007/s40735-021-00501-y

Thakur A, Kumar A (2022) Recent advances on rapid detection and remediation of environmental pollutants utilizing nanomaterials-based (bio) sensors. Sci Total Environ 834(January):155219. https://doi.org/10.1016/j.scitotenv.2022.155219

Thakur A, Kumar A (2023a) Computational insights into the corrosion inhibition potential of some pyridine derivatives: a DFT approach. Eur J Chem 14(2):246–253. https://doi.org/10.5155/eurjchem.14.2.246-253.2408

Thakur A, Kumar A (2023b) Ecotoxicity analysis and risk assessment of nanomaterials for the environmental remediation. Macromol Symp 410(1):1–23. https://doi.org/10.1002/masy.202100438

Thakur A, Kumar A (2023c) Recent trends in nanostructured carbon-based electrochemical sensors for the detection and remediation of persistent toxic substances in real-time analysis. Mater Res Express 10:034001. https://doi.org/10.1088/2053-1591/acbd1a

Thakur RC, Sharma R, Kaur V, Kaur H (2019) Molecular interactions of Niacin in binary aqueous solutions of KCl and $FeCl_3$ at different temperatures: volumetric approach. Think India J 22(37):1604–1613

Thakur RC, Kaur R, Kaur H (2020) Transport properties of thiaminehydrochloridein binary aqueous mixtures of galactose. Eur J Mol Clin Med 7(07):2020

Thakur A, Kaya S, Kumar A (2021) Recent innovations in nano container-based self-healing coatings in the construction industry. Curr Nanosci 18(2):203–216. https://doi.org/10.2174/1573413717666210216120741

Thakur A, Kaya S, Abousalem AS, Kumar A (2022a) Experimental, DFT and MC simulation analysis of Vicia Sativa weed aerial extract as sustainable and eco-benign corrosion inhibitor for mild steel in acidic environment. Sustain Chem Pharm 29(July):100785. https://doi.org/10.1016/j.scp.2022.100785

Thakur A, Kaya S, Abousalem AS, Sharma S, Ganjoo R, Assad H, Kumar A (2022b) Computational and experimental studies on the corrosion inhibition performance of an aerial extract of Cnicus Benedictus weed on the acidic corrosion of mild steel. Process Saf Environ Prot 161:801–818. https://doi.org/10.1016/j.psep.2022.03.082

Thakur A, Kumar A, Kaya S, Marzouki R, Zhang F, Guo L (2022c) Recent advancements in surface modification, characterization and functionalization for enhancing the biocompatibility and corrosion resistance of biomedical implants. Coatings 12(10):1459. https://doi.org/10.3390/coatings12101459

Thakur A, Kumar A, Kaya S, Vo DVN, Sharma A (2022d) Suppressing inhibitory compounds by nanomaterials for highly efficient biofuel production: a review. Fuel 312(September 2021):122934. https://doi.org/10.1016/j.fuel.2021.122934

Thakur A, Kumar A, Sharma S, Ganjoo R, Assad H (2022e) Computational and experimental studies on the efficiency of *Sonchus arvensis* as green corrosion inhibitor for mild steel in 0.5 M HCl solution. Mater Today Proc 66:609–621. https://doi.org/10.1016/j.matpr.2022.06.479

Thakur A, Sharma S, Ganjoo R, Assad H, Kumar A (2022f) Anti-corrosive potential of the sustainable corrosion inhibitors based on biomass waste: a review on preceding and perspective research. J Phys Conf Ser 2267(1):012079. https://doi.org/10.1088/1742-6596/2267/1/012079

Thakur A, Savaş K, Kumar A (2023a) Recent trends in the characterization and application progress of nano-modified coatings in corrosion mitigation of metals and alloys. Appl Sci 13:730. https://doi.org/10.3390/app13020730

Thakur RC, Sharma A, Sharma R, Kaur H (2023b) A comparative analysis of volumetric, viscometric and conductometric properties of triethylmethylammonium tetrafluoroborate (TEMABF4) and tetraethylammonium tetrafluoroborate (TEABF4) in pure propylene carbonate (PC) and binary aqueous propylene carbonate. J Mol Liq 374:121244. https://doi.org/10.1016/j.molliq.2023.121244

Verma C, Thakur A, Ganjoo R, Sharma S, Assad H (2023) Coordination bonding and corrosion inhibition potential of nitrogen-rich heterocycles: azoles and triazines as specific examples. Coord Chem Rev 488(November 2022):215177. https://doi.org/10.1016/j.ccr.2023.215177

Villaseñor J, Capilla P, Rodrigo MA, Cañizares P, Fernández FJ (2013) Operation of a horizontal subsurface flow constructed wetland—microbial fuel cell treating wastewater under different organic loading rates. Water Res 47(17):6731–6738. https://doi.org/10.1016/j.watres.2013.09.005

Wang X, Tian Y (2021) Seasonal variations of pollutants removal and microbial activity in integrated constructed wetland–microbial fuel cell systems. Water Reuse 11(2):312–328. https://doi.org/10.2166/wrd.2021.094

Wang J, Song X, Wang Y, Bai J, Bai H, Yan D, Cao Y, Li Y, Yu Z, Dong G (2017a) Bioelectricity generation, contaminant removal and bacterial community distribution as affected by substrate material size and aquatic macrophyte in constructed wetland–microbial fuel cell. Biores Technol 245:372–378. https://doi.org/10.1016/j.biortech.2017.08.191

Wang J, Song X, Wang Y, Bai J, Li M, Dong G, Lin F, Lv Y, Yan D (2017b) Bioenergy generation and rhizodegradation as affected by microbial community distribution in a coupled constructed wetland–microbial fuel cell system associated with three macrophytes. Sci Total Environ 607–608:53–62. https://doi.org/10.1016/j.scitotenv.2017.06.243

Wang Y, Zhao Y, Xu L, Wang W, Doherty L, Tang C, Ren B, Zhao J (2017c) Constructed wetland integrated microbial fuel cell system: looking back, moving forward. Water Sci Technol 76(2):471–477. https://doi.org/10.2166/wst.2017.190

Wang G, Guo Y, Cai J, Wen H, Mao Z, Zhang H, Wang X, Ma L, Zhu M (2019a) Electricity production and the analysis of the anode microbial community in a constructed wetland–microbial fuel cell. RSC Adv 9(37):21460–21472. https://doi.org/10.1039/c8ra10130b

Wang X, Tian Y, Liu H, Zhao X, Peng S (2019b) Optimizing the performance of organics and nutrient removal in constructed wetland–microbial fuel cell systems. Sci Total Environ 653:860–871. https://doi.org/10.1016/j.scitotenv.2018.11.005

Wang X, Tian Y, Liu H, Zhao X, Wu Q (2019c) Effects of influent COD/TN ratio on nitrogen removal in integrated constructed wetland–microbial fuel cell systems. Biores Technol 271(September):492–495. https://doi.org/10.1016/j.biortech.2018.09.039

Wang W, Zhang Y, Li M, Wei X, Wang Y, Liu L, Wang H, Shen S (2020) Operation mechanism of constructed wetland–microbial fuel cells for wastewater treatment and electricity generation: a review. Biores Technol 314(May):123808. https://doi.org/10.1016/j.biortech.2020.123808

Wang L, Xu D, Zhang Q, Liu T, Tao Z (2022) Simultaneous removal of heavy metals and bioelectricity generation in microbial fuel cell coupled with constructed wetland: an optimization study

on substrate and plant types. Environ Sci Pollut Res 29(1):768–778. https://doi.org/10.1007/s11356-021-15688-3
Wang X, Zhang Y, Zhang C, Li J, Xue M, Xia W (2023) Influence of electrodes configuration on hydraulic characteristics of constructed wetland–microbial fuel cell systems using graphite rods and plates as electrodes. Sustainability (Switzerland). https://doi.org/10.3390/su15086397
Xu L, Zhao Y, Doherty L, Hu Y, Hao X (2016) Promoting the bio-cathode formation of a constructed wetland–microbial fuel cell by using powder activated carbon modified alum sludge in anode chamber. Sci Rep 6(May):1–9. https://doi.org/10.1038/srep26514
Xu F, Cao F, Kong Q, Zhou L, Yuan Q, Zhu Y, Wang Q, da Du Y, de Wang Z (2018) Electricity production and evolution of microbial community in the constructed wetland–microbial fuel cell. Chem Eng J 339(217):479–486. https://doi.org/10.1016/j.cej.2018.02.003
Xu F, Ouyang D, Rene ER, Ng HY, Guo L, Zhu Y, Zhou L, Yuan Q, Miao M, Wang Q, Kong Q (2019) Electricity production enhancement in a constructed wetland–microbial fuel cell system for treating saline wastewater. Bioresour Technol 288(March):121462. https://doi.org/10.1016/j.biortech.2019.121462
Xu F, Zhu Y, Wang Y, Chen H, Zhang Y, Hao D, Qi X, da Du Y, Wang B, Wang Q, Zhao C, Kong Q (2020) Coupling iron pretreatment with a constructed wetland–microbial fuel cell to improve wastewater purification and bioelectricity generation. J Clean Prod 276:123301. https://doi.org/10.1016/j.jclepro.2020.123301
Xu L, Yu W, Graham N, Zhao Y (2021) Revisiting the bioelectrochemical system based biosensor for organic sensing and the prospect on constructed wetland–microbial fuel cell. Chemosphere 264:128532. https://doi.org/10.1016/j.chemosphere.2020.128532
Yadav AK, Dash P, Mohanty A, Abbassi R, Mishra BK (2012) Performance assessment of innovative constructed wetland–microbial fuel cell for electricity production and dye removal. Ecol Eng 47:126–131. https://doi.org/10.1016/j.ecoleng.2012.06.029
Yadav AK, Srivastava P, Kumar N, Abbassi R, Mishra BK (2018) Constructed wetland–microbial fuel cell: an emerging integrated technology for potential industrial wastewater treatment and bio-electricity generation. Construct Wetlands Ind Wastew Treat. https://doi.org/10.1002/9781119268376.ch22
Yadav A, Jadhav DA, Ghangrekar MM, Mitra A (2022) Effectiveness of constructed wetland integrated with microbial fuel cell for domestic wastewater treatment and to facilitate power generation. Environ Sci Pollut Res 29(34):51117–51129. https://doi.org/10.1007/s11356-021-17517-z
Yang Y, Zhao Y, Tang C, Xu L, Morgan D, Liu R (2020) Role of macrophyte species in constructed wetland–microbial fuel cell for simultaneous wastewater treatment and bioenergy generation. Chem Eng J. https://doi.org/10.1016/j.cej.2019.123708
Yang Y, Zhao Y, Tang C, Liu R, Chen T (2021a) Dual role of macrophytes in constructed wetland–microbial fuel cells using pyrrhotite as cathode material: a comparative assessment. Chemosphere 263:128354. https://doi.org/10.1016/j.chemosphere.2020.128354
Yang Y, Zhao Y, Tang C, Mao Y, Chen T, Hu Y (2021b) Novel pyrrhotite and alum sludge as substrates in a two-tiered constructed wetland–microbial fuel cell. J Clean Prod 293:126087. https://doi.org/10.1016/j.jclepro.2021.126087
Yaqoob AA, Khatoon A, Setapar SHM, Umar K, Parveen T, Ibrahim MNM, Ahmad A, Rafatullah M (2020) Outlook on the role of microbial fuel cells in remediation of environmental pollutants with electricity generation. Catalysts. https://doi.org/10.3390/catal10080819
Zhang K, Wu X, Chen J, Wang W, Luo H, Chen W, Ma D, An X, Wei Z (2021) The role and related microbial processes of Mn-dependent anaerobic methane oxidation in reducing methane emissions from constructed wetland–microbial fuel cell. J Environ Manag 294(January):112935. https://doi.org/10.1016/j.jenvman.2021.112935
Zhang Y, Liu F, Lin Y, Sun L, Guo X, Yang S, He J (2022) Enhanced swine wastewater treatment by constructed wetland—microbial fuel cell systems. Water (Switzerland) 14(23):1–16. https://doi.org/10.3390/w14233930

Zhong F, Yu C, Chen Y, Wu X, Wu J, Liu G, Zhang J, Deng Z, Cheng S (2020) Nutrient removal process and cathodic microbial community composition in integrated vertical-flow constructed wetland–microbial fuel cells filled with different substrates. Front Microbiol 11(August):1–12. https://doi.org/10.3389/fmicb.2020.01896
Zhou Y, Xu D, Xiao E, Xu D, Xu P, Zhang X, Zhou Q, He F, Wu Z (2018) Relationship between electrogenic performance and physiological change of four wetland plants in constructed wetland–microbial fuel cells during non-growing seasons. J Environ Sci (China) 70:54–62. https://doi.org/10.1016/j.jes.2017.11.008

Application of Other MFC-Derived Technologies (e.g. MEC, PMFC, AMFC, SMFC) to Mitigate Water Pollution and Resource Recovery

Anand Dave

Abstract A technology known as a microbial fuel cell (MFC) incorporates the metabolic mechanisms of microorganisms to produce energy. MFCs may be utilised in water treatment facilities for wastewater because they may extract contaminants from the wastewater and turn its organic matter into power. By oxidising organic materials and dissolving contaminants, the microorganisms within the anode electrode produce electrons that go to the cathode compartment via an electrical connection. As an outcome of this process, clean water is also produced, which may be recycled or returned to the environment. Since Given that MFCs can generate electricity fromthe organic matter in wastewater, they extend a more efficient signifies of energy to typical treatment of wastewater plants by Efficiently monitoring the energy needs of the treatment facilities. The energy requirements of wastewater treatment plants can increase treatment costs All factors were considering and enhance greenhouse gas emissions. By enhancing energy efficiency and reducing operational costs and lowering greenhouse gas emissions as well as MFC emissionsin treatment facilities for wastewater can improve the stability of the processes involved in wastewater treatment. But since MFC The study is still in its early phases, there is still a lot to learn about how to develop it to a commercial scale. The fundamental structure and kinds of MFCs, as well as the materials and membrane used in their construction, their functioning mechanism, and important process components that affect how successful they are at work, are all covered in detail in this chapter. This chapter discusses the use of the challenges facing the broader use of this technology for sustainable wastewater treatment.

Graphical Abstract

A. Dave (✉)
Department of Microbiology, Seventh Day Adventist Arts and Science College, Ahmedabad 380008, India
e-mail: ananddave16@gmail.com

R. Selvasembian et al. (eds.), *Emerging Trends in Microbial Electrochemical Technologies for Sustainable Mitigation of Water Resources Contamination*,
https://doi.org/10.1007/978-3-031-74636-9_10

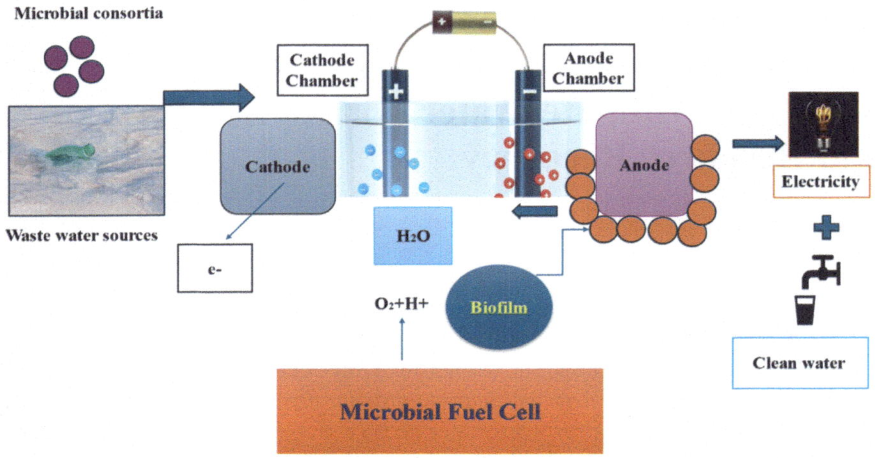

Keywords Microbial fuel cells · Greenhouse gas · Wastewater treatment · Microorganisms

1 Introduction

The industrial revolution led to a boom in urbanisation, which in turn caused an exponential rise in global energy consumption. Non-renewable energy sources are used was mandated by the increase in energy demand. However, the emphasis has switched to renewable sources that support sustainability because of the adverse effects they cause to the ecosystemand gradual depletion. These days, renewable energy sources that transform chemical energy in bonds into electrical energy are called fuel cells. The special class of biological fuel cells (BFCs) known as microbial fuel cells (MFCs) has demonstrated promise in the production of bioelectricity, the treatment of wastewater, and the bioremediation of heavy metals. Through a series of electrochemical processes, it transforms the chemical energy contained in organic or inorganic substance into electrical energy. MFC primarily consists of anodic and cathodic chambers divided by a salt bridge or proton exchange membrane (PEM). As biocatalysts, microorganisms oxidise the organic substrate at the anode while securing protons and electrons. Protons enter the cathodic chamber through the PEM, while electrons are guided there using an external circuit. Thus, protons and electrons combine with oxygen in the cathodic chamber to form a successful reduction process that yields water (Rabaey and Verstraete 2005; Enamala et al. 2020). Exoelectrogens are microorganisms that are employed as biocatalysts because of their remarkable properties. Since they may transport electrons to the anode surface and catalyse the electron acceptor (oxygen) reduction process at the cathode (Enamala et al. 2020; Jatoi et al. 2020). Due to several issues, including a short lifespan, poor output rate (Clauwaert et al. 2008), membrane fouling (Daud et al. 2019), high cost, and restricted efficiency, MFCs have not yet reached their optimal performance (Do et al. 2018; He et al. 2017). Because organic wastewater is abundant throughout the year and rich in nutrients, it is one of the best substrates for bioremediation. It is made up of wastewater from industrial, municipal, and other sources, which is regarded as the best way to gather energy to produce electricity (Tharali et al. 2016a, b). The main drawbacks of conventional wastewater treatment methods were their high energy and operating costs as well as their negative environmental effects. Traditional wastewater treatment methods are estimated to cost around 3% of the world's power consumption, with effluent disposal, or the disposal of sludge, accounting for 50% of the entire cost of wastewater treatment (Saba et al. 2017; Ye et al. 2019). Furthermore, the inadequate treatment of traditional wastewater leads to the release of hazardous dissolved compounds such as ammonia and phosphates, as well as greenhouse gases (Li and Li 2014). Microbial fuel cells hold the answer to addressing these issues. It is essential because it breaks down the organic content in wastewater and lowers the chemical oxygen demand (COD), which eliminates the need for effluent disposal and promotes environmental sustainability, low energy use, and cost. According to several studies, MFC removes COD with an effectiveness of between 80 and 90 percent (Catal et al. 2008; Gadhamshetty et al. 2013; Cui et al. 2014; Ye et al. 2019). Power output has been achieved by the treatment of wastewater from several sources (Hu 2008; Aldrovandi et al. 2009; Sangeetha and Muthukumar 2011; Thung et al. 2015).

1.1 Microbial Fuel Cells (MFCs)

Microorganisms can produce electrochemically active compounds in an MFC, which is a device that operates on the principle of electrochemical energy conversion. According to Allen and Bennetto (1993), these materials react on the anode to generate a biofilm that functions as a biocatalyst to produce energy. It is composed of two glass, plexiglass, or polycarbonate chambers with electrodes made of carbon paper, graphite, or carbon cloth, among other materials. Nitrogen gas is utilised to keep the anaerobic state in the anodic chamber, an anaerobic portion of the MFC. In this chamber, wastewater containing a variety of microorganisms will be treated electrochemically. The other is an aerobic component called a cathodic chamber, which is made up of an oxygen-based high potential electron acceptor electrode. Usually, the two chambers are kept apart by a proton exchange membrane composed of Ultrex and Nafion. Because they let cations flow through them, they serve as an efficient membrane for PEM fuel cells (Zhou et al. 2018). Using MFC to generate bioelectricity from wastewater One efficient way to store renewable energy is through wastewater. Microbial fuel cells are used to produce power at a reasonable cost and clean wastewater at the same time. Wastewater comes from food processing, animal husbandry, or human sources and is used as a substrate. In a typical MFC, a biochemical conversion system recovers the electrical energy potential of organic contaminants found in wastewater in the form of carbonaceous molecules, producing electricity (Tatinclaux et al. 2018). MFC is the perfect technology for producing bioelectricity because of its renewable biomass feature. Wastewater contains microorganisms that function as biocatalysts, oxidising organic or inorganic materials. In the process of creating and growing energy, they metabolise organic materials, producing protons, electrons, and CO_2. The process of microbial reduction generates electrons that are drawn to the anode. They bind to the anodic surface by a biological catalytic process facilitated by cytochromes or redox-active proteins. An external circuittypically a copper wirewill direct electrons towards the cathode. Protons that the microbes make will go via PEM in the direction of the cathode in the meanwhile. Electrons will be reduced to pure and clean water (H2O) by interacting with a strong oxidising agent like oxygen, causing protons to generate H2 (hydrogen gas) and hydroxyl (OH-) in the cathode chamber (Hamelers et al. 2010). Another oxidising agent that isn't as frequently utilised in the cathodic chamber is ferricyanide (Ucar et al. 2017). Platinum and palladium, two non-precious metals, catalyse the reduction of oxygen at the cathode, respectively (Cheng et al. 2006; Dombrovskis et al. 2016; Kumar et al. 2015). The equations that show the total reaction as well as the individual reactions occurring at the anodic and cathodic chambers are listed below.

$$\text{Anodic reaction: } CH_3COO - + H_2O\,2CO_2 + 8H + + 8e$$
$$\text{Cathodic reaction : } 8H + + 8e - + 2O_2 4\,H_2 0$$
$$\text{Overall reaction : } CH_3COO - + 2O_2\,2H_2O + 2CO_2$$

On a commercial scale, the water generated by MFCs may also be used to drive the turbines that produce bioelectricity. Water that has been stored is transported via tunnels and lands on the turbines. The generator is then turned by the turbine shaft to generate energy. Penstock is the name for the pressure pipe that carries water to the turbine. When the turbine needs to be shut down and evacuated for maintenance, a valve that can be closed must be kept at the top of the penstock during installations (Paish 2002).

1.2 Electron Transfer Mechanisms

Proteins and other biomolecules found in bacteria function as intermediaries in the transport of electrons between the bacteria and the electrodes. Extracellular electron transfer occurs via two primary mechanisms: indirect electron transfer, also known as Mediated Electron Transfer (MET), and direct electron transfer, also known as mediator-less electron transfer (DET) (Kumar et al. 2018). Extensive research has been conducted on the extracellular electron transfer methods of two bacteria,

Geobacter spp. and *Shewanella* spp. (Kumar et al. 2016). Microorganisms that participate in mediated electron transfer MFC lack surface proteins that are electrochemically active and may transfer electrons to the anode. As a result, to move these electrons between the electrode and the microbes, they need substances known as mediators or electro-active metabolites. Higher concentrations of these mediators make the transmitted electrons more available. However, mediators can be natural or manufactured, such as sulphate or sulphide, depending on the microorganism involved (Rahimnejad et al. 2015). Neutral red is a metaphor of a synthetic mediator. After passing through the plasma wall and lipid membrane of the outer cell, they begin to release electrons, which are often absorbed by oxygen and other intermediates. Subsequently, the decreased mediator exits the electron-loaded cell, leaving the transported electrons on the surface of the anode electrode. Consequently, a negative charge is acquired by the anode electrode. After releasing the electrons, the mediator reverts to its original oxidised state and is prepared to begin the process again. On the other hand, direct electron transfer is less expensive and non-toxic as it doesn't require a mediator. Without external electron carriers, mediator-less MFCs operate.

1.3 Mediator-Less Electron Transfer

Without the use of a shuttle or mediator, electrons are transferred directly from the microorganisms to the electrodes in a mediator-less or direct electron transfer method. The bacterial cell membrane and electrode must come into physical touch for this to work (Lovley 2006). Direct electron transmission is accomplished by conductive pili, self-assembled appendages known as nanowires, or outer membrane cytochromes (OMCs) (Reguera 2018). The outer surface of bacterial cells contains C-type cytochromes, which function as redox proteins and aid in the passage of electrons to terminal electron acceptors. An electrode was used as the only electron acceptor in research by Holmes et al. (2006) on *Geobacter sulfurreducens,* and it was shown that this led to a 19-fold increase in OMC transcripts, which encoded the OMCs needed for metal reduction. concluded that complex networks of OMCs seen in Shewanella and Geobacter participate in extracellular electron transfer, which raises the current densities in MFCs. The development of conductive pili, or nanowires, in exoelectrogenic bacteria that enable the microbes to transport electrons to a solid electron acceptor, is another method of generating direct electron transfer (Reguera et al. 2006). These nanowires enable electron transfer to terminal electron acceptors (TEAs) that are physically separated from the microorganisms by being connected to cytochromes found on bacterial membranes (Reguera 2018). Nanowires are an assembly of several cytochromes and outer membrane proteins, in contrast to flagella and pilli, which are homopolymers made up of a single component (Sure et al. 2016). As of right now, three different types of nanowires have been identified: type IV pili-based nanowires (Li and Li 2014); periplasmic protein- and outer membrane cytochrome-based nanowires (e.g., MtrC, Omc A) (Pirbadian et al. 2014). *G. sulfurreducens* and Synechocystis are the primary hosts of type IV nanowires, whereas Shewanella oneidensis is the host of nanowires containing MtrC and Omc A (Sure et al. 2016).

1.4 Microorganisms in MFCs

Because they carry out the biochemical processes in a microbial fuel cell, microorganisms are essential to the energy generation process. MFCs use two different kinds of microbes, such as oxidising and reducing electrodes. While electrode oxidising microbes are those that oxidise organic components and then transfer them to an electrode, electrode reducing microorganisms perform the reduction of electrodes and electron transport towards the anode (Fan and Xue 2016). Electrons can be transferred by pili on the outside of bacteria, proteins in the outer membrane, or a mediator molecule in the solution. On the other hand, microorganisms known as electrode oxidising bacteria are those that take electrons from the cathode and use them to break down other materials like carbon dioxide and acetate. The ability of the microorganisms to transmit electrons from substrate to anode must be taken into consideration while choosing them (Nawaz et al. 2020). Pure or mixed cultures of microorganisms can be employed in MFCs to generate power (Shehab et al. 2017). Most of the microbial population employed in MFC technology are pure cultivated species like Shewanella (Gorby et al. 2006) and Geobacter (Lovley et al. 1993). We still must find several more microbes that can contribute significantly to the production of power. Several kinds of pure-cultured microorganisms have been used in microbial fuel cells to produce electricity. Among the powerful eletricigens are: (1) Archaebacteria, a kind of bacteria that can survive harsh environments such high salt and temperature. Under some circumstances, they are necessary for producing power (Cao et al. 2019). (2) Acidobacteria, a broad class of acidophilic bacteria found in a range of habitats. A wide range of substrates may be consumed by them (Dedysh and Damsté 2018). (3) Cyanobacteria are photosynthetic microorganisms that are beneficial to the environment and are used in MFCs. Using light as a source of energy, they oxidise water to create electricity (Indrama et al. 2019). (4) The main class of bacteria utilised in MFCs that produce energy is called proteobacteria. Electrons may be transferred straight to the electrodes by most of its stresses. Strong electricigens are found in several species that fall within the end classifications of proteobacteria. Eukaryotes have not been extensively studied in MFCs. Because of their quick development rate, yeasts are among the most advantageous eukaryotic microorganisms for generating energy. It still produces less energy than certain types of bacteria, though. By creating strains of it that produce more energy, advancements have been achieved (Cao et al. 2019). Additionally, microorganisms belonging to the genera Clostridium (Park and Zeikus 2000), Shewanella (Kim et al. 2002), Geobacter (Reguera et al. 2006), and Pseudomonas are deserving exoelectrogens. To perform the oxidation of substrate, bacteria create biofilms and align themselves near the anode, where they use the anode as an electron receiver. To stick to the anode, these electroactive bacteria in a biofilm create internal components that form a matrix. According to

Marcus and Rittmann (2008), it is made up of carbohydrates, extracellular proteins, and bacterial cells that assist the bacteria in moving electrons to the anode surface. The electrostatic interaction of organic molecules with negative charges in nutritional buffer with bacterial waste also leads to the production of biofilms. Furthermore, the extracellular polymeric materials that the microbes create are a crucial component of biofilms. These (EPSs) are very large molecules that affect several physicochemical processes, such as electron transmission (Gunes 2021; Marsili et al. 2008).

1.5 MFC Design and Configurations Applied to Wastewater Treatment

The reactor's architecture affects MFC performance in addition to the biocatalysts. The electrodes, PEM or salt bridge, and external circuit (resistor) are the three main components of the fuel cell. The output of power, the cost, and the cell's efficiency will all be directly impacted by the design. For MFC to be designed efficiently, several parameters are needed, including size, shape, kind of electrode, and direction of stacking (Gul et al. 2021). Different MFC types exist, depending on the reactor architecture. The various MFC setups are covered in depth in the sections that follow. The MFC model can have one or two chambers. It can also be utilised in combination (Pant et al. 2010). Table 1 displays the power outputs of various MFC setups. The main reasons double chamber MFCs are employed are their high energy output, two independent chambers, and ion exchange membranes. Despite being a development from the double chamber, the single chamber has more positive outcomes (Call and Logan 2008).

Amounts influencing MFC output One of the most promising technical advancements for effectively treating wastewater and producing power at the same time is the microbial fuel cell (Bose et al. 2019). Many parameters, including the composition of the electrode and membrane, temperature, pH, substrate loading rate, biofilm, and external resistance, affect MFC performance and the amount of power it generates. It is necessary to optimise these elements to increase the energy production.

1.6 Anode

The anode compartment, often known as the MFC's heart, is one of the most important components of the microbial fuel cell device since it is where electrons are produced. In the anode compartment, oxidation of organic contaminants produces protons together with electrons, biofilm formation, and hydrogen creation in addition to electron production. The choice of an appropriate anode material is essential to improving MFC performance. High electrical conductivity, microbiological compatibility, chemical stability, good surface area, and cheap cost are all desirable qualities

Table 1 Possibility of producing bioelectricity in connection with MFC kinds and configurations

Types	Electrodes	Advantages	Disadvantages	Power density	References
Single chamber	A graphite fibre anode. Cathode: flexible carbon fabric coupled with PEM	Lower setup costs and increased power output	Unsuitable for industrial manufacturing	1.5 W/m^3	He et al. (2017)
Double chamber	Carbon electrodes	Ussed to increase the output of power	Cathodic chamber solution has to be changed on a regular basis	26 mW/m^2	Aelterman et al. (2006) and Rodrigo et al. (2007)
Flat plate	Anode made of carbon paper and cathode made of carbon fabric	Proton and electron diffusion rates are high. little separation between electrodes	Regular usage causes damage to membranes. oxygen permeability of a membrane	687.7 mW/m^2	Tran et al. (2019) and Min and Logan (2004)
Up flow	Composite made of graphite and cement	Greater COD removal is achieved by increased fluid recirculation, which also operates continuously, is simple to scale up, and has a high loading rate	Long distance between the cathode and anode, costly as fluid pumping requires a lot of energy	93 mW/m^2	Subha et al. (2019) and Arvaniti and Fountoulakis (2021)
Stacked	Electrodes made of carbon cloth	MFC's flexibility	Since they have not been thoroughly examined, a few variables, including height, form, and flow direction, must be noted	2.28 W/m^3 (series) 179.9 W/m^3 (parallel)	Zhang et al. (2017) and Zhang et al. (2017)

for the material to be used (Jaiswal et al. 2020). It is essential that the anode material be compatible with microorganisms. Numerous materials exhibit high electrical conductivity, including gold, copper, stainless steel, and platinum. As an alternative, carbonaceous anode materials that are extremely compatible and inexpensive have been used. These materials include carbon fibre, carbon cloth, carbon paper, carbon brush, carbon mesh, carbon felt, carbon rods, and carbon nanotubes. Because it is fibrous, carbon brush has a larger surface area than other carbon compounds. Carbon rods have a tiny surface area, are reasonably priced, and exhibit substantial electrical conductivity. It is essential that the anode material be compatible with microorganisms. Numerous materials, including gold, copper, stainless steel, and platinum, have high electrical conductivity; however, their use as anode materials is restricted due to their high cost and the fact that antimicrobial substances like cisplatin, which is derived from platinum, inhibit cellular growth (Munoz-Cupa et al. 2020; Slate et al. 2019).

1.7 Cathode

Protons and electrons produced at the anode of an MFC are collected by the cathode electrode. In a double chamber MFC, the cathode is divided by a dividing membrane, whereas in a single chamber MFC, the electrons created are received by an air cathode that is not separated by a membrane. When the electrons reach the cathode, they are lowered by the cathode's electron acceptors (Flimban et al. 2018). Because of its higher redox potential, oxygen is a widely desired electron acceptor. Its restricted utilisation stems from its poor electrode contact and higher power consumption relative to its supply. As an alternative, other electron acceptors have been used in place of oxygen (Ucar et al. 2017). Tetrachloroethane, sodium bromide, ferric ions, hexavalent chromium, perchlorate, nitrite, dichloroethane, potassium ferricyanide, and fumarate are a few of them (Jaiswal et al. 2020). Furthermore, catalysts like platinum are needed by the cathode for improved performance. But the method is expensive since these catalysts are expensive, have low sustainability, and take longer to prepare (Chaturvedi and Verma 2016). As an alternative, biocathodes have been developed, which use microorganisms as catalysts to take protons and electrons produced at the anode and carry out the reduction process. Biocathodes include a few enhancements as their use in the working process can produce useful materials and help get rid of extraneous chemical components. Moreover, oxygen from an external source is not required for biocathodes that use algae as a catalyst (Jung and Pandit 2019). The utilisation of microbes as biocatalysts in biocathodes is advantageous due to their cost-effectiveness, sustainability, ability to function efficiently at neutral pH, and potential for self-regeneration. These characteristics improve the reduction action at the cathode and augment the power generation potential of MFC (Nitisoravut et al. 2017). Other benefits include their capacity to be applied in biocathodes.

1.8 Temperature

Since temperature controls a variety of kinetic and thermodynamic variables, it has a significant effect on MFC performance. It has been discovered that temperature directly affects how well MFCs produce bioelectricity. According to Jatoi et al.

(2020), an increase in temperature causes an increase in power density output and a decrease in MFC resistance. The existence of microorganisms in the MFC's operating mechanism highlights the importance of temperature in comparison to conventional fuel cells. While certain psychrophiles, thermophiles, and hyperthermophilles are also electroactive microorganisms, the majority of electricigens are mesophilic bacteria (Butti et al. 2016). According to a study (Liu et al. 2011), microorganisms show their peak activity between 30 and 45 °C, which results in higher electric power generation in MFC. Temperature variations outside the ideal range may lead to inadequate biofilm growth, which lowers the MFC's capacity to produce bioelectricity. Moreover, temperature deviations beyond the ideal range can potentially physically harm microorganisms and change their physiological functions. The temperature at which various types of bacteria presents may form a biofilm varies; once the required temperature is reached, the bacterial population will proceed with its metabolic processes as needed (Jatoi et al. 2020; Song et al. 2017). Increased microbial growth and biofilm development that include enhanced electrochemical activity may be the cause of the improved voltage output seen during operation at elevated temperature ranges (Tee et al. 2017). A study looked at how the denitrification of wastewater treatment and the ensuing production of energy using dual-chamber MFC were affected by temperature ranges between 25 and 45 °C. Wang et al. (2018) stated that the maximum power production for electricity was seen at 35 °C, which resulted in the highest current density and coulombic efficiency of $8.12 \pm 0.04\%$ and $6.51\ \mathrm{A\ m^{-3}}$, respectively. Research using a dual-chamber MFC was conducted to examine the effects of various factors, such as temperature, on power productivity. According to Tremouli et al. (2017), it was found that raising the temperature from 24 to 35 °C increased power density and produced the highest MFC efficiency.

1.9 Biofilm

The basic component of MFC for treating wastewater and producing power afterward are microorganisms. Anodic biofilm in microbial fuel cells is very important as the growth and characteristics of the biofilm greatly influence MFC performance. The growth of biofilm at the anode is correlated with the production of bioelectricity in MFC. As the percentage of bacterial cells that function as catalysts in a thick biofilm is sufficiently enhanced, a rise in biofilm thickness and density encourages better substrate utilisation (Arbianti et al. 2018; Choudhury et al. 2017). Power generation in MFC increases because of the anode's polarisation resistance decreasing due to an increase in biofilm development (Baranitharan et al. 2015). Since most bacteria have a negative surface charge and are hydrophilic by nature, several enhancements have been made to facilitate the production of biofilms, such as the use of hydrophilic, positively charged anodes (Kalathil et al. 2017). It has been observed that using an enhanced hybrid biofilm that contains bacterial multiwalled carbon nanotubes can increase the power generation potential of MFCs. It was shown that hybrid biofilm significantly reduced the MFC operation's startup time by 53.8%. In comparison to the naturally produced biofilm, the hybrid biofilm exhibited better adsorption qualities and improved resistance to sudden changes in substrate concentration, which led to an increase in current density (Zhang et al. 2017). Electrode-based microbial modification's main goal is to produce microbial biofilm with high conductivity, activity, and penetrating power. By genetically modifying microorganisms to increase their electrogenic potential, MFCs that use wastewater can produce more electricity. Genetically modified microorganisms take the place of conventional microbial strains. The choice of an appropriate genetically modified strain is essential to guaranteeing prolonged MFC function. The formation of a biofilm on the air cathode in a single chamber microbial fuel cell as a result of exposure to scattered organic matter and the lack of a separating membrane has a negative effect on the air cathode's activity. The growth of biofilms on air cathodes obstructs the active sites of catalyst coatings, which can occasionally negatively impact catalyst performance due to extracellular materials. This can thus have an impact on the oxygen reduction potential. Furthermore, anodic microbial biofilm is competing with biofilm formation at the cathode for substrate utilisation. Aerobic bacterial growth that forms a biofilm impedes protons and charged particles from reaching the catalyst and stops hydroxide from diffusing from the cathode, which would otherwise raise the electrode's pH. Additionally, aerobic biofilms use up some oxygen, which influences the cathode's reduction process. Various strategies have been put out to avoid biofouling.

1.10 MFC Based Integrated Process

According to Xu et al. (2017), hybrid technologies based on MFC have recently been presented. These include desalination MFC (DS-MFC), built wetland MFC (CW-MFC), membrane-bioreactor MFC (MBR-MFC), sediment MFC (SMFC), and MFC-based denitrification systems. Of them, MBR-MFC is proficient in producing energy from renewable sources, and CW-MFC, SMFC, and DS-MFC offer unique potential for treating wastewater. The main obstacle, though, is to update them more widely while lowering their cost while maximising productivity. To make these hybrid technologies economically and commercially feasible for usage in the near future, research must be done to improve their stability, performance, and power output (Nawez et al. 2020).

1.11 Prospects

Because MFCs have so many advantages, including environmental sustainability, they are essential to the production of energy and the treatment of wastewater. To cut outrageous costs and power intake, the ideal design hasn't yet been created. Based on this analysis, it can be concluded that large-scale wastewater removal and high-power production give double chambered MFCs an advantage over single chambered

MFCs. Single chambered MFCs, on the other hand, show promising small-scale results since they need less internal resistance, which lowers the cost and increases efficiency. By stacking MFCs in parallel and series connections in vertical, horizontal, cylindrical, and flat-plate topologies, power generation is improved, bigger volumes result from bigger reactors, which then reduce power proportionately because of an increase in internal resistance. To overcome this issue, though, the surface area may be increased by connecting more MFCs in series. When combined, higher power outputs can be obtained than with separate electrodes or configurations. There isn't yet a set design that maximises wastewater treatment and power generation at the same time. Since every configuration has several benefits and drawbacks, scaling up a single or combinatory design can be difficult. There are still a lot of variables to be taken into account for MFC scalability in the future, such as electrode material, stacking orientation, material selection, reactor endurance, and cost-effectiveness. To reach to an optimal design, MFC performance must improve. This may be done by achieving a few goal targets. Increasing the electrogenecity of microorganisms, choosing the right electrode material, optimising operational parameters, efficiently recovering byproducts, minimising capital costs, and maximising power output are some of the goals. (vi) Putting MFC-based hybrid and integrated technologies to the market. Microorganisms play a critical role in the anode's electron transport and substrate breakdown. Mixed microflorae oxidise complex organics in wastewater more effectively than pure cultures. To boost the electrogenic activity of bacteria, researchers are genetically altering the microorganisms in future studies. improving the elimination of organic compounds and electron transport, which in turn improves MFC performance. The electrode material serves as a place for biofilm adhesion as well as an acceptor of electrons. It is crucial to choose the right electrode material. Pt electrodes are generally suitable for cathode and anode applications. Nonetheless, carbon-based compounds, such as graphite and PVC, are often used and have a strong attraction for microbes. Further investigation for electrode materials is needed for long-standing working of MFCs. To achieve optimum power output and COD elimination, operational factors such as pH, kind of microorganisms, temperature, and O2 flow rate are optimised. As a result, the large-scale process cost is directly impacted by the electrode material and MFC setup. The MFC arrangement lowers operating costs while the electrode material raises capital costs. The extraction of high-value byproducts such as heavy metals increases the viability of using MFCs for wastewater treatment. It is also possible to use other byproducts as energy sources, such as H2 and CH4. In the future, MFCs might be used to recover high-value byproducts more effectively and lessen the need for chemicals to precipitate these products. Heavy metal recovery, which yields high-value byproducts, substantially lowers the operational costs of MFCs used to clean wastewater. Activated sludge treatment is less expensive, but the capital cost is thirty times greater. Finding a balance between operating and capital costs is crucial for achieving sustainable wastewater treatment and bioelectricity production at the same time. Since the input power is substantially greater and the output power is insufficient to satisfy the needs of wastewater treatment alone, power output can be increased. More research is needed to boost energy efficiency by accumulating energy for treatment by storing it in batteries and capacitors. The development of cost-effective technologies is crucial to produce sustainable power and the breakdown of substrate for wastewater treatment. By creating new technologies or combining many MFC-based technologies already in use, waste and byproducts would be taken advantage of as additional energy sources. This lessens the environmental burden by minimising the carbon impact. Combining these technologies commercially would add up the entire output of renewable energy, making it economically feasible because renewable energy produced from a single technique is expensive. Significant research is being conducted on novel materials, MFC designs, and scaling strategies to bridge the gap between wastewater treatment and bioelectricity generating production and commercialization.

2 Conclusion

From this point of view, we suggest the idea of an MFC-centered hybrid process as an illustration to encourage the real-world use of MFCs and enhance the wastewater treatment process' sustainability. An analysis is conducted on the potential and constraints of MFC technology in fulfilling the various sustainability requirements. We describe a potential concept of a wastewater hybrid system centred on MFCs refining, which integrates MFC with membrane filtration, struvite precipitation, algae treatment, and AD among other treatment methods, in an appropriate manner. Problems and alternatives for developing the use of such hybrid processes and for practically scaling up MFCs are examined. The paper just offers a preliminary investigation into the potential of using MFCs as a sustainable wastewater treatment technology and to direct their advancement, given the paucity of available data about the practical application of MFCs and the many uncertainties surrounding future technological and economic development. More statistical data must be established before a quantitative assessment about MFCs' sustainability, and their hybrid processes can be made; additionally, field testing must be done to confirm the viability of the proposed applications.

Acknowledgements The author acknowledges the support of financial support of Department of Biosciences, Manipal University Jaipur in the form of DST-FIST project (DST/2022/1012) from Govt. of India, is gratefully acknowledged."

Declarations

Conflict of interest No potential conflict of interest was reported by the author.

References

Aelterman P, Rabaey K, Pham HT, Boon N, Verstraete W (2006) Continuous electricity generation at high voltages and currents using stacked microbial fuel cells. Environ Sci Technol 40(10):3388–3394. https://doi.org/10.1021/es0525511

Aldrovandi A, Marsili E, Stante L, Paganin P, Tabacchioni S, Giordano A (2009) Sustainable power production in a membrane-less and mediator-less synthetic wastewater microbial fuel cell. Bioresour Technol 100(13):3252–3260. https://doi.org/10.1016/J.BIORTECH.2009.01.041

Allen RM, Bennetto HP (1993) Microbial fuel-cells. Appl Biochem Biotechnol 39(1):27–40. https://doi.org/10.1007/BF02918975

Arbianti R, Utami TS, Leondo V, Putri SA, Hermansyah H (2018) Effect of biofilm and selective mixed culture on microbial fuel cell for the treatment of tempeh industrial wastewater. IOP Conf Ser Mater Sci Eng. https://doi.org/10.1088/1757-899X/316/1/012073

Arvaniti I, Fountoulakis MS (2021) Use of a graphite-cement composite as electrode material in up-flow constructed wetland-microbial fuel cell for greywater treatment and bioelectricity generation. J Environ Chem Eng 9(3):105158. https://doi.org/10.1016/j.jece.2021.105158

Baranitharan E, Khan MR, Prasad DMR, Teo WFA, Tan GYA, Jose R (2015) Effect of biofilm formation on the performance of microbial fuel cell for the treatment of palm oil mill effluent. Bioprocess Biosyst Eng 38(1):15–24. https://doi.org/10.1007/s00449-014-1239-9

Bose D, Sridharan S, Dhawan H, Vijay P, Gopinath M (2019) Biomass derived activated carbon cathode performance for sustainable power generation from Microbial Fuel Cells. Fuel 236:325–337. https://doi.org/10.1016/j.fuel.2018.09.002

Butti SK, Velvizhi G, Sulonen ML, Haavisto JM, Koroglu EO, Cetinkaya AY, Mohan SV (2016) Microbial electrochemical technologies with the perspective of harnessing bioenergy: maneuvering towards upscaling. Renew Sustain Energy Rev 53:462–476. https://doi.org/10.1016/j.rser.2015.08.058

Call D, Logan BE (2008) Hydrogen production in a single chamber microbial electrolysis cell lacking a membrane. Environ Sci Technol 42(9):3401–3406. https://doi.org/10.1021/es8001822

Cao Y, Mu H, Liu W, Zhang R, Guo J, Xian M, Liu H (2019) Electricigens in the anode of microbial fuel cells: pure cultures versus mixed communities. Microb Cell Factor 18(1):1–14. https://doi.org/10.1186/s12934-019-1087-z

Catal T, Li K, Bermek H, Liu H (2008) Electricity production from twelve monosaccharides using microbial fuel cells. J Power Sources 175(1):196–200. https://doi.org/10.1016/J.JPOWSOUR.2007.09.083

Chaturvedi V, Verma P (2016) Microbial fuel cell: a green approach for the utilization of waste for the generation of bioelectricity. Bioresour Bioprocess 3(1):1–14. https://doi.org/10.1186/s40643-016-0116-6

Cheng S, Liu H, Logan BE (2006) Increased performance of single-chamber microbial fuel cells using an improved cathode structure. Electrochem Commun 8(3):489–494. https://doi.org/10.1016/j.elecom.2006.01.010

Choudhury P, Prasad Uday US, Bandyopadhyay TK, Ray RN, Bhunia B (2017) Performance improvement of microbial fuel cell (MFC) using suitable electrode and bioengineered organisms: a review. Bioengineered 8(5):471–487. https://doi.org/10.1080/21655979.2016.1267883

Clauwaert P, Aelterman P, De Schamphelaire L, Carballa M, Rabaey K, Verstraete W (2008) Minimizing losses in bio-electrochemical systems: the road to applications. Appl Microbiol Biotechnol 79(6):901–913. https://doi.org/10.1007/s00253-008-1522-2

Cui Y, Rashid N, Hu N, Rehman MSU, Han JI (2014) Electricity generation and microalgae cultivation in microbial fuel cell using microalgae-enriched anode and bio-cathode. Energy Convers Manag 79:674–680. https://doi.org/10.1016/J.ENCONMAN.2013.12.032

Daud SM, Daud WRW, Bakar MHA, Kim BH, Somalu MR, Jahim JM, Muchtar A, Ghasemi M (2019) A comparison of long-term fouling performance by zirconia ceramic filter and cation exchange in microbial fuel cells. Int Biodeterior Biodegrad 136:63–70. https://doi.org/10.1016/j.ibiod.2018.11.001

Dedysh SN, Damsté JSS (2018) Acidobacteria. eLS. https://doi.org/10.1002/9780470015902.a0027685

Do MH, Ngo HH, Guo WS, Liu Y, Chang SW, Nguyen DD, Nghiem LB, Ni BJ (2018) Challenges in the application of microbial fuel cells to wastewater treatment and energy production: a mini review. Sci Total Environ 639:910–920. https://doi.org/10.1016/j.scitotenv.2018.05.136

Dombrovskis JK, Palmqvist AEC (2016) Recent progress in synthesis, characterization and evaluation of non-precious metal catalysts for the oxygen reduction reaction. Fuel Cells 16(1):4–22. https://doi.org/10.1002/fuce.201500123

Enamala MK, Dixit R, Tangellapally A, Singh M, Dinakarrao SMP, Chavali M, Chandrasekhar K (2020) Photosynthetic microorganisms (algae) mediated bioelectricity generation in microbial fuel cell: concise review. Environ Technol Innov 19:100959. https://doi.org/10.1016/j.eti.2020.100959

Fan LP, Xue S (2016) Overview on electricigens for microbial fuel cell. Open Biotechnol J. https://doi.org/10.2174/1874070701610010398

Flimban SG, Kim T, Ismail IMI, Oh SE (2018) Overview of microbial fuel cell (MFC) recent advancement from fundamentals to applications: MFC designs, major elements, and scalability. https://doi.org/10.20944/preprints201810.0763.v1

Gadhamshetty V, Belanger D, Gardiner CJ, Cummings A, Hynes A (2013) Evaluation of Laminaria-based microbial fuel cells (LbMs) for electricity production. Bioresour Technol 127:378–385. https://doi.org/10.1016/J.BIORTECH.2012.09.079

Gorby YA, Yanina S, McLean JS, Rosso KM, Moyles D, Dohnalkova A, Beveridge TJ, Chang IS, Kim BH, Kim KS et al (2006) Electrically conductive bacterial nanowires produced by *Shewanella oneidensis* strain MR-1 and other microorganisms. Proc Natl Acad Sci USA 103(30):11358–11363. https://doi.org/10.1073/pnas.0604517103

Gul H, Raza W, Lee J, Azam M, Ashraf M, Kim KH (2021) Progress in microbial fuel cell technology for wastewater treatment and energy harvesting. Chemosphere. https://doi.org/10.1016/j.chemosphere.2021.130828

Gunes B (2021) A critical review on biofilm-based reactor systems for enhanced syngas fermentation processes. Renew Sustain Energy Rev 143:110950. https://doi.org/10.1016/j.rser.2021.110950

Hamelers HV, Ter Heijne A, Sleutels TH, Jeremiasse AW, Strik DP, Buisman CJ (2010) New applications and performance of bioelectrochemical systems. App Microbiol. Biotechnol. 85(6):1673–1685. https://doi.org/10.1007/s00253-009-2357-1

He L, Du P, Chen Y, Lu H, Cheng X, Chang B, Wang Z (2017) Advances in microbial fuel cells for wastewater treatment. Renew Sustain Energy Rev 71:388–403. https://doi.org/10.1016/j.rser.2016.12.069

Holmes DE, Chaudhuri SK, Nevin KP, Mehta T, Methé BA, Liu A, Lovley DR (2006) Microarray and genetic analysis of electron transfer to electrodes in *Geobacter sulfurreducens*. Environ Microbiol 8(10):1805–1815. https://doi.org/10.1111/j.1462-2920.2006.01065.x

Hu Z (2008) Electricity generation by a baffle-chamber membraneless microbial fuel cell. J Power Sources 1(179):27–33. https://doi.org/10.1016/J.JPOWSOUR.2007.12.094

Indrama T, Tiwari ON, Bandyopadhyay TK, Mondal A, Bhunia B (2019) Cyanobacteria: a biocatalyst in microbial fuel cell for sustainable electricity generation. Waste Sustain Energy 2019:125–140

Jaiswal KK, Kumar V, Vlaskin MS, Sharma N, Rautela I, Nanda M, Aora N, Singh A, Chauhan PK (2020) Microalgae fuel cell for wastewater treatment: recent advances and challenges. J Water Process Eng 38:101549. https://doi.org/10.1016/j.jwpe.2020.101549

Jatoi AS, Akhter F, Mazari SA, Sabzoi N, Aziz S, Soomro SA, Mubarak NM, Baloch H, Memon AQ, Ahmed S (2020) Advanced microbial fuel cell for waste water treatment—a review. Environ Sci Pollut Res. https://doi.org/10.1007/s11356-020-11691-2

Jung SP, Pandit S (2019) Important factors influencing microbial fuel cell performance. Microb Electrochem Technol. https://doi.org/10.1016/B978-0-444-64052-9.00015-7

Kalathil S, Patil SA, Pant D (2017) Microbial fuel cells: electrode materials. Encyclop Interfac Chem Surf Sci Electrochem. https://doi.org/10.1016/B978-0-12-409547-2.13459-6

Kim HJ, Park HS, Hyun MS, Chang IS, Kim M, Kim BH (2002) A mediator-less microbial fuel cell using a metal reducing bacterium, *Shewanella putrefaciens*. Enzym Microb Technol 30(2):145–152. https://doi.org/10.1016/S0141-0229(01)00478-1

Kumar R, Singh L, Wahid ZA, Din MFM (2015) Exoelectrogens in microbial fuel cells toward bioelectricity generation: a review. Int J Energy Res 39(8):1048–1067. https://doi.org/10.1002/er.3305

Kumar R, Singh L, Zularisam AW (2016) Exoelectrogens: recent advances in molecular drivers involved in extracellular electron transfer and strategies used to improve it for microbial fuel cell applications. Renew Sustain Energy Rev 56:1322–1336. https://doi.org/10.1016/j.rser.2015.12.029

Kumar R, Singh L, Zularisam AW, Hai FI (2018) Microbial fuel cell is emerging as a versatile technology: a review on its possible applications, challenges and strategies to improve the performances. Int J Energy Res 42(2):369–394. https://doi.org/10.1002/er.3780

Li Y, Li H (2014) Type IV pili of Acidithiobacillus ferrooxidans can transfer electrons from extracellular electron donors. J Basic Microbiol 54(3):226–231. https://doi.org/10.1002/jobm.201200300

Liu Y, Climent V, Berná A, Feliu JM (2011) Effect of temperature on the catalytic ability of electrochemically active biofilm as anode catalyst in microbial fuel cells. Electroanalysis 23(2):387–394. https://doi.org/10.1002/elan.201000499

Lovley DR (2006) Microbial fuel cells: novel microbial physiologies and engineering approaches. Curr Opin Biotechnol 17(3):327–332. https://doi.org/10.1016/j.copbio.2006.04.006

Lovley DR, Giovannoni SJ, White DC, Champine JE, Phillips EJP, Gorby YA, Goodwin S (1993) Geobacter metallireducens gen. nov. sp. nov., a microorganism capable of coupling the complete oxidation of organic compounds to the reduction of iron and other metals. Arch Microbiol 159(4):336–344. https://doi.org/10.1007/BF00290916

Marcus AK, Rittmann B (2008) New insights into fuel cell that uses bacteria to generate electricity from waste. News Archive of the Biodesign Institute, Arizona State University

Marsili E, Baron DB, Shikhare ID, Coursolle D, Gralnick JA, Bond DR (2008) Shewanella secretes flavins that mediate extracellular electron transfer. Proc Natl Acad Sci USA 105(10):3968–3973. https://doi.org/10.1073/pnas.0710525105

Min B, Logan BE (2004) Continuous electricity generation from domestic wastewater and organic substrates in a flat plate microbial fuel cell. Environ Sci Technol 38(21):5809–5814. https://doi.org/10.1021/es0491026

Munoz-Cupa C, Hu Y, Xu CC, Bassi A (2020) An overview of microbial fuel cell usage in wastewater treatment, resource recovery and energy production. Sci Total Environ. https://doi.org/10.1016/j.scitotenv.2020.142429

Nawaz A, Hafeez A, Abbas SZ, Haq IU, Mukhtar H, Rafatullah M (2020) A state-of-the-art review on electron transfer mechanisms, characteristics, applications and recent advancements in microbial fuel cells technology. Green Chem Lett Rev 13(4):365–381. https://doi.org/10.1080/17518253.2020.1854871

Nitisoravut R, Thanh CN, Regmi R (2017) Microbial fuel cells: advances in electrode modifications for improvement of system performance. Int J Green Energy 14(8):712–723. https://doi.org/10.1080/15435075.2017.1326049

Paish O (2002) Small hydro power: technology and current status. Renew Sustain Energy Rev 6(6):537–556. https://doi.org/10.1016/S1364-0321(02)00006-0

Pant D, Van Bogaert G, Diels L, Vanbroekhoven K (2010) A review of the substrates used in microbial fuel cells (MFCs) for sustainable energy production. Bioresour Technol 101(6):1533–1543. https://doi.org/10.1016/j.biortech.2009.10.017

Park DH, Zeikus JG (2000) Electricity generation in microbial fuel cells using neutral red as an electronophore. Appl Environ Microbiol 66(4):1292–1297. https://doi.org/10.1128/AEM.66.4.1292-1297.2000

Pirbadian S, Barchinger SE, Leung KM, Byun HS, Jangir Y, Bouhenni RA, ElNaggar MY (2014) Shewanella oneidensis MR-1 nanowires are outer membrane and periplasmic extensions of the extracellular electron transport components. Proc Natl Acad Sci USA 111(35):12883–12888. https://doi.org/10.1073/pnas.1410551111

Rabaey K, Verstraete W (2005) Microbial fuel cells: novel biotechnology for energy generation. Trends Biotechnol 23:291–298. https://doi.org/10.1016/j.tibtech.2005.04.008

Rahimnejad M, Adhami A, Darvari S, Zirepour A, Oh SE (2015) Microbial fuel cell as new technology for bioelectricity generation: a review. Alex Eng J 54(3):745–756. https://doi.org/10.1016/j.aej.2015.03.031

Reguera G (2018) Microbial nanowires and electroactive biofilms. FEMS Microbiol Ecol 94(7):fiy086. https://doi.org/10.1093/femsec/fy086

Reguera G, Nevin KP, Nicoll JS, Covalla SF, Woodard TL, Lovley DR (2006) Biofilm and nanowire production leads to increased current in Geobacter sulfurreducens fuel cells. Appl Environ Microbiol 72(11):7345–7348. https://doi.org/10.1128/AEM.0144

Rodrigo MA, Canizares P, Lobato J, Paz R, Sáez C, Linares JJ (2007) Production of electricity from the treatment of urban waste water using a microbial fuel cell. J Power Sources 169(1):198–204. https://doi.org/10.1016/j.jpowsour.2007.01.054

Saba B, Christy AD, Yu Z, Co AC (2017) Sustainable power generation from bacterio-algal microbial fuel cells (MFCs): an overview. Renew Sustain Energy Rev 73:75–84. https://doi.org/10.1016/J.RSER.2017.01.115

Sangeetha T, Muthukumar M (2011) Catholyte performance as an influencing factor on electricity production in a dual-chambered microbial fuel cell employing food processing wastewater. Energy Sources Part A Rec Util Environ Eff 33(16):1514–1522. https://doi.org/10.1080/15567030903397966

Shehab NA, Ortiz-Medina JF, Katuri KP, Hari AR, Amy G, Logan BE, Saikaly PE (2017) Enrichment of extremophilic exoelectrogens in microbial electrolysis cells using Red Sea brine pools as inocula. Bioresour Technol 239:82–86. https://doi.org/10.1016/j.biortech.2017.04.122

Slate AJ, Whitehead KA, Brownson DA, Banks CE (2019) Microbial fuel cells: an overview of current technology. Renew Sustain Energy Rev 101:60–81. https://doi.org/10.1016/j.rser.2018.09.044

Song Y, An J, Chae KJ (2017) Effect of temperature variation on the performance of microbial fuel cells. Energy Technol 5(12):2163–2167. https://doi.org/10.1002/ente.201700277

Subha C, Kavitha S, Abisheka S, Tamilarasan K, Arulazhagan P, Banu JR (2019) Bioelectricity generation and effect studies from organic rich chocolaterie wastewater using continuous upflow anaerobic microbial fuel cell. Fuel 251:224–232. https://doi.org/10.1016/j.fuel.2019.04.052

Sure S, Ackland ML, Torriero AA, Adholeya A, Kochar M (2016) Microbial nanowires: an electrifying tale. Microbiology 162(12):2017–2028. https://doi.org/10.1099/mic.0.000382

Tatinclaux M, Gregoire K, Leininger A, Biffinger JC, Tender L, Ramirez M, Kjellerup BV (2018) Electricity generation from wastewater using a floating air cathode microbial fuel cell. Water Energy Nexus 1(2):97–103. https://doi.org/10.1016/j.wen.2018.09.001

Tee PF, Abdullah MO, Tan IA, Amin MA, Nolasco-Hipolito C, Bujang K (2017) Effects of temperature on wastewater treatment in an affordable microbial fuel cell-adsorption hybrid system. J Environ Chem Eng 5(1):178–188. https://doi.org/10.1016/j.jece.2016.11.040

Tharali AD, Sain N, Osborne WJ (2016a) Microbial fuel cells in bioelectricity production. Front Life Sci 9(4):252–266. https://doi.org/10.1080/21553769.2016.1230787

Tharali AD, Sain N, Jabez Osborne W (2016b) Microbial fuel cells in bioelectricity production. Front Life Sci 9(4):252–266. https://doi.org/10.1080/21553769.2016.1230787

Thung W-E, Ong S-A, Ho L-N, Wong Y-S, Oon Y-L, Oon Y-S, Lehl HK (2015) Simultaneous wastewater treatment and power generation with innovative design of an upflow membraneless microbial fuel cell. Water Air Soil Pollut 226(5):1–7. https://doi.org/10.1007/S11270-015-2410-X

Tran TV, Lee IC, Kim K (2019) Electricity production characterization of a sediment microbial fuel cell using different thermo-treated flat carbon cloth electrodes. Int J Hydrog Energy 44(60):32192–32200. https://doi.org/10.1016/j.ijhydene.2019.10.076

Tremouli A, Martinos M, Lyberatos G (2017) The effects of salinity, pH and temperature on the performance of a microbial fuel cell. Waste Biomass Valoriz 8(6):2037–2043. https://doi.org/10.1007/s12649-016-9712-0

Ucar D, Zhang Y, Angelidaki I (2017) An overview of electron acceptors in microbial fuel cells. Front Microbiol 8:643. https://doi.org/10.3389/fmicb.2017.00643

Wang S, Zhao J, Liu S, Zhao R, Hu B (2018) Effect of temperature on nitrogen removal and electricity generation of a dual-chamber microbial fuel cell. Water Air Soil Pollut 229(8):1–13. https://doi.org/10.1007/s11270-018-3840-z

Xu L, Zhao Y, Fan C, Fan Z, Zhao F (2017) First study to explore the feasibility of applying microbial fuel cells into constructed wetlands for COD monitoring. Bioresour Technol 243:846–854. https://doi.org/10.1016/j.biortech.2017.06.179

Ye Y, Ngo HH, Guo W, Chang SW, Nguyen DD, Liu Y, Ni B, Zhang X (2019) Microbial fuel cell for nutrient recovery and electricity generation from municipal wastewater under different ammonium concentrations. Bioresour Technol 292:121992. https://doi.org/10.1016/J.BIORTECH.2019.121992

Zhang P, Liu J, Qu Y, Zhang J, Zhong Y, Feng Y (2017) Enhanced performance of microbial fuel cell with a bacteria/multi-walled carbon nanotube hybrid biofilm. J Power Sources 361:318–325. https://doi.org/10.1016/j.jpowsour.2017.06.069

Zhou Y, Zhu N, Guo W, Wang Y, Huang X, Wu P, Xian J (2018) Simultaneous electricity production and antibiotics removal by microbial fuel cells. J Environ Manag 217:565–572. https://doi.org/10.1016/j.jenvman.2018.04.013

Integrating Microbial Electrochemical Technologies with Other Biological Systems for the Remediation of Water Pollution and Resource Recovery

Nur Izyan Wan Azelee, Akmali Mokhter, Nurrulhidayah Salamun, Rohaida Che Man, Zaitul Iffa Abd Rasid, Noor Hidayah Abd Rahman, and Ida Madiha Yusoff

Abstract The increasing presence of emerging contaminants in water resources poses a significant threat to human health and the environment. Numerous biological systems have been employed for addressing the emerging contaminants however, the efficacy and cost effectiveness of the systems remained intricate. The most recent trend in wastewater treatment is utilizing microbial electrochemical technologies (MET) where microbial activity is harnessed to generate energy while simultaneously treating the wastewater. Moreover, several studies have been performed and evaluated their contaminants removal from diverse water resources. This chapter explores the prevailing biological systems utilized for remediation of water pollutions. Three types of innovative integrations are highlighted and discussed in the chapter which

N. I. W. Azelee (✉) · N. H. A. Rahman
Faculty of Chemical and Energy Engineering, Universiti Teknologi Malaysia (UTM), 81310 Johor Bahru, Johor, Malaysia
e-mail: nur.izyan@utm.my

N. I. W. Azelee · I. M. Yusoff
Institute of Bioproduct Development (IBD), Universiti Teknologi Malaysia, UTM, 81310 Johor Bahru, Johor, Malaysia

A. Mokhter · N. Salamun
Department of Chemistry, Faculty of Science, Universiti Teknologi Malaysia (UTM), Johor Bahru, Malaysia

A. Mokhter
Advanced Membrane Technology Research Centre (AMTEC), School of Chemical and Energy Engineering, Faculty of Engineering, Universiti Teknologi Malaysia, Johor, Malaysia

R. C. Man
Faculty of Chemical and Process Engineering Technology, Universiti Malaysia Pahang, Lebuhraya Tun Razak, 26300 Gambang, Pahang, Malaysia

Z. I. A. Rasid
UTM Research Ethics Committee, Department of Vice-Chancellor (Research and Innovation), Universiti Teknologi Malaysia, 81310 Johor Bahru, Johor, Malaysia

R. Selvasembian et al. (eds.), *Emerging Trends in Microbial Electrochemical Technologies for Sustainable Mitigation of Water Resources Contamination*,
https://doi.org/10.1007/978-3-031-74636-9_11

are integrated microbial fuel cell technology (MFC), integrated microbial electrolysis cell technology (MEC) and integrated microbial desalination cell technologies (MDC). While highlighting these innovative technologies, this chapter also addresses the challenges faced especially in the real implementation such as scalability, cost efficacy, overall system optimizations and long-term durability of the integrated technology. This chapter offers insights into new innovative technologies for sustainable water remediation and resource recovery.

Keywords Microbial electrochemical technologies · Microbial desalination cells · Microbial fuel cells · Microbial electrolysis cell

1 Introduction

Water pollution is an escalating global issue with far-reaching consequences for the environment and public health (Lin FY et al. 2022, Lin L et al. 2022). The increased demand for water brought on by population growth, industrialization, and agricultural development might be satisfied by limiting water pollution and improving wastewater treatment (Mateo-Sagasta et al. 2017). In recent years, the increasing challenges associated with water resource contamination have spurred a growing interest in innovative technologies capable of addressing these environmental concerns. The complexity of water contamination issues has been heightened by the emergence of novel toxins introduced by various industries. These contaminants, which range from the residues of agricultural processes to the residues of cosmeceutical products, are difficult to classify and require advanced remediation techniques (Khan et al. 2022).

Most electrochemical water treatment techniques employ electricity as the primary reactant to facilitate the treatment procedure (Sillanpää and Shestakova 2017). To date, researchers are continuously exploring innovative approaches to overcome the problems posed by conventional technologies. Conventional electrochemical technologies have long been employed in wastewater treatment, relying on chemical processes for pollutant removal (Nagarajan et al. 2023). However, the shortcomings of these traditional methods, including high energy consumption, dependence on expensive catalysts, and the generation of environmentally harmful by-products, have prompted researchers to explore alternative approaches. Among the emerging trends in this field, microbial electrochemical technologies (METs) have garnered significant attention for their potential to revolutionize wastewater treatment processes, representing a paradigm shift from conventional electrochemical technologies (Mohammadi et al. 2023).

The first documentation of microorganisms' ability to produce electricity through their metabolic activities can be attributed to the English botanist Potter in 1911. Utilizing platinum electrodes in a galvanic cell, Potter measured the electric potential of both *Saccharomyces cerevisiae* and *Escherichia coli,* attributing the observed phenomenon to the breakdown of organic compounds under specific conditions

(Hernandez and Osma 2020). In the past 20 years, microbial electrochemical technologies (METs) have witnessed a significant surge in scientific publications across various fields, including water treatment, chemical synthesis or catalysis, energy production, and sensor applications (Holtmann and Harnisch 2018; Mohan et al. 2019).

Microbial electrochemical technologies represent a promising intersection of microbiology and electrochemistry. It refers to the utilization of microbial-driven electrochemical processes for the production of complex organic compounds, environmental solutions and energy generation (Ramírez-Vargas et al. 2018). The two predominant approaches for generating bioelectricity involve microbial desalination cells (MDCs) and microbial fuel cells (MFCs), which can utilize a wide range of soluble or dissolved complex organic molecules. This substrate encompasses solid wastes, wastewaters, and renewable biomass, serving as the anolyte (Zhang et al. 2016). The fundamental principle of microbial electrochemical technology (MET) depends on the capacity of microorganisms to convert chemical energy into electrical energy. According to Schröder et al. (2015), microbial electrochemistry entails the exploration and application of interactions between living microbial cells and electrodes, facilitated through either capacitive or Faraday interactions. In this mechanism, electrons function as the intermediary currency, connecting the microbial cell to the electrical interface through inherent mechanisms of the respiratory process, as mentioned by Schoepp-Cothenet et al. (2013). These technologies leverage the unique abilities of microorganisms to catalyze electrochemical reactions, offering a novel approach to treat wastewater while simultaneously producing energy (Domínguez-Garay and Esteve-Núñez 2018). Microbial electrochemical systems present a multifaceted solution to water contamination, deriving benefits for both the environment and energy utilization through intricate interactions between microorganisms and electrodes. Since 2012, numerous studies have explored the technical feasibility and advantages of integrating MET with constructed wetlands (CW) for both wastewater treatment and energy production. Constructed wetlands are engineered systems that utilize plants, microorganisms, and various processes for wastewater treatment (Yakar et al. 2018).

A key strength of microbial electrochemical technologies lies in their ability to utilize naturally occurring microorganisms to catalyze electrochemical reactions. Microbes, such as certain species of bacteria, can directly transfer electrons to electrodes, driving redox reactions that facilitate the breakdown of organic pollutants (Hernandez and Osma 2020). This biological-electrochemical synergy not only minimizes the need for external energy inputs but also reduces the reliance on synthetic chemicals, contributing to a more sustainable approach to wastewater treatment. Furthermore, microbial electrochemical technologies offer versatility in treating a wide range of contaminants. Sonawane et al. (2022) have demonstrated the effectiveness of microbial fuel cells and microbial electrolysis cells in removing diverse pollutants, including heavy metals and pharmaceutical residues, showcasing the potential for broad-spectrum applications in water treatment.

One notable aspect of this integration involves combining METs with natural microbial found in diverse environments. Natural environments harbor diverse microbial populations with unique metabolic capabilities. By integrating METs with these communities, researchers aim to capitalize on the synergies between engineered electrochemical processes and the inherent abilities of native microorganisms. This collaborative approach holds potential for improving the efficiency and adaptability of water remediation processes, as well as facilitating resource recovery (Shi et al. 2022). Apart from that, integration with anaerobic digestion systems can enhance the breakdown of complex organic compounds. The by-products of METs, such as hydrogen and methane, can serve as substrates for anaerobic bacteria, promoting the production of biogas and aiding in the removal of persistent pollutants. Besides that, the combination of METs with plant-based systems, known as phytoremediation, has shown significant promise. Phytoremediation, an emerging technology, employs plants and their associated rhizosphere microorganisms to eliminate, transform, or contain hazardous chemicals present in soils, sediments, groundwater, surface water, and even the atmosphere (Nedjimi 2021). Plants can actively absorb and accumulate pollutants from the environment, a process that can be augmented by the electrochemical stimuli provided by METs (Zhang et al. 2016). The synergy between plant-driven processes and microbial electrochemical reactions creates a powerful tool for removing contaminants and promoting sustainable water management (Yang et al. 2023).

METs offer numerous advantages in addressing water pollution and facilitating resource recovery. A significant advantage is their energy-efficient design, harnessing the metabolic activities of microorganisms to simultaneously generate electricity during the remediation process (Butti et al. 2016). This dual functionality enables both wastewater treatment and sustainable energy production. METs demonstrate versatility in effectively addressing a wide range of pollutants, including organic contaminants and heavy metals. Integration with natural microbial communities enhances system efficiency by leveraging the diverse metabolic capabilities of indigenous microorganisms (Das et al. 2023). Furthermore, METs contribute to environmental sustainability by reducing sludge production, a common by-product in traditional wastewater treatment. The collaborative synergy between METs and phytoremediation, where plants assist in pollutant removal, enhances the overall effectiveness of water treatment (Khan et al. 2022). Additionally, METs play a role in resource recovery by capturing energy in the form of electricity and extracting valuable chemicals or nutrients from wastewater, aligning with the principles of a circular economy (Butti et al. 2016).

Despite their potential, METs encounter challenges that limit their widespread adoption and effectiveness. Low power densities, stemming from the need for enhanced electron transfer efficiency and cathode reactions, pose a fundamental limitation (Zhang et al. 2016; Butti et al. 2016). For instance, the scalability of microbial electrochemical technologies remains a key concern, with many existing systems designed for laboratory-scale applications (Patil et al. 2012). Additionally, the narrow substrate specificity of some microbial communities restricts the range of pollutants that can be efficiently treated. These challenges have prompted researchers to explore strategies for enhancing the performance and applicability of METs, and one promising avenue is the integration with other biological systems.

The shift towards microbial electrochemical technologies in wastewater treatment signifies a progressive approach to address the pressing challenges of water resource contamination. By understanding the advantages these technologies offer over conventional methods, we pave the way for a more sustainable and effective solution to safeguard our water ecosystems. The decentralized nature of METs allows for flexible application in various settings. This interdisciplinary approach harnesses the strengths of various biological processes to create efficient, environmentally friendly solutions for water remediation and resource management. Ongoing research in the field continues to unlock new possibilities, positioning METs as a promising and dynamic technology for the future of water treatment and resource recovery. This chapter aims to provide an overview of the recent research trends and future outlook of METs in wastewater treatment, delving into their applications, challenges, and potential advancements. It will explore the evolving landscape of microbial electrochemical technologies, discussing their role in addressing diverse water pollutants and contributing to sustainable wastewater management practices.

2 Prevailing Biological Systems for Emerging Contaminants and Current Treatments

The biological system for emerging contaminant is a growing concern worldwide, as it contributes to environmental pollution which harmful for human being. The source of contaminant mainly from cosmeceutical, nutraceutical, pharmaceutical product surfactant and agricultural industry. Various studies and research have been done to prevent the contamination and minimize its negative effects human well-being. Clean and safe water has become a basic necessity in human life. Water contamination, especially with foodborne pathogens, contribute to advancement of water treatment technology. Continued development and research in water treatment is crucial for safety. Moreover, safe and clean water aligned to Sustainable Development Goal (SDG) with related to topic SDG 6: Clean Water and Sanitation. The main objective of SDG 6 is to provide sustainable water and sanitation to all part of the world. The polluted water might be treated by enhancing bacterial removal, preventing biofouling, and improving the disinfection capacity.

Traditionally, water treatment methods including filtration, and ultraviolet irradiation has been successfully in preventing the waterborne disease. However, emerging contaminant from various industries contribute to long term health and environmental negative side effects (Shahnaz and Hayder 2023). Cosmeceutical, nutraceuticals, pharmaceuticals products produce unwanted and harmful substances contribute to microbial contamination, radiological contamination, physical contamination, chemical contamination, and allergen contamination. Agricultural processing contributes to the pesticide contamination. Pesticides are commonly used to prevent pest infestation, leading to a decrease in plant productivity. Additionally, the pesticides also important for protecting livestock from external parasites like ticks and fleas. The application of pesticides in agricultural field has potential to be washed out by rain or water irrigation. The contaminants caused by pesticides eventually flow into nearby water sources such as lakes, rivers, and ocean. This process contributes to water contamination. For drinking supplies, pesticides have potential to leach through soil and groundwater. Long term exposure to pesticide in drinking water has potential to various chronic disease. It may cause carcinogenic effect, neurogenerative disease and lastly, contributed to the disruption to endocrine system.

A study has been conducted by Tahmasebi et al. (2023) focusing on pesticides removal from water sources. The pesticides enter the water system through spray drift, leaking of poison container and washing the equipment. Tahmasebi et al. (2023) studied on function of granular and biological activated carbon column in a water treatment plant. This advance water treatment systems enhanced pesticides removal from water system. The suggested method is claimed to reduce the processing cost and not contributed to other toxic side product harmful to the environment. The efficiency of water treatment can be optimized by manipulating several treatment parameters such as pollutant concentration, water flow rate, processing temperature, porosity and particle size of contaminant. Activated carbon has been widely used to reduce the undesirable odour produced by organic compound and pesticides sources. The common methods to removes organic matter, taste, and odour of water is ozonation. The study found that application of activated carbon is more efficient compared to traditional method such as flocculation, filtration, ozonation, coagulation and chlorination. Additionally, chlorination, the most common methods worldwide, has been proven to be able to remove pesticides at high concentrations. Household activities also contributed to water pollution. A study conducted by Asefa et al. (2023) on household water treatment found that there were no any household used membrane filter as a household water treatment methods in their research areas. Therefore, it is important to investigate the most efficient method to protect the water system from any contamination. According to Rahman et al. (2023), batch electrocoagulation systems have the potential to preserve peat water in Sarawak. The study demonstrated that the electric voltage and treatment time significantly influence chemical oxygen demand, turbidity, total dissolved solids, total suspended solid, and reduction of colour.

According to Shahnaz and Hayder (2023) in their review on potential of graphene's antibacterial potential for advanced and sustainability solution in water treatment. The graphene is a two-dimensional carbon material with specific properties. This material is well-known for various applications including water treatment. The graphene-based material offers various strength including two-dimensional structure, high mechanical strength, excellent electrical conductivity, chemical stability, inherent antibacterial properties, tailorable surface chemistry, large surface area, photocatalytic activity, broad-spectrum effectiveness and enhanced disinfection efficiency for potential advanced water treatment. Lin et al. (2023) investigated on water treatment operation by comparing deep neural network model and artificial model. This study is important for predicting the coagulant amount and settled water turbidity. On the other hands, Rosadi et al. (2023) studied on fluorescence-based indicator to investigate the effectiveness and safeness of drinking water. These listed methods have their pros and cons which needs to be balanced with their efficacy and operating costs. Another latest technology recently been introduced for wastewater treatment is microbial electrochemical technologies. The incorporation of microbial electrochemical technologies in real applications can be strengthen by integrating with other biological systems and is discussed in the following sections.

3 Integrated Microbial Fuel Cell Technologies for Wastewater Remediation

Microbial fuel cell (MFC) is a system that is known as bioelectrochemical process consisting of electrons that uses biochemical reactions which is catalyzed by bacteria. They are widely used in the applications that aim for the generation of electricity. There are a lot of advantages of using MFC in technologies thanks to its green technology aspect that does not use any additional chemical additives. MFC technology is very promising in the wastewater remediation field. There are a lot of industries who are looking forward to this kind of technology especially for the industries related to the alimentation, biological and pharmaceutical productions. This is because of the prospectus of MFC itself that aims not only for the wastewater treatment from the industries, but also have a great tendency to generate a renewable source of energy from the process.

3.1 Microbial Fuel Cell (MFC) Concept

It is believed that one of the most important aspects in MFC is the substrate where it is the source of nutrients for the microorganisms. According to Kloch and Toczyłowska-Mamińska (2020), the electricity generation will depend on this biological aspect. Usually in MFC, the electrons need to be released by bacteria from the oxidation process at the anode electron, which named as negative terminal. The electrons next will be transferred to the positive terminal which is the cathode electron with the help of conductive materials. In this terminal, the electrons will be combined with oxygen and will be diffused through a medium called proton exchange membrane (PEM). Currently there are different and various models of MFC with different designs. There is no optimal MFC design that will fit for all applications. Several changes and modifications need to be done and the investigations need to be performed in getting to know how the model can enhance the performance of MFC. For example, there can be different materials for electrodes (anode and cathode) and the membranes used as a medium of diffusion can vary according to applications. The selective membranes used sometime might be the types of PEM or cation exchange membranes (CEM) or anion exchange membranes (AEM). The best selection of materials for the membrane's development will certainly give impact and enhance on their overall performance.

3.2 *Various MFC Configuration for Different Applications*

Microbial fuel cells have shown an important increase in the wastewater treatment industries since it is capable to show an excellent result in removing contaminants like chemical oxygen demand (COD) up to 90% efficiency (Jatoi et al. 2021). With the second objective in energy harvesting, MFC possesses great features like thermophilic metabolism and high temperature biodegradability that can lead to the generation of higher power densities (Guan et al. 2019). It can be seen that various types of wastewaters have been treated using MFC method by different group of researchers. As for example, we can find the MFC used in swine wastewater treatment, food production wastewater, and domestic wastewater (Mubarak et al. 2023). As for microbial fuel cell configuration, different systems or designs of MFC can be found based on different applications and materials developed by different researchers. Generally, MFC consists of 2 electrodes which are anode and cathode that will perform the oxidation and reduction of sepsis involved in the wastewater and a membrane that will act as a separator of 2 compartments. During the MFC process, electrons and protons will be produced on the anode by oxidizing the microbes present in wastewater treated which will be then transferred to the cathode. At the cathode side, electrons and protons received will reduce the oxygen content. The MFC performance can be tested by varying certain parameters like the materials used for the anode and cathode development, membranes used as separator (anion or cation exchange membranes), presence/added of microbes, pH, temperature, and concentration of the wastewater subject to the treatment. Some of the research works related to MFC for different applications, configuration of MFC, the focus of the MFC treatment and the performance of the studies were gathered in the Table 1

3.3 *MFC Limitation in Wastewater Treatment and Future Works*

MFC as a wastewater treatment is very promising as this technique promotes the green technology aspect and eco-friendly by using microbes from nature for tackling the pollution issues and as one of the alternatives for generating a renewable energy and less dependency to the current non-renewable fossil fuels consumption. It has a great potential to be explored in the aspect mentioned above. However, the limitations that can be discussed further on this technology is on the new energy

Table 1 Summary of different MFC applications, configurations, focus of study and performances

Application	Types of MFC	Focus	Performance	Researchers
Domestic wastewater	MFC-based biosensor (graphite rods and gravel)	Chemical oxygen demand assessment	Significant linear relationship between COD concentrations and MFC energy E_{cell} (contact time > 10 h)	Corbella et al. (2019)
	Double-chamber MFC (graphite rods and microorganisms from cow dung + acetate)	Harvest energy and production of methane gas	90 mV energy produced + up to 85.65 m^3/day off methane production with 0.8 g acetate added	
Swine wastewater	Single-chambered MFC (anode—reduced graphene oxide and cathode—platinum-coated carbon cloth) in continuous flow mode	COD removal and energy generation	COD removal up to 52% (530 mg L^{-1} day^{-1}) + maximum electricity generation up to 2.2 Wh m^{-3}	Goto and Yashida (2019)
	Double-chambered MFC (swine wastewater containing sulfonamide antibiotics) using cation exchange membranes (CEM) with anode (graphite) and cathode (carbon-fiber brush) in continuous mode	COD and nutrients removal	More than 90% of COD, 66.6 +/− 1.4% and 32.1 +/− 2.8% of ammonium and phosphates, respectively removed from swine wastewater	Cheng et al. (2021)
Food production wastewater	Two-parallel MFC reactors (fed with undiluted dairy wastewater) used in continuous mode for 2.5 months	Organic matter removal and energy harvesting	Maximum power density recovered over 27 W m^{-3} and up to 95% of COD removal	Cecconet et al. (2018)
	Two glass containers for anode and cathode chambers containing *Saccharomyces cerevisiae.* Anode and cathode from graphite plates with salt bridge used	COD removal and electricity generation at different temperature and pH	COD removal—92% Maximum electrical power generated up to 850 mV and 28 μA for voltage and current, respectively	Al-saned et al. (2021)
Pharmaceutical wastewater	Paraboloid configuration (2 cups for graphite-based anode and cathode)	COD and TDS removal, energy generation	Maximum power and current generation—2.01 W m^{-3} 168 mA m^{-2} + 81% COD removal and 36% of TDS removal	Rashid et al. (2021)

(continued)

Table 1 (continued)

Application	Types of MFC	Focus	Performance	Researchers
	Photobioreactor with microalgae *Scenedesmus abundans*, separated by PEM and electrodes made up of graphite rods	Biomass and electricity production + COD removal	COD removal 77 +/− 3%	Nayak and Ghosh (2019)
			Maximum voltage and power density produced—745.96 mV and 838.68 mW m^{-2}, respectively	

yield harvested. It might be low energy generated from the continuous long duration process. Sometimes, the electrodes materials used also can be expensive that may contribute to its limitation. Current challenges and future perspectives of MFC in wastewater treatment need to be highlighted to rend this technology more pronounced in the environmental field especially on its performance aspect like COD removal efficiency as has been highlighted by Hassan et al. (2023). Most of them achieve more than 90% of COD removal besides ensuring bioenergy production for the benefit of human beings.

4 Integrated Microbial Electrolysis Cell Technologies for Wastewater Remediation

4.1 Principles and Mechanism of Microbial Electrolysis Cell (MEC)

Anaerobic biological processes known as microbial electrolysis cells (MEC) transform organic compounds in wastewater into sustainable energy sources in the form of hydrogen (H_2) and methane (CH_4). MEC are bioelectrochemical reactors that convert chemical energy stored in organic compounds into electrical energy, or H_2 through catalytic oxidation by the action of microorganisms. These microorganisms act as biocatalysts, which are responsible for extracting electrons and protons through the degradation of organic content and releasing carbon dioxide as a by-product. The electrons produced by the oxidation are then transferred to the anode surface and further conducted to the cathode chamber via an external circuit. To be specific, the cell will convert to an MEC if the cathode chamber becomes anaerobic, in which the transferred protons will be reduced to H_2 as the product. This reduction is a thermodynamically non-spontaneous reaction; thus, an external energy input is essential. MECs are a new alternative strategy for wastewater treatment instead of using centralised water systems. In MEC systems, the growth of microorganisms at the anode decreases the activation potential of a given redox reaction, thus increasing the voltage efficiency and product yield of MECs. The microorganisms that adhere

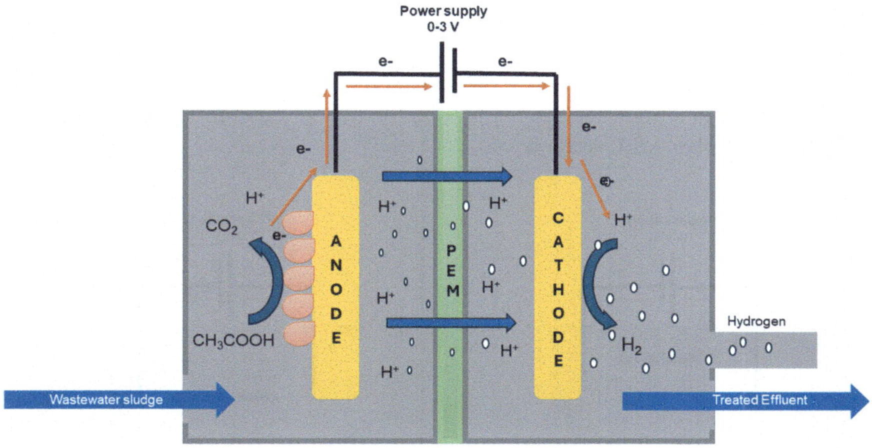

Fig. 1 Schematic mechanism of a microbial electrolysis cells (MEC) for wastewater treatment for hydrogen production (Fudge et al. 2021)

to the anode produce biofilm on the anode surface, which initiates the conversion of chemical energy from wastewater sludge into electrical energy. The generated electrical energy is applied to exhibit the production of valuable products such as H_2 and CH_4 at the cathode. In addition, the microorganisms continue to exchange electrons with the electrode to maintain their cellular growth and function (Sleutels et al. 2012). MEC systems require a separator between the bioanode and cathode if they are designed for reduction of hydrogen ion to hydrogen and CH_4, as shown in Fig. 1.

4.2 *Microbial Electrolysis Cell Architecture*

Basically, MEC design consists of two chambers, which are the anode and cathode. They are separated from each other by an ion-exchange membrane. Initially, the MEC design comprised H-type cells with gas collection ports to the cathode (Kadier et al. 2016b). However, the design was modified to improve the operation of the MEC system. The first modification to the MEC design is the addition of a double chamber to the cell. Unfortunately, the presence of a membrane enlarges the distance between electrodes, which can cause high overpotentials and also lead to a pH imbalance between two chambers, thus increasing the loss of voltage. To tackle the problem occurring in double chamber MEC design, a membraneless single chamber is needed to recover the net current output of the MEC system (Call and Logan 2008). However, this configuration still has a major drawback, such as the possibility of hydrogen crossover to the anode, which is later utilized by the methanogenic bacteria, which reduces the overall yield of the MEC system. The comparison configuration system of double and single of MEC system are illustrated as shown in Fig. 2. Overall, both configurations have their own advantages and disadvantages, as summarized in Table 2. A comparative analysis of both modifications has shown that a single chamber results in a higher hydrogen production rate and current density compared to a double chamber MEC system.

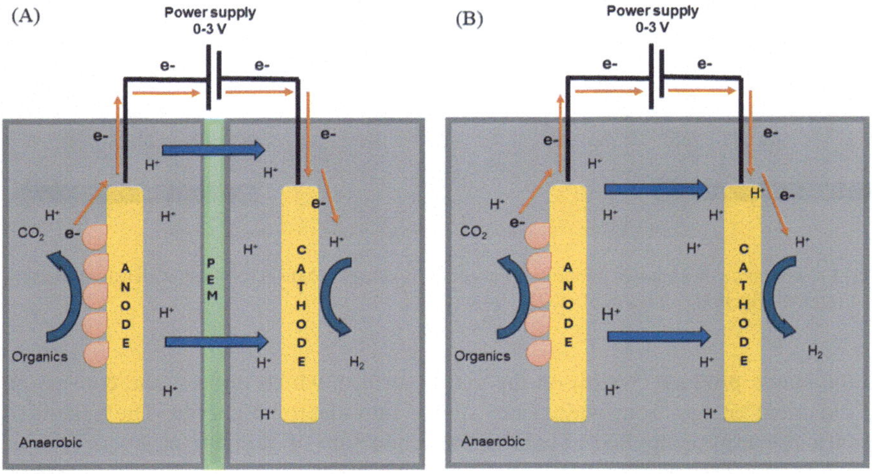

Fig. 2 **A** double chamber MEC system with proton-exchange membrane (PEM), and **B** single chamber MEC system

Table 2 Advantages and disadvantages of different configurations in MEC design

	Advantages	Disadvantages
Double chamber MEC	Prevention of back diffusion of H_2 from cathode to anode	System resistance is increased due to large distance of electrodes
	Purity of H_2 is maintained	Large increment in pH gradient cause potential losses
		High operation cost for H_2 production
Single chamber MEC	Reduce potential losses	Hydrogenotrophic bacteria produce methane
	High current density and H_2 production rate	Less H_2 purity
	pH gradients impact decreases	

4.3 Integration of MEC with Other Technologies

4.3.1 Integration with Anaerobic Digestion (AD)

Anaerobic digestion (AD) is a process of converting organic compounds into biogas via syntrophic microbial interaction, such as the conversion of biomass into biofuel with more energy efficiency. This technology reclaims waste pollution by recovering energy contents from waste and reducing waste quantity. However, there are some issues that should be taken into consideration as this technology has a long start-up, needs specific nutrients, and is highly susceptible to changes in operation conditions. These drawbacks hinder the benefit of AD. Consequently, the integration of MEC with AD is a recognised electromethanogenesis process that can help improve the overall performance of AD. At the same time, the integration of MEC with AD can reduce instability, increase biogas yield by AD, and increase CH_4 generation and purity in MEC system. For example, if MEC is designed for CH_4 production, the biocathode is surrounded by methanogenic microorganisms that are used to produce CH_4 from H_2 and CO_2 at the cathode of the cells. When wastewater flows into the MEC-AD under anaerobic conditions, electricity is supplied to the electrodes with precise control to produce biogas, and at the same time, the wastewater is treated to prevent organic contaminants (Zakaria and Ranjan Dhar 2021). The electrogens present in the anode chamber oxidize the organic matter to produce electricity.

Furthermore, integrated MEC-AD has been categorized into "in-situ" and "ex-situ" systems (Fig. 3). The in-situ MEC-AD is built in a single reactor without a separator between anode and cathode, with the enrichment of electroactive bacteria (EAB) or electrotropic methanogenic archaea (EMA). The EAB oxidizes organic acid and transfers electrons to the electrode by extracellular electron transfer, while the EMA consumes electrons to generate CH_4 via electromethanogenesis. In addition, the MEC-AD enriches hydrogenetrophic methanogens by providing more H_2 for electrodes or bulk solutions to increase growth rate, methane production rate, and resilience toward environmental change (Huang et al. 2022).

The second configuration of MEC-AD processes is known as ex-situ MEC-AD which operates in separate reactors. In this process, AD is performed as a pre-fermented phase of the feedstock, which improves the biodegradability of organic compounds prior to MEC processes as a strategy to increase current density, chemical oxygen demand (COD) removal rate, and the coulombic efficiency of the MEC system. For example, the production of H_2 and CH_4 from palm oil mill effluent by implementing two-stages of dark fermentation and mesophilic MECs to achieve high total energy recovery and COD removal efficiency (Krishnan et al. 2019). Other studies also show an improvement in organic phosphorus release and increased H_2 recovery from waste activated sludge (WAS) via MEC-AD approach (Zhou et al. 2019). The improved degradation of propionate was facilitated by syntrophic interactions between fermentative microorganisms, exoelectrogens, and hydrogenotrophic methanogens. From these studies, it has been proven that ex-situ MEC-AD could

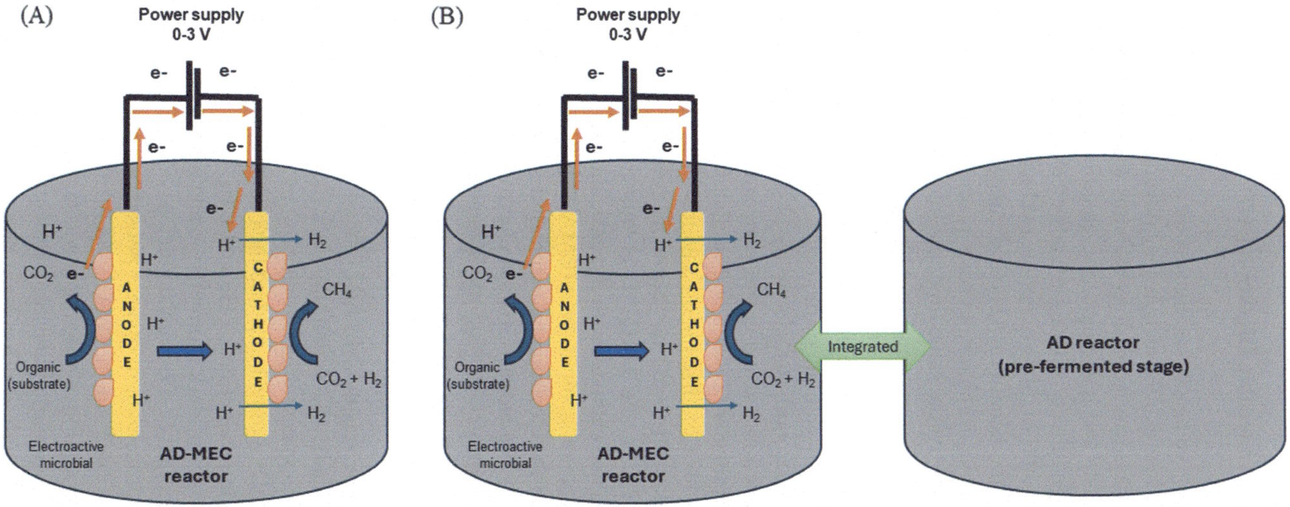

Fig. 3 Schematic diagram of microbial electrolysis cell-anaerobic digestion (MEC-AD) **A** in-situ MEC-AD, and **B** ex-situ MEC-AD to produce methane (Wang et al. 2022a, b)

produce a better methane yield by MEC-AD system than that of the individual MEC system.

4.3.2 Integration with Membrane Bioreactor

There are some challenges that should be taken into consideration in the process of MEC for wastewater treatment, such as low efficiency for practical application and the need for additional energy input, which increases the processing cost. Another treatment process, the membrane bioreactor, has been chosen to integrate with MEC to achieve high processing efficiency and an affordable cost of wastewater treatment. The integration of MEC with a membrane bioreactor or filtration is intended to enhance the quality of mill effluent so that it can be reused. The integration of a single chamber of MEC with electrically conductive and catalytic microfiltration hollow fibers (HFM) called an anaerobic electrochemical membrane bioreactor (EMBR) (Katuri et al. 2014, 2018). In this system, HFM has a dual function as the membrane material for the filtration of the treated water and the cathode for H_2 evolution. Through this system, the treated water is allowed to pass through the microfiltration pores of the electrically porous hollow fiber cathode while blocking the entrance of bacteria. Moreover, HFM designs with small radial dimensions increase the electrode surface-area-to-reactor volume ratio, which is needed for large scale applications of MEC (Katuri et al. 2014, 2018). In this system, HFM has a dual function as the membrane material for the filtration of the treated water and the cathode for H_2 evolution. Through this system, the treated water is allowed to pass through the microfiltration pores of the electrically porous hollow fiber cathode while blocking the entrance of bacteria. Moreover, HFM designs with small radial dimensions increase the electrode surface-area-to-reactor volume ratio, which is needed for large scale applications of MEC.

On the other hand, the use of nickel-based HFM improves the recovery of CH_4-rich biogas in the EMBR system. This is due to the mitigation of biofouling in the Ni-based EMBR system by gas bubble formation on the cathode surface via in-situ cleaning (Katuri et al. 2014). Additionally, the energy input to the MEC-MBR system exhibits different impacts on microorganisms, such as the fact that the generated electric current in this system stimulates microbial growth while simultaneously degrading pollutants, and the excessive use of electricity may constrain microbial growth. Therefore, in the MEC-MBR system, different voltage supplies affect COD removal efficiency. In previous studies, applying the MEC-MBR system for wastewater treatment without voltage supply resulted in only 52.6% of COD removal efficiency, however, the removal efficiency increased up to 70.6% when the voltage was set at 0.6 V (Ding et al. 2018). It seems that the COD removal efficiency decreases when the voltage supply increases, which is due to the fact that high voltage or electricity can exhibit severe plasma rupture, lower metabolic activity, and decrease microbial growth rates (Chen et al. 2016a, b). A new insight has emerged for an improved membrane efficacy such as ceramic. The low internal resistance of the ceramic-based membrane coupled with microbial fuel cell lead to a significant fouling control and electricity utilization (Wang et al. 2022a, b).

5 Integrated Microbial Desalination Cell Technologies for Wastewater Remediation

Drinkable water can be produced from seawater from a process called desalination where salts from seawater are removed or filtered (Al-Othman et al. 2019). Electrodialysis, reverse osmosis, and multistage flash are examples of traditional desalination techniques. Large amounts of energy are needed for these conventional processes, and most of that energy originates from fossil fuels (Sophia and Gohil 2018). Microbial desalination cells (MDCs) offer future potential benefits for producing biogas, desalinating water, treating wastewater sustainably, and producing power.

MDC is a promising approach that may effectively address the challenges of wastewater remediation, saltwater desalination, and eco-energy production simultaneously (Al-Mamun et al. 2018). By use of the bio-electrochemical oxidation of organic compounds found in wastewater, this approach produces electrical energy. In the MDC system, the power generated is utilised to facilitate ion mobility. Ionic migration produces useful by-products in addition to desalination. The MDC is an enhanced iteration of bio-electrochemical systems (BES) whereby the anodic and cathodic parameters of BES can be fabricated with novel electrode materials and catalysts. The process involves the direct degradation of organic molecules to produce electrical energy, with the help of anode-respiring bacteria (ARB) that move electrons outside the cell. Consequently, they have the capability to function independently (Kumar et al. 2022).

Compared to conventional desalination techniques, MDC shows a notable increase in energy output (mostly in the biohydrogen form) throughout the operation, ranging from 180 to 231% (Wang and Ren 2013). The desalination unit together with the anode and cathode, placed in between, are the parts of the MDC. The cation and anion exchange membranes are included into this unit. In this configuration, the anode chamber breaks down the organic compounds in wastewater while producing electricity. The central portion of the setup carries out the process of removing salt from saline water, while the cathode component of the MDC setup completes the electrical circuit (Al-Mamun et al. 2018; Luo et al. 2012).

The main findings show that desalination was effectively completed by the MDC system without requiring external power, as well as wastewater clean-up and power recovery without the need of intermediary processes (Odunlami et al. 2023). The practical uses of this technology faced many technological hurdles, including the need to maintain the pH levels in both cathodic and anodic fluids, dealing with increased internal resistance, using catalysts on electrodes, and addressing issues related to membrane fouling and durability.

Researchers and scientists use MDC systems with additional methods such as electrodialysis, reverse osmosis (RO), and chemical production cells to overcome the limitations of the MDC technology. Combining a variety of treatment techniques can improve a solution and provide the most economical and energy-efficient result. Shivakumar and Razaviarani (2021) study looked into how MDC systems and RO technologies may work together. Based on the research, once the drawbacks of MDC technologies are addressed, it shows great promise as a pretreatment method before using reverse osmosis (RO). This is because it effectively eliminates significant amounts of pollutants that hinder the process of desalination in the RO plant, while also requiring low amount of energy and achieving high removal of salinity.

Integrating forward osmosis membrane (FOM) into microbial desalination cells (MDC) may provide an additional benefit for water purification. A FOM has been used in place of the anion exchange membrane (AEM) in the conventional MDC to improve its performance. This combination form of FOM and MDC is known as osmotic MDC (Os-MDC) (Chen et al. 2016a, b; Santoro et al. 2017). FOM minimises the ions flow from the desalination chamber to the electrode-containing chambers while facilitating the movement of water from the anode to the desalination chamber during the process. In the end, this increases the process's overall efficiency (Ragab et al. 2019). The creation of water flow in this MDC module is a result of the difference in chemical potential between the draw and feed solutions (Zhang et al. 2012). The water flow helps regulate the pH of the anode, which promotes the optimal development of microorganisms and facilitates the efficient degradation of organic materials. Instead of the gradient in salt concentration observed in a traditional osmotic unit, the main driving factor in osmotic (MDCs) is the difference in electrical potential (Cath and Elimelech 2006). Os-MDC has many key benefits, including its ability to effectively treat large volumes of wastewater, generate power, and dilute salty water (Salameh et al. 2020). The primary disadvantages of the approach are reduction of efficiency, diminished ionic separation, membrane fouling, and potential membrane contamination by bacteria, organic compounds, and viruses (Chen et al. 2016a, b; Santoro et al. 2017).

Lan et al. (2019) conducted a study on wastewater remediation using a combination of a MDC and a chemical production cell (MDCC) together with the Fenton process. The MDCC system has four distinct chambers, including the anode, acid-production, desalination, and cathode chambers. The effluent generated from an electroplating procedure, including substantial quantities of hazardous heavy metals like zinc and nickel, poses a threat to biological systems. Additionally, it has a significant concentration of refractory organics that exhibit resistance to biodegradation.

The research investigated the effectiveness of combining microbial desalination cell (MDC) with microbial electrolysis cell (MEC), known as MDC-MEC, for the removal of Pb (Koomson et al. 2022). The study found a removal efficiency of up to 90%. A high cathodic reduction potential was attained and maintained by the MDC-MEC system, which made it easier to reduce Pb^{2+} by precipitating it at the MEC cathode. The system demonstrated 550 mA m^{-2} of current and 166.08 mW m^{-2} of power, respectively. An anode, an anion exchange membrane, a desalination chamber, a cation exchange membrane, and a cathode make up the basic components of MDC-MEC. A supplementary bipolar membrane (BPM) was inserted between the anion exchange membrane (AEM)/cation exchange membrane (CEM) and the anode/cathode to hinder the movement of anions/phosphate buffer. This setup also generates valuable acid/alkali within this compartment. In a study by Zhu and Logan (2014) on MDC-MEC, serpentine was dissolved using the acid produced in the acid production unit, while ambient carbon dioxide was absorbed using the alkali produced. Compared to conventional procedures, the acid created in MDC-MEC increased the dissolving efficiency of calcium and magnesium ions by a factor of 145 and 20, respectively. Moreover, it has been shown that the fed batch mode permits the alkali solution in MDC-MEC to absorb 24 mg of carbon dioxide, which precipitates as calcium and magnesium carbonates over the course of 24 h. Throughout this procedure, 94% of the COD has been eliminated, achieving a desalination efficiency of 22%.

A bipolar membrane (BPM) integrated with MDC (BPM-MDC) consists of both a cation exchange membrane (CEM) and an anion exchange membrane (AEM). The transitional layer of the BPM-MDC facilitates the dissociation of water into hydrogen and hydroxyl ions when an electric voltage is applied (Gujjala et al. 2022). The BPM-MDC is composed of a four-compartment configuration that enables the salt conversion into acid and base. It is positioned near to the anodic chamber. The BPM-MDC typically comprises both CEM and AEM that are physically attached together using unit operations and heat pressing. A number of parameters influence BPM performance, including as the membrane's endurance, the voltage drops during water splitting, the permeability selectivity, and the electrical resistance, all of which should be minimal. Mehanna et al. (2010) found that enhancing the ion exchange capacity of BPM resulted in a 1.26-fold improvement of efficiency. BPM is susceptible to fouling as a result of extended exposure in the anodic chamber. In general, the benefit of integrated BPM-MDC is the improvement in chemical synthesis, especially in the context of acid/base reactions. However, a drawback is the need for external power to facilitate the splitting of water.

A separator-coupled stacked circulation MDC is another type of integrated MDC. It consists of one cathode chamber, one anode chamber, one concentrate compartment and two desalination chambers (Gujjala et al. 2022). The AEM and CEM were separated the chambers. A graphite rod functioned as the anode. Meanwhile the activated carbon parts were added on to the anode chamber. According to Morel et al. (2012), the graphite rod makes it easier for electrons to bridge the external circuit, while the activated carbon particles inside the chamber encourage the growth of active bacteria. A piece of glass fibre and a carbon cloth were incorporated in the air cathode, with the carbon cloth facing the water. Glass fibre serves as a partition. A titanium wire is connected to the glass fibre to link the air cathode to the external circuit (Chen et al. 2012). To enable the flow of electrodes in this integrated unit, a channel links to electrode chambers for the electrode circulation. Moreover, when in operation, the electrolyte without any buffer substance flows between the anode chamber and cathode chamber, originating from a distinct electrolyte chamber. The use of separator linked stacking circulation MDC resulted in many benefits, including the reduction of pH imbalance, enhancement of coulombic efficiency, prevention of biofouling on the cathode, and the ability to operate the system smoothly for extended periods of time (Saeed et al. 2015).

An ion-exchange resin coupled MDC (R-MDC) is referred to the chamber of MDC that is contained with an anion-exchange and cation-exchange resins. The three chambers of the R-MDC were the same as in the traditional MDC: the anode chamber, the cathode chamber, and the desalination chamber. The distillation chamber is divided from the anode and cathode rooms by the AEM and CEM. Carbon felt is placed within the chambers, and graphite rods function as both anodes and cathodes (Zhang et al. 2012). When low-salt water is used to desalinate in standard MDC, the ohmic resistance in the cell goes up because there is less salt in the water. The ion-exchange resins used in the desalination chamber of R-MDC serve as effective ionic conductors because of their high conductivity. Therefore, ion-exchange resins help to maintain a consistent electrical resistance and decrease the amount of energy used in the process of desalination (Saeed et al. 2015). The benefits of the integrated R-MDC include a higher rate of desalination, reduced the utilization of energy, consistent ohmic resistance, and improved efficiency of charge transfer (Sophia et al. 2016).

To create a long-lasting photosynthetic MDC, algae is used to act as a biocathode and this integrated unit is called as PMDC. Algae aid in the production of oxygen and the elimination of chemical oxygen demand (COD) on-site (Koomson et al. 2022). In the reactor, algae behave as an electron acceptor. The presence of light during the light and dark cycles affects the efficiency of PMDC (Tawalbeh et al. 2020). Algae provides an oxygen for receiving the electron and offers the biomass that is well-suited for applications of biofuel. Additionally, algae also have a vital function in lessen the emissions of carbon dioxide from the wastewater treatment plant. All the nutrients are reuse simultaneously and serve a meaningful purpose. PMDC enables the concurrent processes of desalination, wastewater remediation, and the algal cultivation (Sevda et al. 2015). Algae are photosynthetic organisms that rely on light for the formation of biomass. Nevertheless, the provision of external light seems to be economically unviable. Therefore, the provision of illumination via the establishment of regular light and dark patterns would provide a viable option. In their study, Kokabian et al. (2018) examined the PMDC efficiency in different operation methods: static, continuous, and photobioreactor. They found that the static mode of operation yielded the highest power density of 753.75 mW/m^3 and voltage of 167 mV. At a concentration of 35 g/L, the removal rates for total dissolved solids (TDS), salts, and organics were reported to be 32%, 27%, and 31%, respectively. Meanwhile, in the continuous operation, the rates of COD and salt removal were 45.6% and 22.7%, respectively. The static mode of operation was shown to be more advantageous for generating electricity among the three modes of operation. However, the continuous-MDC and photobioreactor-MDC are appropriate for algal biomass production.

6 Challenges Faced and Future Prospect

Integrated microbial electrochemical technologies (METs) are attracting significant interest because of their distinctive and effective approach to managing diverse wastewaters containing emerging contaminants (ECs). In these systems, electroactive microorganisms oxidize the contaminants, transforming them into electricity and valuable byproducts such as H_2, H_2O_2, and CH_4. In summary, MFCs focus on bioelectricity generation from organic matter, MECs are designed for hydrogen production through water electrolysis, and MDCs combine microbial metabolism with desalination processes for simultaneous power generation and water desalination. However, there are still numerous challenges require attentions. Most of the ongoing research operates at laboratory scale, with current system density and wastewater remediation capabilities remaining insufficient for successful upscaling. Ensuring consistent performance, scalability, and cost-effectiveness while maintaining efficiency and stability at larger scales requires extensive research and development. Considering the overall setup expenses of METs, approximately 60% of the cost is associated with the membrane and its maintenances. There is a pressing need to significantly decrease these costs to broaden the practical application of METs on a larger scale (Do et al. 2018).

Membrane that is suitable to be applied for MET should have characteristics such as good proton conductivity, foul resistance over its operational lifetime, minimal loss of substrate, low in thickness, impermeability to positive ions (Na^+ and Mg^{2+}) and O_2, excellent scalability, mechanical properties, remarkable stability, resistance to the passage of gases, such as H_2 and N_2, low internal resistance, and importantly, be cost-effective. Till this date, the most applied membrane is Nafion membrane (a polymer-based proton exchange membrane) have shown superior characteristics with regard to coulombic efficiency in METs regardless of the significant costs involved. The effort to develop membranes with both a more affordable cost and attractive performance to substitute Nafion is still in developmental phases. Current studies show that hybrid membranes such as biopolymer chitosan-carbon nanotubes nanocomposites, graphene oxide-sulfonated poly ether ether ketone (GO-SPEEK), polysulfone-SPEEK (PS-SPEEK), have proven to achieve higher efficiency than Nafion membranes (Srinophakun et al. 2017; Ayyaru and Ahn 2021).

The durability and environmental impact of membranes used are often overlooked in studies, posing challenges for sustainability and biodegradability. While various studies explore the effects of membranes as separators on METs, significant research gap is apparent, including optimization for maximum output, membrane resilience against biofouling and analysis of long-term durability. Further research is needed for practical applications, with an emphasis on the simultaneous recovery and reuse of valuable resources while treating wastewater. Resolving these research gaps can contribute to the broad expansion of METs development for various environmental and cleaner energy purposes. Recently, researchers have been investigating membrane-less METs to address fouling issues and costs for manufacturing in wastewater treatment. However, this type of membrane face challenges such as substrate and oxygen crossover between anodic and cathodic chambers due to lack of partition. This shows that although membranes are not essential in METs, their presence significantly enhances efficacy despite the associated higher capital and operational costs. Membranes play a crucial role in MET designs, ensuring continuous operational cycles and enhancing the effectiveness of bioelectrochemical performance.

The effectiveness of METs also relies significantly on anodic parameters and cathodic parameters. When considering anodic factors, it is crucial that the chosen anode material possesses necessary attributes such as biocompatibility, to support the adhesion and growth of electroactive microorganisms. Maintaining a stable and active microbial biofilm on the electrode surface is essential for sustained MET performance. Anode material with high specific surface area and porosity are desirable characteristics to promote a larger contact area for microbial attachment and enhance mass transfer, allowing efficient exchange of electrons and substrates between microorganisms and electrodes. Apart from that, the selected anode materials must be chemically stable in the wastewater or electrolyte solution. The harsh conditions in wastewater, including varying pH and the presence of contaminants, can affect the stability of the electrodes. Choosing materials that resist corrosion and degradation is crucial for long-term performance. On the cathodic side, limitations arise due to poor reduction kinetics and a slow rate of electron acceptor reduction reactions at the cathode, acting as significant constraints on the overall product output from METs. Cathode materials should exhibit good catalytic activity for the reduction reactions occurring at the cathode. Common reduction reactions include the reduction of oxygen to water or the production of hydrogen gas. Suitable catalysts can enhance reaction kinetics and overall efficiency. Pt-based catalysts are commonly utilized on the cathode to improve cathode reaction kinetics. However, these materials are expensive and prone to issues such as biotoxicity and biofouling (Rozendal et al. 2008). Over the past years, many studies have suggested alternative materials for cathode that offer both cost-effectiveness and high catalytic activity. These materials encompass various categories, including those with carbon structures (Lin et al. 2022a, b), Nitrogen-modified carbon (Modi et al. 2016), polymers with good conductivity (Singh et al. 2016), among others.

Despite the materials of membrane as separators, electrodes and catalyst compositions, operating parameters also has significant effects on the efficiencies of METs. Variations in operating conditions, such as pH and temperature can affect microbial activity and electrochemical performance. Finding the right set of conditions for a specific microbial electrochemical system is a key challenge. Temperature plays a crucial role in METs, as an increase within a specific range enhances the electron transport and activity of electricigens. While most MFCs works at room temperature, the optimal temperature for both degradation and production of electricity varies among electricigens (Palanisamy et al. 2019). In term of pH, studies have shown that MFCs is preferable at lower pH due to a better electrolyte conductivity, reduction of activation energy of reaction and improvement in cathodic reaction kinetics (Dai et al. 2016). However, the transport of cation species other than protons will lead to an increment in pH at the cathode chamber due to proton consumption by the cathodic reaction. However, this issue can be solved by addition of external chemicals but may result in secondary pollution (Rozendal et al. 2006; Ramya and Kumar 2022). Applying a proper buffer is recommended to maintain the desired pH while ensuring optimal MFC performance. The performance optimization of MFC part can be improved by various challenges such as methanogenic substrate loss, architecture of the reactor, membrane fouling, enhancing mediated electron transfer. Studies have demonstrated that lower temperatures and a proper pH control effectively inhibit methanogen activity, leading to increased purity and rates of H_2 production in MECs (Muddasar et al. 2022; Kadier et al. 2016a; Lu et al. 2011). Additionally, lower pH levels have been shown to enhance electrolyte conductivity, thereby improving H_2 production. The performance of MDCs also significantly influenced by pH and temperature. Optimal pH conditions are crucial for promoting microbial activity in the anode chamber and ensuring efficient electrochemical reactions. Deviations from the preferred pH range can impact microbial viability and hinder overall MDCs efficiency. Additionally, pH imbalances between the anode and cathode chambers, often caused by ion exchange membranes, can affect electrochemical reactions and ion transport. Temperature plays a vital role in microbial activity, with higher temperatures generally enhancing organic matter degradation and electron transfer.

Extreme temperatures can have adverse effects. The optimization of pH and temperature is essential for maximizing the desalination efficiency of MDCs, and research focuses on identifying specific ranges that promote microbial activity and electrochemical performance in these systems (Safwat et al. 2023). Improving the performance of MFCs, MECs and MDCs typically relies on incorporating technological advancements and advanced materials for electrodes and membrane as separators. Nevertheless, many of these METs face a bottleneck due to high costs, undermining their economic and environmental viability when scaled up. Consequently, further research is imperative, with a focus on developing novel materials and technologies such as electro-fenton process, microbial electro synthesis cell and treatment systems for different pollutants that are both cost-effective and efficient, while also evaluating their longevity during extended operational periods.

References

Al-Mamun A, Ahmad W, Baawain MS, Khadem M, Dhar BR (2018) A review of microbial desalination cell technology: configurations, optimization and applications. J Clean Prod 183:458–480

Al-Othman A, Darwish NN, Qasim M, Tawalbeh M, Darwish NA, Hilal N (2019) Nuclear desalination: a state-of-the-art review. Desalination 457:39–61

Al-Saned AJO, Kitafa BA, Badday AS (2021) Microbial fuel cells (MFC) in the treatment of dairy wastewater. IOP Conf Ser Mater Sci Eng 1067(1):012073

Asefa L, Ashenafi A, Dhengesu D, Roba H, Lemma H (2023) Household water treatment practice and associated factors among rural *Kebeles* (villages) in west Guji zone, southern Ethiopia: community based cross-sectional study. Clin Epidemiol Global Health 22:101311. https://doi.org/10.1016/j.cegh.2023.101311

Ayyaru S, Ahn YH (2021) Enhanced performance of sulfonated GO in SPEEK proton-exchange membrane for microbial fuel-cell application. J Environ Eng 147(2):04020153

Butti SK, Velvizhi G, Sulonen ML, Haavisto JM, Koroglu EO, Cetinkaya AY, Mohan SV (2016) Microbial electrochemical technologies with the perspective of harnessing bioenergy: maneuvering towards upscaling. Renew Sustain Energy Rev 53:462–476

Call D, Logan BE (2008) Hydrogen production in a single chamber microbial electrolysis cell lacking a membrane. Environ Sci Technol 42(9):3401–3406

Cath TAC, Elimelech M (2006) Forward osmosis: principles, applications, and recent developments. J Membr Sci 281(1–2):70–87

Cecconet D, Molognoni D, Callegari A, Capodaglio AG (2018) Agro-food industry wastewater treatment with microbial fuel cells: energetic recovery issues. Int J Hydrog Energy 43(1):500–511

Chen X, Liang P, Wei Z, Zhang X, Huang X (2012) Sustainable water desalination and electricity generation in a separator coupled stacked microbial desalination cell with buffer free electrolyte circulation. Biores Technol 119:88–93

Chen X, Sun H, Liang P, Zhang X, Huang X (2016a) Optimization of membrane stack configuration in enlarged microbial desalination cells for efficient water desalination. J Power Sources 324:79–85

Chen Y, Yu B, Yin C, Zhang C, Dai X, Yuan H, Zhu N (2016b) Biostimulation by direct voltage to enhance anaerobic digestion of waste activated sludge. RSC Adv 6(2):1581–1588

Cheng D, Ngo HH, Guo W, Chang SW, Nguyen DD, Nguyen QA, Liang S (2021) Improving sulfonamide antibiotics removal from swine wastewater by supplying a new pomelo peel derived biochar in an anaerobic membrane bioreactor. Biores Technol 319:124160

Corbella C, Hartl M, Fernandez-gatell M, Puigagut J (2019) MFC-based biosensor for domestic wastewater COD assessment in constructed wetlands. Sci Total Environ 660:218–226

Dai H, Yang H, Liu X, Zhao Y, Liang Z (2016) Performance of sodium bromate as cathodic electron acceptor in microbial fuel cell. Biores Technol 202:220–225

Das A, Das S, Das N, Pandey P, Ingti B, Panchenko V, Pandey P (2023) Advancements and innovations in harnessing microbial processes for enhanced biogas production from waste materials. Agriculture 13(9):1689

Ding A, Fan Q, Cheng R, Sun G, Zhang M, Wu D (2018) Impacts of applied voltage on microbial electrolysis cell-anaerobic membrane bioreactor (MEC-AnMBR) and its membrane fouling mitigation mechanism. Chem Eng J 333:630–635

Do MH, Ngo HH, Guo WS, Liu Y, Chang SW, Nguyen DD, Ni BJ (2018) Challenges in the application of microbial fuel cells to wastewater treatment and energy production: a mini review. Sci Total Environ 639:910–920

Domínguez-Garay A, Esteve-Núñez A (2018) Designing strategies for operating microbial electrochemical systems to clean up polluted soils under non-flooded conditions. Bioelectrochemistry 124:142–148

Fudge T, Bulmer I, Bowman K, Pathmakanthan S, Gambier W, Dehouche Z, Constantinou A (2021) Microbial electrolysis cells for decentralised wastewater treatment: the next steps. Water 13(4):445

Goto Y, Yoshida N (2019) Scaling up microbial fuel cells for treating swine wastewater. Water 11(9):1803

Guan YF, Zhang F, Huang BC, Yu HQ (2019) Enhancing electricity generation of microbial fuel cell for wastewater treatment using nitrogen-doped carbon dots-supported carbon paper anode. J Clean Prod 229:412–419

Gujjala LKS, Dutta D, Sharma P, Kundu D, Vo DVN, Kumar S (2022) A state-of-the-art review on microbial desalination cells. Chemosphere 288:132386

Hassan M, Kanwal S, Singh RS, SA MA, Anwar M, Zhao C (2023) Current challenges and future perspectives associated with configuration of microbial fuel cell for simultaneous energy generation and wastewater treatment. Int J Hydrog Energy 50:323–350

Hernandez CA, Osma JF (2020) Microbial electrochemical systems: deriving future trends from historical perspectives and characterization strategies. Front Environ Sci 8:44

Holtmann D, Harnisch F (2018) Electrification of biotechnology: Quo vadis? In: Holtmann D, Harnisch F (eds) Bioelectrosynthesis. Advances in biochemical engineering/biotechnology, vol 167. Springer, Cham, pp 395–411. https://doi.org/10.1007/10_2018_75

Huang Q, Liu Y, Dhar BR (2022) A multifaceted screening of applied voltages for electro-assisted anaerobic digestion of blackwater: significance of temperature, hydrolysis/acidogenesis, electrode corrosion, and energy efficiencies. Biores Technol 360:127533

Jatoi AS, Akhter F, Mazari SA, Sabzoi N, Aziz S, Soomro SA, Ahmed S (2021) Advanced microbial fuel cell for waste water treatment—a review. Environ Sci Pollut Res 28:5005–5019

Kadier A, Kalil MS, Abdeshahian P, Chandrasekhar K, Mohamed A, Azman NF, Hamid AA (2016a) Recent advances and emerging challenges in microbial electrolysis cells (MECs) for microbial production of hydrogen and value-added chemicals. Renew Sustain Energy Rev 61:501–525

Kadier A, Simayi Y, Abdeshahian P, Azman NF, Chandrasekhar K, Kalil MS (2016b) A comprehensive review of microbial electrolysis cells (MEC) reactor designs and configurations for sustainable hydrogen gas production. Alex Eng J 55(1):427–443

Katuri KP, Werner CM, Jimenez-Sandoval RJ, Chen W, Jeon S, Logan BE, Saikaly PE (2014) A novel anaerobic electrochemical membrane bioreactor (AnEMBR) with conductive hollow-fiber membrane for treatment of low-organic strength solutions. Environ Sci Technol 48(21):12833–12841

Katuri KP, Kalathil S, Ragab AA, Bian B, Alqahtani MF, Pant D, Saikaly PE (2018) Dual-function electrocatalytic and macroporous hollow-fiber cathode for converting waste streams to valuable resources using microbial electrochemical systems. Adv Mater 30(26):1707072

Khan S, Naushad M, Govarthanan M, Iqbal J, Alfadul SM (2022) Emerging contaminants of high concern for the environment: current trends and future research. Environ Res 207:112609

Kloch M, Toczyłowska-Mamińska R (2020) Toward optimization of wood industry wastewater treatment in microbial fuel cells—mixed wastewaters approach. Energies 13(1):263

Kokabian B, Smith R, Brooks JP, Gude VG (2018) Bioelectricity production in photosynthetic microbial desalination cells under different flow configurations. J Ind Eng Chem 58:131–139

Koomson DA, Huang J, Li G, Miwornunyuie N, Darkwah WK (2022) Performance of recirculatory microbial desalination cell-microbial electrolysis cell coupled system with different catholytes. Renew Energy 189:1375–1382

Krishnan S, Din MFM, Taib SM, Nasrullah M, Sakinah M, Wahid ZA, Singh L (2019) Accelerated two-stage bioprocess for hydrogen and methane production from palm oil mill effluent using continuous stirred tank reactor and microbial electrolysis cell. J Clean Prod 229:84–93

Kumar V, Dwivedi SK, Oh S (2022) A review on microbial-integrated techniques as promising cleaner option for removal of chromium, cadmium and lead from industrial wastewater. J Water Process Eng 47:102727

Lan J, Ren Y, Lu Y, Liu G, Luo H, Zhang R (2019) Combined microbial desalination and chemical-production cell with Fenton process for treatment of electroplating wastewater nanofiltration concentrate. Chem Eng J 359:1139–1149

Lin FY, Lin YY, Li HT, Ni CS, Liu CI, Guan CY, Chen HY (2022) Trapa natans husk-derived carbon as a sustainable electrode material for plant microbial fuel cells. Appl Energy 325:119807

Lin L, Yang H, Xu X (2022) Effects of water pollution on human health and disease heterogeneity: a review. Front Environ Sci 10:880246

Lin S, Kim J, Hua C, Kang S, Park MH (2023) Comparing artificial and deep neutral network models for prediction of coagulant amount and settled water turbidity: lesson learned from big data in water treatment operation. J Water Process Eng 54:103949. https://doi.org/10.1016/j.jwpe.2023.103949

Lu L, Ren N, Zhao X, Wang H, Wu D, Xing D (2011) Hydrogen production, methanogen inhibition and microbial community structures in psychrophilic single-chamber microbial electrolysis cells. Energy Environ Sci 4(4):1329–1336

Luo H, Xu P, Jenkins PE, Ren Z (2012) Ionic composition and transport mechanisms in microbial desalination cells. J Membr Sci 409:16–23

Mateo-Sagasta J, Zadeh SM, Turral H, Burke J (2017) Water pollution from agriculture: a global review. Executive summary. Published by Food and Agriculture Organization of the United Nations Rome, 2017 and the International Water Management Institute on behalf of the Water Land and Ecosystems research program Colombo, 2017

Mehanna M, Kiely PD, Call DF, Logan BE (2010) Microbial electrodialysis cell for simultaneous water desalination and hydrogen gas production. Environ Sci Technol 44(24):9578–9583

Modi A, Singh S, Verma N (2016) In situ nitrogen-doping of nickel nanoparticle-dispersed carbon nanofiber-based electrodes: its positive effects on the performance of a microbial fuel cell. Electrochim Acta 190:620–627

Mohammadi F, Vakili-Nezhaad GR, Al-Rawahi N, Gholipour S (2023) Microbial electrochemical systems for bioelectricity generation: current state and future directions. Results Eng 2023:101619

Mohan SV, Sravan JS, Butti SK, Krishna KV, Modestra JA, Velvizhi G et al (2019) Microbial electrochemical technology: emerging and sustainable platform. In: Mohan SV, Varjani S, Pandey A (eds) Microbial electrochemical technology. Elsevier, Amsterdam, pp 3–18. https://doi.org/10.1016/B978-0-444-64052-9.00001-7

Morel A, Zuo K, Xia X, Wei J, Luo X, Liang P, Huang X (2012) Microbial desalination cells packed with ion-exchange resin to enhance water desalination rate. Biores Technol 118:43–48

Mubarak NM, Sattar A, Mazari SA, Nizamuddin S (eds) (2023) Advanced nanomaterials and nanocomposites for bioelectrochemical systems. Elsevier, London

Muddasar M, Liaquat R, Aslam A, Ur Rahman MZ, Abdullah A, Khoja AH, Bahadar A (2022) Performance efficiency comparison of microbial electrolysis cells for sustainable production of biohydrogen—a comprehensive review. Int J Energy Res 46(5):5625–5645

Nagarajan AM, Subramanian A, Gobinathan KP, Mohanakrishna G, Sivagami K (2023) Electrochemical-based approaches for the treatment of pharmaceuticals and personal care products in wastewater. J Environ Manag 344:118385

Nayak JK, Ghosh UK (2019) Post treatment of microalgae treated pharmaceutical wastewater in photosynthetic microbial fuel cell (PMFC) and biodiesel production. Biomass Bioenerg 131:105415

Nedjimi B (2021) Phytoremediation: a sustainable environmental technology for heavy metals decontamination. SN Appl Sci 3(3):286

Odunlami O, Vershima DA, Tagbo CV, Ogunlade S, Nkongho S (2023) Microbial desalination cell technique—a review. S Afr J Chem Eng 46(1):312–329

Palanisamy G, Jung HY, Sadhasivam T, Kurkuri MD, Kim SC, Roh SH (2019) A comprehensive review on microbial fuel cell technologies: processes, utilization, and advanced developments in electrodes and membranes. J Clean Prod 221:598–621

Patil SA, Hägerhäll C, Gorton L (2012) Electron transfer mechanisms between microorganisms and electrodes in bioelectrochemical systems. Bioanal Rev 4:159–192

Ragab M, Elawwad A, Abdel-Halim H (2019) Evaluating the performance of microbial desalination cells subjected to different operating temperatures. Desalination 462:56–66

Rahman NA, Jol CJ, Linus AA, Ming CK, Arif P, Baharuddin N, Borhan SW, Jalal AA, Hamid DFAAA (2023) Treatment of tropical peat water in Sarawak peatlands nature reserve by utilising a batch electrocoagulation system. Sustain Chem Environ 4:100043. https://doi.org/10.1016/j.scenv.2023.100043

Ramírez-Vargas CA, Prado A, Arias CA, Carvalho PN, Esteve-Núñez A, Brix H (2018) Microbial electrochemical technologies for wastewater treatment: principles and evolution from microbial fuel cells to bioelectrochemical-based constructed wetlands. Water 10(9):1128

Ramya M, Kumar PS (2022) A review on recent advancements in bioenergy production using microbial fuel cells. Chemosphere 288:132512

Rashid T, Sher F, Hazafa A, Hashmi RQ, Zafar A, Rasheed T, Hussain S (2021) Design and feasibility study of novel paraboloid graphite based microbial fuel cell for bioelectrogenesis and pharmaceutical wastewater treatment. J Environ Chem Eng 9(1):104502
Rosadi MY, Maysaroh S, Sagita ND, Anggreini S, Desmiarti R, Deng Z, Li F (2023) Flourescence-based indicator predict the performance of conventional drinking water treatment processes: evaluation based on the changes in the composition of dissolved organic matter. Chemosphere 337:139410. https://doi.org/10.1016/j.chemosphere.2023.139410
Rozendal RA, Hamelers HV, Buisman CJ (2006) Effects of membrane cation transport on pH and microbial fuel cell performance. Environ Sci Technol 40(17):5206–5211
Rozendal RA, Hamelers HV, Rabaey K, Keller J, Buisman CJ (2008) Towards practical implementation of bioelectrochemical wastewater treatment. Trends Biotechnol 26(8):450–459
Saeed HM, Husseini GA, Yousef S, Saif J, Al-Asheh S, Fara AA, Aidan A (2015) Microbial desalination cell technology: a review and a case study. Desalination 359:1–13
Safwat SM, Meshref MNA, Salama M, Elawwad A (2023) Influence of co-substrate existence, temperature, pH, and salt concentration on phenol removal, desalination, and power generation using microbial desalination cells. Int J Environ Sci Technol 20(10):10695–10712
Salameh T, Tawalbeh M, Al-Shannag M, Saidan M, Melhem KB, Alkasrawi M (2020) Energy saving in the process of bioethanol production from renewable paper mill sludge. Energy 196:117085
Santoro C, Abad FB, Serov A, Kodali M, Howe KJ, Soavi F, Atanassov P (2017) Supercapacitive microbial desalination cells: new class of power generating devices for reduction of salinity content. Appl Energy 208:25–36
Schoepp-Cothenet B, van Lis R, Atteia A, Baymann F, Capowiez L, Ducluzeau A-L et al (2013) On the universal core of bioenergetics. Biochim Biophys Acta (BBA) Bioenerget 1827:79–93. https://doi.org/10.1016/j.bbabio.2012.09.005
Schröder U, Harnisch F, Angenent LT (2015) Microbial electrochemistry and technology: terminology and classification. Energy Environ Sci 8:513–519
Sevda S, Yuan H, He Z, Abu-Reesh IM (2015) Microbial desalination cells as a versatile technology: functions, optimization and prospective. Desalination 371:9–17
Shahnaz T, Hayder G (2023) Exploring graphene's antibacterial potential for advanced and sustainable solutions in water treatment. J Water Process Eng 56:104530. https://doi.org/10.1016/j.jwpe.2023.104530
Shi P, Wang H, Feng M, Cheng H, Yang Q, Yan Y, Zhang M (2022) The coupling response between different bacterial metabolic functions in water and sediment improve the ability to mitigate climate change. Water 14(8):1203
Shivakumar T, Razaviarani V (2021) An integrated approach to enhance the desalination process: coupling reverse osmosis process with microbial desalination cells in the UAE. Water Supply 21(3):1127–1143
Sillanpää M, Shestakova M (2017) Emerging and combined electrochemical methods. Electrochem Water Treat Methods Fundam Methods Full Scale Appl 2017:131–225
Singh S, Modi A, Verma N (2016) Enhanced power generation using a novel polymer-coated nanoparticles dispersed-carbon micro-nanofibers-based air-cathode in a membrane-less single chamber microbial fuel cell. Int J Hydrog Energy 41(2):1237–1247
Sleutels TH, Ter Heijne A, Buisman CJ, Hamelers HV (2012) Bioelectrochemical systems: an outlook for practical applications. Chemsuschem 5(6):1012–1019
Sonawane JM, Mahadevan R, Pandey A, Greener J (2022) Recent progress in microbial fuel cells using substrates from diverse sources. Heliyon. https://doi.org/10.1016/j.heliyon.2022.e12353
Sophia AC, Bhalambaal VM, Lima EC, Thirunavoukkarasu M (2016) Microbial desalination cell technology: contribution to sustainable waste water treatment process, current status and future applications. J Environ Chem Eng 4(3):3468–3478
Sophia C, Gohil JM (2018) Microbial desalination cell technology: functions and future prospects. In: Progress and recent trends in microbial fuel cells. Elsevier, London, pp 399–422. ISBN 978-0-444-64017-8

Srinophakun P, Thanapimmetha A, Plangsri S, Vetchayakunchai S, Saisriyoot M (2017) Application of modified chitosan membrane for microbial fuel cell: roles of proton carrier site and positive charge. J Clean Prod 142:1274–1282

Tahmasebi AA, Beni AS, Azhdarpoor A, Moeini Z (2023) The application of granular and biological activated carbon columns in removal of organochlorine and organophosphorus pesticides in a water treatment plant. J Water Process Eng 56:104383. https://doi.org/10.1016/j.jwpe.2023.104383

Tawalbeh M, Al-Othman A, Singh K, Douba I, Kabakebji D, Alkasrawi M (2020) Microbial desalination cells for water purification and power generation: a critical review. Energy 209:118493

Wang H, Ren ZJ (2013) A comprehensive review of microbial electrochemical systems as a platform technology. Biotechnol Adv 31:1796–1807. https://doi.org/10.1016/j.biotechadv.2013.10.001

Wang L, Wu Y, You Z, Bao H, Zhang L, Wang J (2022a) Electrochemical impedance spectroscopy (EIS) reveals the role of microbial fuel cell-ceramic membrane bioreactor (MFC-CMBR): electricity utilization and membrane fouling. Water Res 222:118854

Wang XT, Zhang YF, Wang B, Wang S, Xing X, Xu XJ, Chen C (2022b) Enhancement of methane production from waste activated sludge using hybrid microbial electrolysis cells-anaerobic digestion (MEC-AD) process—a review. Biores Technol 346:126641

Yakar A, Türe C, Türker OC, Vymazal J, Saz Ç (2018) Impacts of various filtration media on wastewater treatment and bioelectric production in up-flow constructed wetland combined with microbial fuel cell (UCW-MFC). Ecol Eng 117:120–132

Yang K, Abu-Reesh IM, He Z (2023) Enhancing organic contaminant degradation through integrating advanced oxidation processes with microbial electrochemical systems. J Hazardous Mater Lett 4:100075

Zakaria BS, Dhar BR (2021) An intermittent power supply scheme to minimize electrical energy input in a microbial electrolysis cell assisted anaerobic digester. Biores Technol 319:124109

Zhang F, Chen M, Zhang Y, Zeng RJ (2012) Microbial desalination cells with ion exchange resin packed to enhance desalination at low salt concentration. J Membr Sci 417:28–33

Zhang Q, Hu J, Lee DJ (2016) Microbial fuel cells as pollutant treatment units: research updates. Biores Technol 217:121–128

Zhou Y, Zhang J, Zhang Z, Wang P, Xia S (2019) PH dependent of the waste activated sludge reduction by short-time aerobic digestion (STAD) process. Sci Total Environ 649:1307–1313

Zhu X, Logan BE (2014) Copper anode corrosion affects power generation in microbial fuel cells. J Chem Technol Biotechnol 89(3):471–474

Scaling-Up Attempts of Microbial Electrochemical Technologies for the Removal of Water Pollutants

Mihir Kumar, Shubham Rameshchandra Boda, Aryansh Sandeepbhai Shah, Sanjay Kumar, and Ila Jogesh Ramala Sarkar

Abstract Microbial fuel cell (MFC) technology is an encourage ingsustainable technology, with enormous possible technology for wastewater handling. MFCs are strategies accomplished of producing electricity through the use of microbes as biocatalyst and electro-chemical responses. Current progresses in MFC technology have enhanced the output power. A varied collection of biological-rich wastewater can be processed with MFC. MFC technology is facing stand against about power density, biological exclusion frequency, electrode performance limitation, commercial feasibility, high principle and upkeep cost and difficulty to continue electron motion in an intricate wastewater setting. Prospects occur in MFCs scaling up, MFCs have the ability to increase the capacity of the sustainable wastewater treated unit. This chapter offers serious perceptions about the MFC progresses, uses of MFCs in wastewater handling. The difficulties during scaling-up challenges of MFC technology will be pointed out in this chapter.

Keywords Microbial fuel cell · Wastewater treatment · Micro-organism · Bio-catalyst · Scale-up

1 Introduction to Microbial Fuel Cell

Waste valorisation and the circular economy are required due to the growing need for renewable energy, which causes climate change, worldwide change, and environmental damage. Concerningly, there is an increase in the need for wastewater treatment that is sustainable due to concerns about the environment and public health. More water is consumed as a result of an increasing population and level of life, which also increases the amount of waste water produced.

M. Kumar · S. R. Boda · A. S. Shah · S. Kumar · I. J. R. Sarkar (✉)
Department of Chemical Engineering, Faculty of Technology, Marwadi University, Rajkot 360003, India
e-mail: sarkarila@gmail.com

R. Selvasembian et al. (eds.), *Emerging Trends in Microbial Electrochemical Technologies for Sustainable Mitigation of Water Resources Contamination*,
https://doi.org/10.1007/978-3-031-74636-9_12

Sustainable green technologies were developed in response to the rising need for resources such as clean water, energy recovery, and pollution removal. Therefore, the creation of biological fuel cells, which may address the issues, is a significant advancement in the field of renewable energy.

The 1839 work of William Groves served as the impetus for the invention of fuel cells. The utilisation of biocatalysts and electrochemical processes in biological fuel cells allows for the production of electrical energy from chemical energy. The environment greatly benefits from energy recovery from wastewater treatment.

Waste water is an issue for the environment because of how much is created each day, but it also contains nutrients that can be used for farming. The necessity to comprehend the significance of wastewater and how to benefit from it has risen due to the scarcity of freshwater and nutrient-rich natural resources.

The wastewater treatment process has a number of significant downsides, including a high energy need. Between 2 and 60% of the overall operating costs of a wastewater treatment facility go towards the energy needed for the operation. Wastewater treatment is a good thing since the chemical energy that is stored there is far higher than the energy needed to treat it.

The development of cutting-edge wastewater treatment techniques that enable energy-neutral operations is crucial. These techniques must be able to recover energy as well as nutrients like nitrogen and phosphorus that are present in waste water. These fuel cells are also cutting-edge technology. By employing microorganisms to oxidise different contaminants from the wastewater, they are able to produce electrical energy. Biodegradable chemicals contain energy that can be converted to electrical energy using microbial fuel cells (Logan et al. 2006). In this form of fuel cell, microorganisms in the anode section oxidise the organic-rich materials, releasing electrons, protons, and carbon dioxide as by-products.

The electrodes of a simple MFC are submerged in water and kept apart by a membrane. The anode chamber's resident microbes undergo oxidation, producing electrons and protons. The production of energy is carried out by microbes. From an electron source to an electron acceptor, the electrons are transferred.

In order to produce an electrical current from biomass, anode respiring bacteria that have gathered on the anode perform an oxidation process of organic matter that results in protons (one proton for every electron). As a by-product of oxidation, CO_2 will be created. When producing electricity in the cathode chamber, protons and electrons are spent.

Carbon anode materials such solid graphite, graphite-felt, carbon cloth, and carbon paper are utilised in MFCs, as well as platinized cathodes made of coated graphite, coated carbon paper, and titanium. Microbial fuel cells are a potential green technology that may be used in the process of treating wastewater in addition to the removal of pollutants, organic matter, and nutrients for recovery.

According to several research, microbial fuel cells have been used to clean home wastewater (Shah's), animal manure, brewery wastewater, and waste sludge. MFC is also suitable to the treatment of wastewater, which has additional advantages such as the creation of some bioelectricity during the process as well as the removal and reduction of contaminants. Microbial fuel cells are galvanic cells where the free ions

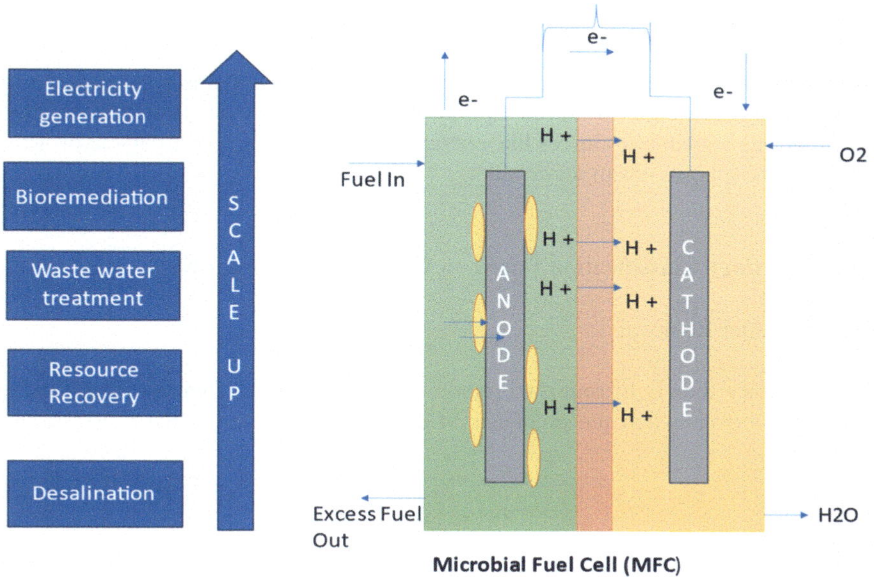

Fig. 1 A schematic of MFC with its various application

(negative) undergo an electrochemical reaction free reaction energy directing to a positive standard cell voltage (Gude 2016) (Fig. 1).

2 Classification of Wastewater Treated by Microbial Fuel Cell

Wastewater treated by MFCs can be categorize based on various parameters. One common classification is based on the source or type of wastewater. In this case, we can classify domestic wastewater treated by MFCs:

Primary classification based on source.

2.1 Domestic Wastewater

This category includes wastewater generated from households. It contains various contaminants such as organic matter, nutrients, pathogens, and sometimes household chemicals.

2.1.1 Secondary Classification Based on Treatment Objective

Wastewater treatment: when the primary goal is to clean and treat the wastewater to meet discharge standards or for reuse purposes. Resource recovery: when the focus is on recovering valuable resources like energy or nutrients (e.g., electricity, hydrogen, nitrogen, phosphorus) from wastewater.

2.1.2 Tertiary Classification Based on Contaminant Removal

Organic Matter Removal

If the primary aim is to remove organic compounds from domestic wastewater. Nutrient removal: when the emphasis is on reducing nutrients like nitrogen and phosphorus.

Pathogen Reduction

In cases where the treatment's primary goal is to reduce the presence of harmful microorganisms.

Domestic wastewater as a feedstock: domestic wastewater, often referred to as sewage or sanitary waste-water, is the wastewater generated from households, including kitchen, bathroom, and laundry wastewater. It contains a mixture of organic matter, nutrients (such as nitrogen and phosphorus), pathogens, and various contaminants.

Microbial Fuel Cells

These bio-electrochemical devices use microorganisms to catalyse the oxidation of organic matter present in wastewater, generating electrical energy in the process. They consist of an anode (where bacteria oxidize organic matter), a cathode (where oxygen reduction occurs), and a membrane or separator to separate the two chambers.

2.1.3 Treatment Process

Domestic wastewater is introduced into the anode chamber of the MFC. Microbes in the anode chamber break down organic matter in the wastewater, releasing these electrons flow through an external circuit to the cathode, creating an electric current. At the cathode, electrons combine with oxygen to form water. The treated wastewater, which has reduced organic content and potentially lower levels of contaminants, exits the anode chamber.

2.1.4 Benefits of Using MFCs for Domestic Wastewater Treatment

Energy Generation

MFCs produce electrical energy during the treatment process, which can be harnessed for various applications.

Reduced Organic Matter

MFCs can effectively remove organic matter from wastewater, reducing its biological oxygen demand (BOD) and chemical oxygen demand (COD). Resource recovery: they can recover valuable resources like electricity, hydrogen gas, and potentially other by-products.

Green Technology

MFCs offer a sustainable and environmentally friendly approach to wastewater treatment.

2.1.5 Challenges and Considerations

Microbial activity: the efficiency of MFCs depends on the microbial community's activity, which can be influenced by factors like temperature, pH, and the composition of wastewater.

Maintenance

MFCs require regular maintenance and monitoring to ensure stable performance.

Scale-Up

Scaling up MFCs for municipal wastewater treatment can be challenging and costly.

Electrode material: the choice of electrode material and design plays a crucial role in MFC performance.

Domestic wastewater treatment using MFCs is an area of ongoing research and development. While it holds promise for small-scale and decentralized wastewater treatment, further advancements are needed to make it more practical and cost-effective for larger municipal wastewater treatment facilitiestop of form.

2.2 *Cattle Manure*

Cattle manure can be used as a source material in MFCs to produce electricity and simultaneously treat the manure. Here's how the process works.

2.2.1 Cattle Manure as a Substrate

Cattle manure is rich in organic matter, including cellulose, lignin, and various organic compounds. This organic matter serves as a substrate for the microorganisms in the MFC's anode chamber.

MFCs are bio-electrochemical devices that use the metabolic activity of microorganisms to generate electricity. They consist of an anode (where oxidation of organic matter occurs), a cathode (where reduction of an electron acceptor, typically oxygen, occurs), and an electrolyte or membrane that separates the two chambers.

2.2.2 Treatment and Electricity Generation Process

Cattle manure is introduced into anode compartment of the MFC. Microorganisms naturally present in manure or inoculated into the system metabolize the organic matter in the manure as part of their normal metabolic processes. During this microbial metabolism, electrons are released as by-products. The released electrons are captured by the anode, creating an electric current. Protons (H^+) are also produced during the metabolic reactions and migrate through the electrolyte or membrane to the cathode.

At the cathode, electrons combine with oxygen or another electron acceptor to form water or other reduced compounds. This flow of electrons from the anode to cathode and across an outer circuit produce electricity.

2.2.3 Challenges and Considerations

The efficiency of MFCs using cattle manure depends on the microbial community's activity, which can be influenced by factors like temperature, pH, and the composition of the manure MFCs require regular maintenance and monitoring to ensure stable performance. Scaling up MFCs for large-scale cattle manure treatment may pose challenges, and the design and configuration of MFC systems need to be carefully considered.

2.3 Brewery Wastewater

Brewery wastewater is generated during the brewing process and contains various organic compounds, including sugars, starches, proteins, and by-products such as yeast cells and hops residues.

2.3.1 Treatment and Electricity Generation Process

Brewery wastewater is introduced into anode chamber of the MFC.

Microorganisms, which can be naturally present in the wastewater or inoculated into the system, metabolize the organic matter in the wastewater. During this metabolic process, microorganisms release electrons as by-products. At the cathode, electrons combine with oxygen or another electron acceptor to form water or other reduced compounds. This flow of electrons from the anode to cathode and across an outer circuit generates power.

2.3.2 Benefits of Using Brewery Wastewater in MFCs

MFCs provide an efficient means to treat brewery wastewater, reducing its organic load and potentially removing contaminants. The microbial breakdown of organic matter in brewery wastewater generates electrical energy, which can be used to offset energy costs or power on-site equipment. MFCs can recover valuable resources such as electricity and potentially other by-products, depending on the specific composition of the wastewater.

2.4 Waste Sludge

Waste sludge typically consists of organic solids, microorganisms, and water. It is a by-product of various processes, such as wastewater treatment, agriculture, and food processing.

MFCs are bio-electrochemical devices that utilize the metabolic activity of microorganisms to convert organic matter into electricity. They comprise an anode (where oxidation of organic matter takes place), a cathode (where reduction of an electron acceptor occurs, usually oxygen), and a separator or electrolyte that separates the two chambers.

2.4.1 Treatment and Electricity Generation Process

Waste sludge is introduced into anode compartment of the MFC. Microorganisms present inside sludge or inoculated into the system metabolize the organic matter in the sludge.

At the cathode, electrons combine with oxygen or another electron acceptor, forming water or other reduced compounds. This flow of electrons from the anode to the cathode through an external circuit generates electricity.

2.4.2 Benefits of Using Waste Sludge in MFCs

MFCs can effectively reduce the organic content of waste sludge, minimizing the volume that needs to be disposed of. the microbial breakdown of organic matter in waste sludge generates electrical energy, which can be used for various applications. Besides electricity, MFCs can potentially recover precious resources like nutrients (e.g., nitrogen, phosphorus) and metals from the sludge.

3 Specified Parameters of Microbial Fuel Cell

3.1 Anodic and Cathodic Reaction

The anode and cathode are positioned in electrolyte through distinct spaces that are divided by a membrane in a MFC. The microbes that produce electrons from the electron transport network are necessary for the microbial fuel cell to function. The microorganism's cell wall, which contains proteins involved in energy production, will allow the electron to leave. The bacterial metabolic activity, where the reduction process occurs and both the electrons and protons are produced, is what causes the establishment of a bio potential at the anodic compartment. And at the cathode side, which is where they are divided by a membrane, this causes the microbial fuel cell to produce bioelectricity.

The electrochemically energetic microbes can provide electrons to an anode (in anodic compartment), which are released by decomposing the biological materials, resulting in creation of an energy source. Exo-electrotrophs are the name given to microorganisms that take electrons.

The following is a representation of an oxidation process carried out in the anodic compartment by electrochemically active bacteria utilising acetate as fuel:

$$2HCO^{3-} + 9H^{+} + 8e^{-} = CH_3COO^{-} + 4H_2O$$

Exo-electrogens are microorganisms that are electrochemically active and capable of giving electrons.

The anode's electrochemically active bacteria create protons, which permeate through the membrane and into the cathodic chamber.

Due to its availability and significant reduction potential, oxygen is typically employed as an oxidant in the cathode chamber. When protons permeate through the membrane into the cathode side, they mix with the oxygen present, resulting in the creation of water.

The oxygen reduction reaction (ORR) is defined as:

$$O_2 + 4H^+ + 4e^- = 2H_20.$$

To be categorised as a microbial fuel cell, a cell must be capable of having its fuel supply oxidised at the anode side (wastewater).

These electrons pass from the anode to the cathode through an outer circuit, resulting in generation of power as a result of potential change connected to the electron flow. Protons created at the anode have the ability to move across the solution and across the polymer electrolyte membrane. Due to its advantageous characteristics, including as strong conductivity to cations, selectively permeable to protons, low internal resistance, and the capacity to withstand extended periods of inactivity without negatively impacting the MFC, polymer electrolyte membranes are the most often employed separators in MFCs.

3.2 *Materials for Microbial Fuel Cell*

An important component of a microbial fuel cell that affects both performance and cost is the electrode material. A substance must possess a variety of qualities in order to function as an electrode. It should be affordable, have advantageous electrochemical characteristics (favourable electron transport), be mechanically stable, and have a wide surface area, which will result in huge current densities.

Because they are more biocompatible than other electrodes and have a relatively cheap cost, carbon-based electrodes are utilised extensively. Precious metals should be avoided in order to keep costs down because their incorporation into electrode materials raises the price of electrodes.

3.2.1 Anode Substance

Graphite foam, Multilayer Graphene Coating on Stainless Steel Electrode, Carbon felt,, Reduced graphene oxide (particles), electrode modified with activated carbon. Carbon fibre, etc.

3.2.2 Cathode Substance

Carbon felt, Carbon foam, Carbon paper, Carbon cloth coated with platinum, Graphite plate, Platinum rod, Glass slide coated with a layer of Cr/Au or Ti/Pt, etc. (Table 1).

4 Microbial Fuel Cell Application in Wastewater Treatment

4.1 Organic Removal

The majority of MFC investigations employed synthetic media like glucose, acetate, sucrose, xylose, and other organic substrates. It is crucial to utilise these kinds of biodegradable organics to understand the biology of the MFC system since they are readily biodegradable by exo-electrogenic bacteria and have a straightforward metabolism. The design and material considerations in organic substrates were primarily made to increase power generation, Gugan and Robina (2017).

The carbon removal from MFCs biodegradable substrates is also discovered to be high (> 90%), demonstrating the high percentages of carbon removal from biodegradable substrates. MFCs are a suitable alternative for treating real wastewaters with high organic loading rates, such as those from food processing, industry, landfill leachate, household effluent, breweries, hospitals, farms, and other facilities. Following waste water treatment, it demonstrates a large reduction in organic matter, Erable et al. (2010a; b).

The most practical use of a microbial fuel cell for potential usage in practical settings is wastewater treatment. The high COD of the microbial fuel cell makes its application in organic wastewater unique.

According to one of the studies, MFC is also capable of effectively treating wastewater, with removal efficiencies of 93% at organic loading rates of 2.9 kg COD per cubic metre per day and 93% at 5.8 kg COD per cubic metre per day, Shimoyama et al. (2008).

4.2 Nitrogen Removal

Most wastewaters include nitrogen, mostly in the form of ammonium and organic nitrogen. With the aid of heterotrophic bacteria, organic nitrogen is transformed into ammonium by the ammonification process.

Table 1 Some of the anode/cathode materials and microbial composition

Anode material	Cathode material	Microbes	Fuel	Energy output	References
Carbon felt	Activated carbon, carbon black and poly(vinylideneflouride) binder	Aerobic mixed community	'High strength' swine wastewater	750 ± 70 mW m^{-2}	Yang et al. (2016)
Carbon cloth	Carbon cloth/carbon felt	Mixed community	–	3.5 A m^{2}	Blanchet (2016)
Graphite foam	Graphite foam	*R. ferrireducens*	–	74 mA m^{2}/445 mV	Chaudhuri and Lovely (2003)
3D few-layer graphene macroporous scaffolds	Glass slide coated with a layer of Cr/Au or Ti/Pt	Mixed community – *Geobacter* spp., enriched	–	11, 2000 W m^{-3}	Ren et al. (2006)
Carbon felt	Graphite felt	Anaerobic mixed community	Sludge from domestic wastewater	7.07 mW m^{-2}	Chen et al. (2013)

4.2.1 Process of Removing Nitrogen in an MFC Reactor

Clauwaert et al. (2007) made the first use of nitrate as an electron acceptor in the cathode of a microbial fuel cell reactor. It has been shown that autotrophic denitrification is possible without the production of H_2. This discovery on nitrogen removal by a microbial fuel cell biocathode has caught the interest of numerous researchers.

4.2.2 Microbiological Communities' Function in an MFC

Proteobacteria make up a large fraction of the bacteria present at the anode and cathode (Chae et al. 2009). The -Proteobacteria are typically found in the biofilm anode. There are microorganisms of this sort that can biodegrade organic substances. Thauera genera was found to make up 70.1% of the total -Proteobacteria in the acetate-enriched microbial fuel cell, according to research by Chae et al. (2009).

4.2.3 Factors that Affect How Much Nitrogen is Removed in MFC

Directly influencing nitrogen removal effectiveness are the Dissolved oxygen DO, pH, and C/N ratio. They have a considerable influence on a microbial fuel cell's efficiency as well. To run a microbial fuel cell for effective, reliable, and long-lasting nitrogen removal, it's also necessary to comprehend some of the other affecting factors, such as temperature and conductivity. In order for nitrification to occur, oxygen is necessary.

4.2.4 Issues with Removing Nitrogen with MFC

Applications of microbial fuel cells for wastewater nitrogen removal outperform traditional systems. However, there are still several obstacles to be addressed before this technology can be used to remove nitrogen in a way that is effective, stable, and sustainable. It is crucial to investigate the best layouts and operating procedures for various wastewater carbon–nitrogen ratios in order to reduce pollution and generate electricity. While removing nitrogen from waste water, the objective is to create neutral energy or positive energy treatment systems.

4.2.5 Conclusion

Microbial fuel cells have attracted a lot of attention in the wastewater treatment industry due to their remarkable benefits over traditional biological approaches. Studies that use microbial fuel cells for simultaneous energy generation and nitrogen and organic removal have produced positive results. A potent way to treat wastewater with a low carbon–nitrogen ratio is to include the short-cut nitrification–denitrification process into microbial fuel cells. This might accomplish 99% nitrogen removal with a carbon–nitrogen ratio of 9.6.

4.3 Phosphorus Removal

A few research looked at how phosphorous may be recovered or removed from wastewaters for use in microbial fuel cells. In recent studies, 70–82% of the phosphorus in large amounts of precipitate containing highly concentrated phosphorous on the surface of the liquid of the air cathode was removed, and struvite precipitation took place on the cathode surface in a microbial fuel cell (with a 70 mL of anodic volume) fed with swine wastewater. Phosphorus is recovered from waste water via struvite crystallisation. The formation of hydroxide and rise in pH are caused by the flow of alkali cations in addition to the proton consumption during oxygen reduction. Struvite is recoverable on the cathode side, although it becomes less soluble as pH rises, investigated the use of microbial fuel cells to extract phosphate from digested sewage sludge.

In this system, the iron phosphate, which contained digested sewage sludge, served as a location for the release of orthophosphate. For the phosphorus to be recovered as struvite, the orthophosphate-containing effluent from the microbial fuel cell required the addition of magnesium and ammonium as well as pH correction. When swine wastewater was used as fuel for an air–cathode microbial fuel cell, many crystals developed on the liquid side of the cathode's surface. A series of chemical characterizations later revealed the crystals to be struvite.

4.4 Metal Removal

One of the most alarming environmental problems is heavy metal contamination. The treatment of heavy metals is crucial due to their resistance, tenacity in the environment, and toxicity.

For the removal of heavy metals from waste water, both single and double chambered microbial fuel cells were used. The anodic chamber served primarily to break down organic matter, serving as both the carbon and electron donor, while the cathodic compartment removes heavy metals. In the part that follows, reduction of metals including chromium, zinc, copper, cobalt, selenite, vanadium, iron, cadmium, and silver is given and studied in both double (abiotic/biocathodes) systems (Table 2).

Table 2 Removal/recovery of metals in double chambered microbial fuel cells with biocathodes

Anode material	Cathode material	Substrate	Metal	Concentration	Pmax	Reference
Graphite plate	Graphite plate submerged in granular graphite	Acetate	Cr (VI)	39.2 mg/L	2.4 W/m^3	Huang et al. (2010)
Carbon fiber felt	Carbon fiber felt	Acetate	Cr (VI)	200 mg/L	529 mW/m^2	Li et al. (2017)
Sheet of graphite felt	Sheet of graphite felt	Glucose	Cr (VI)	20 mg/L	9.7 mW/m^2	Wu et al. (2017)
Graphite plate	Graphite plate	Acetate	Cr (VI)	63 mg/L	55.5 mW/m^2	Tandukar et al. (2009)

5 Challenges and Opportunities

Challenges include scaling up MFCs for practical use, optimizing microbial communities, and addressing cost limitations.

5.1 Challenges

5.1.1 Efficiency

MFCs are relatively inefficient at converting organic matter into electricity compared to other energy sources.

5.1.2 Scaling Up

Scaling MFCs for practical applications can be challenging due to issues like limited power output and the need for large microbial populations.

5.1.3 Maintenance

Maintaining the microbial community's health and stability can be complex and require constant monitoring.

Environmental conditions: MFC performance can be affected by factors like temperature and pH, making them sensitive to environmental fluctuations.

5.2 *Opportunities*

5.2.1 Sustainable Energy

MFCs can generate electricity from organic waste materials, offering a sustainable and environmentally friendly energy source.

5.2.2 Wastewater Treatment

MFCs can simultaneously treat wastewater while producing electricity, reducing the environmental impact of sewage treatment.

5.2.3 Bioremediation

MFCs can assist in the clean-up of contaminated sites by utilizing microorganisms to degrade pollutants.

5.2.4 Research and Innovation

Continued research may lead to improvements in MFC efficiency, making them more viable for various applications.

Overall, microbial fuel cells hold promise for addressing energy and environmental challenges, but their practical implementation requires overcoming several hurdles.

MFCs use of technical challenges including low electricity production, current instability, and high internal resistance.

- There are also some big challenges like:
- Biological matter removal efficiency
- Power density
- High principle and maintenance rate

5.3 *Power Density*

At the MFC, 88.3–98.7% of organic matter was removed with an average power density of 3432–3847 mW m^{-3} depending on the initial COD concentration. Coulombic efficiency (CE) is a frequently used parameter in MFC research to represent substrate conversion, higher current generation could result in a higher CE. However, the maximum power output does not occur with the highest current generation; thus, a high CE does not always mean high energy production. Power density is indeed a significant challenge in microbial fuel cells (MFCs). MFCs generate electricity by harnessing the metabolic activity of microorganisms to oxidize organic matter. However, several factors limit their power density:

1. Microbial Activity: The rate at which microorganisms can oxidize organic matter is relatively slow, leading to limited electron transfer and, consequently, low power output.
2. Electrode Surface Area: Increasing the electrode surface area can enhance power density, but it's often challenging to scale up without increasing cost and complexity.
3. Microbial Efficiency: Not all microorganisms are equally efficient at electron transfer, and optimizing the microbial community is an ongoing challenge.
4. Mass Transport: Effective transport of substrates and products to and from the microorganisms on the electrodes can be limited, affecting MFC performance.

The power density of a microbial fuel cell (MFC) can vary depending on several factors, including the type of microorganisms used, the design of the MFC, and the specific operating conditions. Generally, MFCs have lower power densities compared to other types of fuel cells like hydrogen fuel cells. Typical power densities for MFCs range from a few milliwatts to a few watts per square meter (mW/m^2 to W/m^2) of electrode surface area.

To improve the power density of an MFC, researchers work on optimizing electrode materials, microbial strains, and system configurations. Achieving higher power densities is an ongoing area of research in the field of microbial electrochemical technologies.

5.4 *Biological Matter Removal Efficiency*

To determine the exact removal efficiency for a particular MFC system, you would need to conduct experiments and measurements under specific conditions. The biological matter removal efficiency in a microbial fuel cell (MFC) depends on various factors, including the type of microorganisms used, the design of the MFC, and the characteristics of the wastewater being treated. In general, MFCs can be effective in removing organic matter from wastewater. They utilize microorganisms to break down organic compounds, generating electricity in the process, the organic matter present in the influent, and the operating conditions. Generally, MFCs are designed to treat organic wastewater while generating electricity. The removal efficiency can vary widely, but it's typically in the range of 70–90% for organic matter, such as COD (Chemical Oxygen Demand) removal.

Efficiency can vary, but MFCs have been used successfully to treat organic pollutants, such as in municipal wastewater treatment or in the treatment of certain industrial effluents. The specific removal efficiency will depend on the system's optimization and the nature of the contaminants.

Efficiency can be influenced by factors like hydraulic retention time, electrode materials, and the specific microbial community used. It's important to note that MFCs are not primarily designed for high-efficiency wastewater treatment like conventional treatment plants but rather as a sustainable technology that combines wastewater treatment with electricity generation. The exact efficiency will vary based on the specific MFC setup and goals of the system.

5.5 High Principle and Maintenance Rate

High principles and effective maintenance practices are fundamental to maximizing the electricity generation efficiency and lifespan of microbial fuel cells in various applications, including wastewater treatment, bioenergy production, and environmental monitoring.

5.6 High Principle

5.6.1 Principle of Electron Transfer

MFCs operate based on the principle of electron transfer from microorganisms (usually bacteria) to an electrode. Microbes break down organic matter, releasing electrons in the process. These electrons are transferred to the anode, creating an electric current.

5.6.2 Principle of Redox Reactions

The high principle involves the redox reactions occurring at the anode and cathode. Electrons are released at the anode through microbial metabolism and accepted by an electron acceptor (usually oxygen) at the cathode, forming water. This electron flow generates electricity.

5.6.3 Principle of Microbial Growth

The growth of electroactive microorganisms on the anode's surface is essential. These bacteria form biofilms, allowing efficient electron transfer to the electrode.

5.7 Maintenance Rate

5.7.1 Substrate and Nutrient Supply

Maintaining a continuous supply of organic substrate and nutrients for the microbes is crucial. This ensures a constant source of electrons for electricity generation. Regular monitoring and replenishment may be required.

5.7.2 Electrode Maintenance

The electrodes should be kept clean and free from biofouling. Biofilms can accumulate on the electrode surface, reducing electron transfer efficiency. Cleaning or replacing electrodes may be necessary.

5.7.3 Scaling Up and Scaling Down

Depending on the application, scaling up or down the MFC design and operation may be necessary to achieve the desired power output and efficiency.

5.7.4 Monitoring and Control System

Installing sensors and control systems to monitor key parameters such as voltage, current, pH, and temperature is essential for timely intervention and adjustments.

5.7.5 PH and Temperature Control

Maintaining optimal pH and temperature conditions for the microbial community is important. Microbes have specific pH and temperature ranges at which they thrive and produce electrons efficiently.

5.8 Scaling up of Fuel Cell

Method	Description
Stacking MFCs	Miniature MFCs linked together like Lego bricks. Offers modularity for increased voltage or current
Examples: Wastewater treatment, power generation from scaled stacks	

(continued)

(continued)

Method	Description
Electronic scale-up	Bypasses physical limitations with optimized electronic connections. Maintains power output with separate MFC units
Examples: Promising for maintaining power as the system scales	

5.8.1 Stacking MFCs

Stacking MFCs: A Modular Approach to Scaling Up.

Microbial fuel cells can be scaled up by stacking them, which involves joining several separate MFC units. This strategy has a number of benefits, such as:

Modularity

The simple addition or removal of individual MFC units enables variable system design based on particular power or treatment requirements.

Enhanced Power Output

The total voltage is raised by linking MFCs in series. On the other hand, if they are connected in parallel, the total current will grow.

Increased Efficiency

Stacking, as opposed to a single big MFC, enables better use of available space and resources.

There are two primary MFC stacking configurations:

5.9 Series Connection

The anode of one MFC is linked to the cathode of the following in this setup, creating a chain.

The sum of the voltages from each individual MFC in the stack equals the total voltage output.

When aiming for a greater voltage to power particular applications, this approach is perfect.

One disadvantage is that the MFC in the chain with the lowest output limits the current.

5.9.1 Parallel Connection

In this case, every anode and every cathode is connected to each other.

This arrangement raises the stack's overall current output.

It works well for uses when a higher current is needed, including charging low-power electronics.

5.9.2 Here's a Deeper Look at the Considerations for Stacking MFCs

Membrane Selection

The separator membrane positioned in between each MFC unit is essential for correct stacking. Proton exchange membranes (PEMs), which separate the anode and cathode chambers and permit protons to pass through, are commonly used in this membrane to avoid unintentional solution mixing.

Flow Configuration

Careful design is required for the influent (wastewater) and effluent (treated water) flows through the stacked system. Different flow configurations exist, including co-flow, counter-flow, and single-pass, each with unique benefits and drawbacks for power generation and treatment efficiency.

MFC Design Optimization

The layout of each MFC unit in the stack is also very important. Factors like electrode material, chamber size, and operational parameters like substrate feed rate and oxygen supply need to be optimized for efficient performance in the stacked configuration.

5.9.3 Examples of Stacked MFC Applications

Treatment of Wastewater

Stacked MFCs may produce electricity and handle wastewater at the same time. The microorganisms in the wastewater use the organic stuff as fuel, which helps with treatment and power production.

Bioenergy Production

In larger-scale systems, stacks can be used to produce bioenergy from a variety of organic materials. This provides a sustainable method for producing renewable energy and managing garbage.

5.9.4 While Stacking MFCs is a Promising Approach, Challenges Remain

Internal Resistance

Internal resistance can increase with the number of MFCs in the stack, which can reduce total power output. It is essential to choose low-resistance materials and optimize connections.

Leakage

The biological processes inside the stack may be disrupted and efficiency may be decreased by leakage between compartments. Useful sealing methods must be used.

Long-Term Stability

Ongoing research is necessary to maintain stable operation and microbial activity over extended periods of time. This includes elements like optimal nutrient delivery and biofouling control.

One effective method for increasing the size of microbial fuel cells is to stack them. Stacking MFCs have great promise for bioenergy production, wastewater treatment, and a more sustainable future by tackling current issues and continuously improving designs.

5.9.5 Electronic Scale-Up

An inventive way to get beyond the physical constraints on growing the size of microbial fuel cells (MFCs) is through electronic scale-up. Rather than physically constructing one enormous MFC, this approach focuses on improving the electronic connections between multiple MFC units.

The salient features of electronic scale-up are broken down as follows:

Dispersed MFC Devices

Electronic scale-up maintains the MFCs apart from one other, in contrast to stacking, which connects them physically. This gets rid of problems with constraints on mass transit and inefficiencies inside a big single unit. Because each MFC unit runs on its own, ideal conditions for microbial activity and power generation within each chamber are guaranteed.

Power Management System

The power management system is the key component of electronic scale-up. This system gathers the electricity produced by every single MFC unit in an intelligent manner. In order to guarantee effective power delivery, it then controls and optimizes the electron flow across the whole system.

Within the field of electronic scale-up, there are two main approaches:

Boost Converters

They are electrical circuits that raise the output voltage of every MFC unit. The power management system can raise the voltage from individual units to satisfy the needs of the target application by carefully placing boost converters. Because of this, power output can be increased without the MFCs having to be physically larger.

MPPT Stands for Maximum Power Point Tracking

MFCs, like many other power sources, aren't always running at full capacity. The power management system uses the MPPT approach to dynamically modify the electrical demand on each MFC unit. This maximizes the system's total power output by guaranteeing that every unit runs at its peak power.

Electronic scale-up's benefits.

Breaks Through Physical Barriers

There is no need to construct a huge, labor-intensive MFC. Systems that are scalable and modular and can be readily customized to meet particular power needs are made possible by this approach.

Enhanced Efficiency

Electronic scale-up may be able to attain higher overall efficiency than conventional stacking techniques by maximizing electron flow and guaranteeing that every MFC unit performs at its best.

Flexibility in MFC Design

Different MFC units can have their own unique designs and optimizations made for things like operational parameters and material selection. According to the intended use, customization is possible thanks to this flexibility.

Difficulties with Electronic Scale-Up:

Complexity of the Power Management System

Advanced electronics knowledge is needed to design and install an effective power management system using boost converters and MPPT.

System Control and Optimization

To handle changes in power output from individual MFC units and guarantee overall system stability, the power management system's control algorithms must be highly developed.

Cost Considerations

Compared to more straightforward stacking techniques, the MFC system's total cost may rise due to the additional electronic components needed for the power management system.

6 Future Perceptions

The perception of microbial fuel cells (MFCs) in the future is likely to be positive and evolving. MFCs have the potential to play a significant role in sustainable energy production and wastewater treatment. As technology advances and environmental concerns grow, MFCs may gain more recognition and adoption for their ability to generate electricity from organic matter while reducing pollution. Public perception may become increasingly favourable as these benefits become more apparent, especially if they are integrated into broader renewable energy and environmental sustainability initiatives. However, challenges like scalability and cost-effectiveness may still need to be addressed for widespread acceptance. It also incorporates several prospective for MFC commercialization and scaled-up industrial usage options in the future.

The future perception of microbial fuel cells is quite promising. MFCs are a sustainable technology that uses microorganisms to generate electricity from organic matter, making them an attractive option for renewable energy generation and wastewater treatment. Here are some potential future developments:

1. Increased Efficiency:
   Researchers are working on improving MFC efficiency, both in terms of electricity production and wastewater treatment. This could lead to more widespread adoption in various applications.
2. *Scaling up*:
   Scaling up MFCs for larger applications, such as industrial wastewater treatment or even grid-scale electricity generation, is a focus area. Success in this regard could have a significant impact on sustainable energy production.
3. *Integration into Existing Infrastructure*:
   Integrating MFCs into existing wastewater treatment plants and agricultural systems can help generate energy from waste streams and reduce operational costs.
4. *Biotechnology Advancements*:
   Advances in biotechnology may enable the development of more efficient and robust microbial communities for MFCs, further enhancing their performance.
5. *Hybrid Systems*:
   Combining MFCs with other renewable energy sources like solar or wind could provide a more consistent and reliable energy supply.
6. *Environmental Applications*:
   MFCs may find broader use in environmental monitoring, helping to power remote sensors and data collection devices in ecosystems and water bodies.
7. *Education and Awareness*:
   Increasing public awareness and education about MFCs and their benefits may accelerate their adoption and integration into society.

   - Overall, microbial fuel cells hold great potential in addressing energy and environmental challenges in the future, provided that research and development efforts continue to advance the technology and its applications.

7 Conclusion

The conclusion of a microbial fuel cell (MFC) study or experiment can vary depending on the specific goals and findings of the research. However, a typical conclusion might summarize key points such as:

1. *Energy Generation*:

MFCs have shown promise as a sustainable technology for electricity generation through the microbial conversion of organic matter.

2. *Efficiency*:
   The efficiency of MFCs in converting organic substrates into electricity remains an area of improvement, and further optimization is needed to increase their practical utility.
3. *Environmental Benefits*:
   MFCs offer potential environmental benefits by utilizing waste materials as fuel and potentially treating wastewater simultaneously.
4. *Microbial Communities*:
   The role of microbial communities in MFC performance is significant, and understanding and optimizing these communities can enhance MFC efficiency.
5. *Applications*:
   MFCs have various applications, including wastewater treatment, remote power generation, and sensor devices, but their scalability and cost-effectiveness need further investigation for broader adoption.
6. *Challenges*:
   Challenges such as electrode material selection, scaling up MFCs for practical use, and economic viability need to be addressed to unlock the full potential of this technology.

- In summary, microbial fuel cells hold promise as a sustainable energy and wastewater treatment technology, but further research and development are necessary to overcome existing challenges and make them more practical and efficient for real-world applications.

References

Blanchet E, Erable B, Solan MLD, Bergel A (2016) Two-dimensional carbon cloth and three-dimensional carbon felt perform similarly to form bioanode fed with food waste. Electrochem Commun 66:38–41

Chae KJ, Choi MJ, Lee JW, Kim KY, Kim IS (2009) Effect of different substrates on the performance, bacterial diversity, and bacterial viability in microbial fuel cells. Bioresour Technol 100(14):3518–3525

Chaudhuri SK, Lovely DR (2003) Electricity generation by direct oxidation of glucose in mediatorless microbial fuel cells. Nat Biotechnol 21:1229–1232

Chen CY, Chen TY, Chung YC (2013) A comparison of bioelectricity in microbial fuel cells with aerobic and anaerobic anodes. Environ Technol 35:286–293

Clauwaert P, Rabaey K, Aelterman P, Schamphelaire LD, Pham TH, Boeckx P, Boon N, Verstraete W (2007) Biological denitrification in microbial fuel cells. Environ Sci Technol 41(9):3354–3360

Erable B, Duteanu NM, Ghangrekar MM, Dumas C, Scott K (2010a) Application of electro-active biofifilms. Biofouling 26:57–71

Erable B, Duteanu NM, Ghangrekar MM, Dumas C, Scott K (2010b) Application of electro-active biofifilms. Biofouling 26:57–71

Gude VG (2016) Wastewater treatment in microbial fuel cells—an overview. J Clean Prod 122:287–307

Gugan J, Robina F (2017) Microbial fuel cells and their applications for cost effective water pollution remediation. Proc Natl Acad Sci India B Biol Sci 87:625–635

Huang WE, Li M, Jarvis RM, Goodacre R, Banwart SA (2010) Chapter 5 - Shining Light on the Microbial World: The Application of Raman Microspectroscopy. Adv App Microbiol 70:153–186

Li SW, He H, Zeng RJ, Sheng GP (2017) Chitin degradation and electricity generation by Aeromonas hydrophila in microbial fuel cells. Chemosphere 168:293–299

Logan BE, Hamelers B, Rozendal R (2006) Microbial fuel cells: methodology and technology. Environ Sci Technol 40(17):5181–5192

Ren H, Tian L, Gu Q, Zhu W (2006) Secalonic acid D; A cytotoxic constituent from marine lichen-derived fungusGliocladium sp. T31. Arch Pharmacal Res 29:59–63

Shimoyama T, Komukai S, Yamazawa A, Ueno Y, Logan BE, Watanabe K (2008) Electricity generation from model organic wastewater in a cassette-electrode microbial fuel cell. Appl Microbiol Biotechnol 80:325–330

Tandukar M, Tezel U, Pavlostathis SG (2009) Biological Chromium(VI) Reduction in Microbial Fuel Cell: A Three in One Approach. Proc Water Environ Fed 17:527–535

Wu X, Xiong X, Brunetti G, Yong X, Zhou J, Zhang L, Wei P, Jia H (2017) Effect of MWCNT-modified graphite felts on hexavalent chromium removal in biocathode microbial fuel cells. RSC Adv 7:53932–53940

Yang E, Chae KJ, Alayande AB, Kim KY, Kim IS (2016) Concurrent performance improvement and biofouling mitigation in osmotic microbial fuel cells using a silver nanoparticle-polydopamine coated forward osmosis membrane, J Membr Sci 513:217–225.

Commercialization and Current Market of the Microbial Electrochemical Technologies as Biosensor and Remediation

Rahul Kandpal, Syed Wazed Ali, and Shaikh Ziauddin Ahammad

Abstract Microbial electrochemical technologies (METs) have emerged as promising tools for biosensing and environmental remediation applications. This book chapter provides an in-depth exploration of the commercialization prospects and current market trends of METs in the context of biosensors and remediation. The chapter begins by highlighting the significance of biosensors in biotechnology and the limitations of traditional biosensing approaches. It then delves into the principles and mechanisms underlying METs, focusing on the interplay between microorganisms and electrodes through extracellular electron transfer (EET) mechanisms. The progress made in MET-based biosensors, including advancements in electrode materials, configurations, and microbial consortia, is discussed. Additionally, the chapter showcases the diverse applications of MET-based biosensors, ranging from medical diagnostics to environmental monitoring, through case studies highlighting successful implementations. The market prospects for MET-based biosensors are examined, analyzing key players, market size, and potential growth areas. Furthermore, the chapter explores the market opportunities for MET-based remediation technologies, emphasizing their potential to offer cost-effective and sustainable solutions in the environmental remediation sector. The challenges associated with commercializing METs, such as technological optimization, regulatory requirements, and economic viability, are thoroughly discussed. Finally, the chapter summarizes the

R. Kandpal · S. W. Ali (✉)
School of Interdisciplinary Research (SIRe), Indian Institute of Technology Delhi, Hauz Khas, New Delhi 110016, India
e-mail: wazed@textile.iitd.ac.in

R. Kandpal
e-mail: Rahul.Kandpal@sire.iitd.ac.in

R. Kandpal · S. W. Ali · S. Z. Ahammad (✉)
Department of Textile and Fiber Engineering, Indian Institute of Technology Delhi Hauz Khas, New Delhi 110016, India
e-mail: zia@dbeb.iitd.ac.in

R. Kandpal · S. Z. Ahammad
Department of Biochemical Engineering and Biotechnology, Indian Institute of Technology Delhi, Hauz Khas, New Delhi 110016, India

R. Selvasembian et al. (eds.), *Emerging Trends in Microbial Electrochemical Technologies for Sustainable Mitigation of Water Resources Contamination*,
https://doi.org/10.1007/978-3-031-74636-9_13

commercialization prospects and market trends of METs as biosensors and remediation tools, intending to inspire further research and development in this dynamic field.

Keywords Microbial electrochemical technology · Bioelectrochemical system · Electroactive microbes · Microbial fuel cell · Microbial electrolysis cell · Bioelectrosynthesis

1 Introduction

Microbial electrochemical technologies (METs) use microorganisms to generate electricity or produce valuable chemicals from organic waste or pollutants (Logan and Rabaey 2012; Ghangrekar et al. 2023). They can also be used as biosensors to detect the presence and concentration of various contaminants or analytes in water or soil (Tiquia-Arashiro and Pant 2019; Simoska et al. 2021). METs have many potential applications in environmental remediation, biorefinery, bioenergy, biomedicine, agriculture, biofuel production, biosensor development (Kandpal et al. 2021, 2022c), nutrient recovery, wastewater treatment (Kandpal et al. 2023), and heavy metals remediation from industry effluents (Nara et al. 2020; Simoska et al. 2021). However, despite these promising potentials, the scaling-up and commercialization of METs are still challenging due to technical, economic, and environmental barriers. Some of these challenges include the optimization of electrode materials (Kandpal et al. 2022a; b), membrane properties, reactor design, microbial community structure, and operational parameters (Wang et al. 2020; Chung and Dhar 2021; Jadhav et al. 2022). Moreover, the cost-effectiveness and environmental impact of METs need to be evaluated and compared with conventional technologies (Jadhav et al. 2022). Some recent advances and innovations in METs include the use of 3D printing for fabricating customized electrodes and reactors (Chung and Dhar 2021), the development of microbial electrosynthesis for producing value-added chemicals from CO_2 (Das et al. 2020; Prévoteau et al. 2020; Zhang et al. 2021), and the integration of METs with other renewable energy sources such as solar and wind power (Ganiyu and Martínez-Huitle 2020; Wang et al. 2022).

Over the past few decades, extensive researchas shown in Fig. 1, has been conducted to explore the potential of METs, leading to significant advancements in their design, performance, and scalability (Deng et al. 2023). This chapter aims to provide a comprehensive overview of the commercialization prospects and current market of METs as biosensors and remediation tools, highlighting the advancements, market prospects, challenges, and opportunities that shape their adoption and integration into real-world applications.

The chapter provides a comprehensive overview of METs, elucidating the underlying principles and mechanisms that drive their functionality. METs leverage the synergistic interactions between microorganisms and electrodes to enable the transfer of electrons and catalyze electrochemical reactions. This unique interface

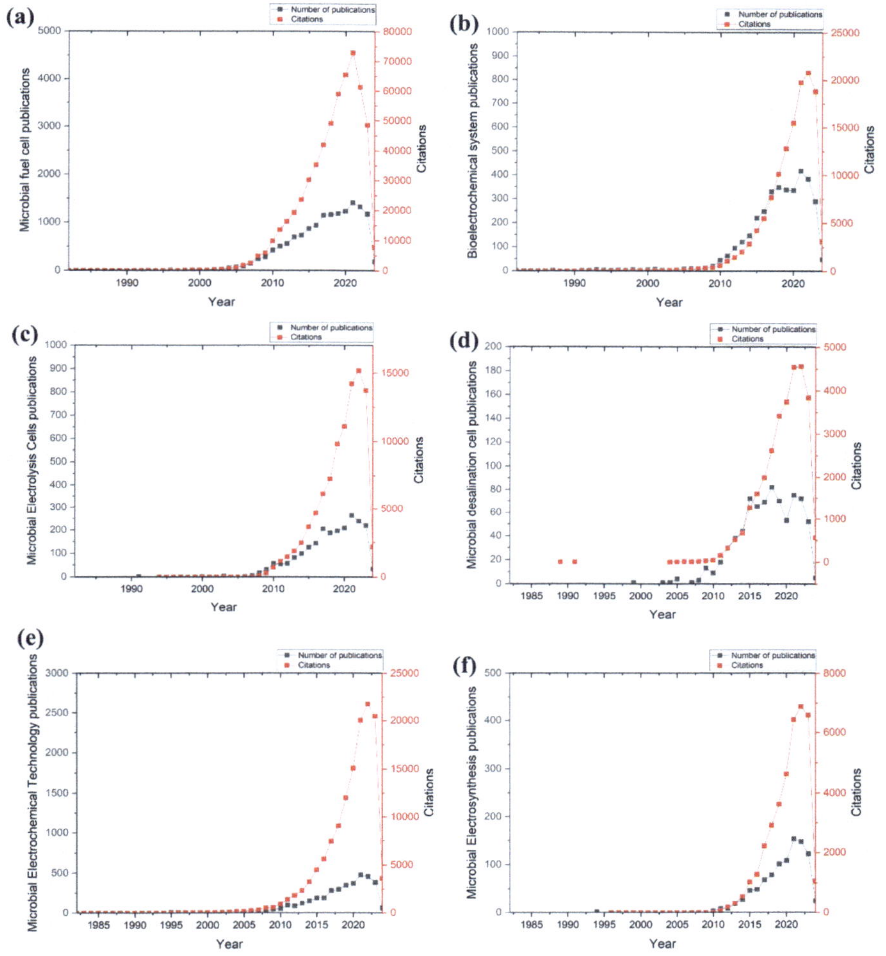

Fig. 1 Quantitative analysis of scientific literature on **a** microbial fuel cell, **b** bioelectrochemical system, **c** microbial electrolysis cells, **d** microbial desalination cell, **e** microbial electrochemical technology, and **f** microbial electrosynthesis from the year 1982 to 2024 (*Source* ISI Web of Science, April 2024)

between biology and electrochemistry allows METs to exhibit distinct capabilities as biosensors and remediation tools, making them attractive for commercialization.

The chapter then delves into the commercialization prospects of MET-based biosensors, exploring their significance in biotechnology. Biosensors based on METs offer a sensitive, specific, and label-free detection platform for various analytes, including toxins, pathogens, and metabolites. The ability of MET-based biosensors to harness microbial communities' inherent selectivity and sensitivity, coupled with the advantages of electrochemical detection, positions them as powerful tools in environmental monitoring, healthcare diagnostics, food safety, and industrial process control. The chapter examines the progress in MET-based biosensors, including developing novel electrode materials, sensor designs, and signal amplification strategies that have paved the way for their commercialization and market adoption.

Furthermore, the chapter explores the commercialization potential of MET-based technologies in environmental remediation. METs offer sustainable and efficient approaches for treating contaminated soil, water, and air. The chapter discusses the progress in MET-based remediation technologies, including the enhancement of performance, integration with other treatment processes, successful scale-up, and cost-reduction efforts (Zhang and Angelidaki 2016). These advancements have propelled the market prospects of METs in environmental remediation, addressing the growing concerns of pollution and the need for sustainable solutions. The chapter also examines the specific applications of METs in soil remediation, groundwater treatment, wastewater management, marine, and aquatic remediation, and air pollution control, highlighting their commercial viability and potential for market expansion.

Alongside the opportunities, the chapter also addresses the challenges hindering the commercialization of METs. These challenges include technological limitations, standardization issues, regulatory frameworks, economic viability, and market acceptance. The chapter highlights the importance of interdisciplinary collaborations among researchers, industry stakeholders, and regulatory bodies to overcome these challenges and drive the commercialization of MET-based technologies. Collaborative efforts enable knowledge exchange, technology transfer, and the establishment of guidelines and regulations that ensure the safe and effective implementation of METs in biosensing and remediation applications.

Finally, the chapter provides an outlook on future commercialization and market development perspectives for METs. It explores emerging trends, potential applications, and untapped market segments that can further drive the adoption and integration of METs. The chapter emphasizes the need for continued research, innovation, and investment in MET-based technologies to unlock their full commercial potential and contribute to a more sustainable and cleaner future.

This book chapter is a comprehensive guide to the commercialization and current market of microbial electrochemical technologies as biosensors and remediation tools. It explores the significance, advancements, market prospects, challenges, and opportunities associated with METs, shedding light on their role in shaping the landscape of biosensing and environmental remediation. By providing insights into the current state and prospects of METs, this chapter aims to facilitate the broader adoption and commercialization of these technologies, contributing to a more sustainable and environmentally conscious society.

The primary objective of this chapter is to present a scholarly examination of the commercialization prospects and current market dynamics concerning microbial electrochemical technologies (METs) employed as biosensors and remediation tools. It endeavors to distill intricate scientific principles into a lucid narrative, elucidating the interplay between MET's fundamental mechanisms, industrial applications, market penetration, and associated challenges. By amalgamating empirical data, theoretical frameworks, and industry perspectives, this chapter seeks to offer an academically rigorous appraisal of METs' potential in addressing contemporary environmental and biotechnological imperatives. Through a rigorous analysis of research findings, technological advancements, regulatory considerations, and market trends, it endeavors to furnish scholarly insights that may inform policy-making, research agendas, and industrial strategies pertinent to METs' integration into broader biosensing and environmental remediation frameworks.

2 Current Market Segments for METs

Some of the current market segments for METs are:

- **Wastewater treatment:** METs can reduce wastewater's organic load and nutrient content while recovering energy or value-added products (Tiquia-Arashiro and Pant 2019). They can also degrade recalcitrant pollutants such as pharmaceuticals, pesticides, and dyes (Anwer et al. 2020; Ghangrekar et al. 2023).
- **Biosensors:** METs can sense the changes in the electrical signals generated by the microbial communities in response to different environmental conditions or stimuli. They can be used to monitor water quality, soil health, bioreactor performance, or disease diagnosis (Tiquia-Arashiro and Pant 2019; Simoska et al. 2021; Ghangrekar et al. 2023).
- **Bioremediation:** METs can enhance the biodegradation of contaminants in soil or groundwater by providing an external electron acceptor or donor. They can also stimulate the growth and activity of beneficial microorganisms that can degrade or transform pollutants (Tiquia-Arashiro and Pant 2019; Wang et al. 2020, 2022).
- **Bioelectrosynthesis:** METs can produce valuable chemicals such as hydrogen, methane, acetate, ethanol, butanol, and organic acids from carbon dioxide or organic waste. They can also synthesize complex molecules such as proteins, enzymes, or pharmaceuticals (Logan and Rabaey 2012; Tiquia-Arashiro and Pant 2019; Ghangrekar et al. 2023).

2.1 Wastewater Treatment

2.1.1 Significance of Wastewater Treatment

Microbial electrochemical technologies (METs) hold great promise for wastewater treatment applications as shown in Fig. 2 which demonstrates continuous flow microbial fuel cells. This section explores the significance of METs in wastewater treatment and highlights their potential for commercialization in this field.

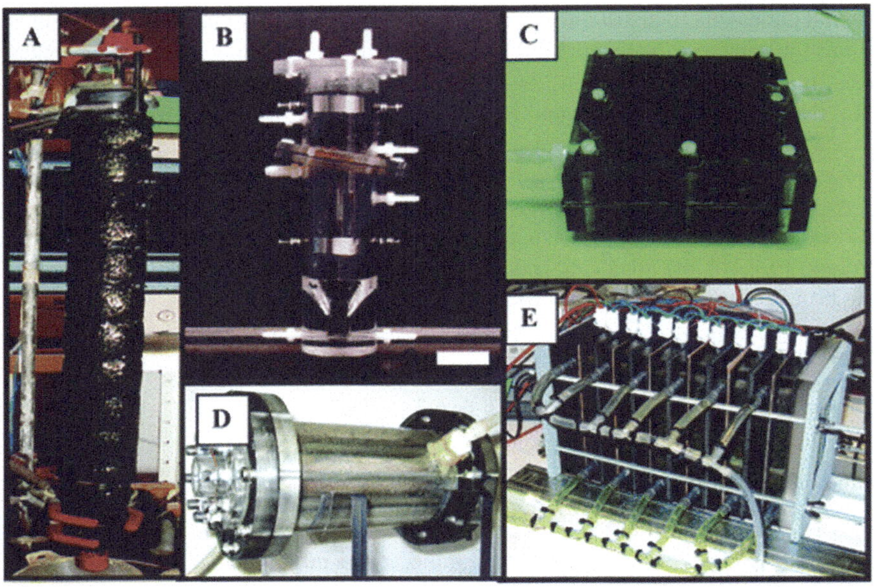

Fig. 2 Continuous flow MFCs. **a** Tubular upflow MFC with outer cathode and ineer graphite bed, **b** tubular upflow MFC with cathode above and anode below, **c** flat plat MFC, **d** single chamber MFC with cathode inside, **e** MFC stack [*Source* Reprinted with permission from Logan et al. (2006) Copyright {2024} American Chemical Society]

Enhanced Treatment Efficiency

METs offer the potential to significantly improve wastewater treatment processes' efficiency. Traditional treatment methods, such as activated sludge systems, can be limited by slow degradation rates, high energy consumption, and the need for costly chemicals. METs can overcome these limitations by harnessing the electroactivity of microorganisms, which enables direct electron transfer to or from solid electrodes. This direct electron transfer pathway enhances the degradation of organic pollutants, accelerates the removal of nutrients (such as nitrogen and phosphorus), and improves overall treatment efficiency (Ramírez-Vargas et al. 2018b; Ghangrekar et al. 2023).

Resource Recovery

Wastewater contains valuable resources such as energy, nutrients, and organic matter (Solon et al. 2019). METs provide an opportunity to recover these resources, transforming wastewater treatment plants into resource recovery centers (Kehrein et al. 2020). For example, the metabolic activity of microorganisms in METs can be harnessed to produce hydrogen gas, methane, or other valuable chemicals from the organic content of wastewater. Additionally, METs can facilitate the recovery of nutrients like phosphorus and nitrogen, which can be utilized as fertilizers or for other applications (Das et al. 2021a). The potential for resource recovery in wastewater treatment using METs presents a compelling economic opportunity (Ghangrekar et al. 2023).

Energy Generation

METs can generate renewable energy during wastewater treatment (Das et al. 2021b). The electrochemical reactions in METs produce electrical current, which can be harvested as usable energy. This energy can be used to power the wastewater treatment process or even fed back into the grid. By integrating METs into wastewater treatment plants, energy self-sufficiency can be achieved, reducing the reliance on external energy sources and enhancing the sustainability of the treatment process (Siddiqui et al. 2021; Das et al. 2021b; Wang et al. 2022; Ghangrekar et al. 2023).

Nutrient Removal

Nutrient pollution, particularly nitrogen and phosphorus, is a significant concern in wastewater treatment. Excessive nutrient discharge can cause eutrophication in receiving water bodies, leading to ecological imbalances and harmful algal blooms. METs offer a potential solution for efficient nutrient removal. The microbial processes within METs can facilitate the conversion of nitrogen into harmless nitrogen gas through processes like denitrification. Additionally, phosphorus can be selectively precipitated and recovered as struvite, a valuable fertilizer. METs can provide a sustainable and cost-effective approach to nutrient removal in wastewater treatment plants (Gupta et al. 2016; Rajesh Banu et al. 2021; Nguyen et al. 2022).

Decentralized and Modular Systems

MET-based wastewater treatment systems can be designed as decentralized and modular units, offering flexibility in implementation. This modular approach allows for the construction of smaller treatment units that can be distributed closer to the point of wastewater generation (Ramírez-Vargas et al. 2018b; Fudge et al. 2021). Decentralized MET systems can reduce the need for extensive sewer networks and centralized treatment plants, enabling cost savings and minimizing environmental impact (Ramírez-Vargas et al. 2018b; Guo et al. 2020; Jadhav et al. 2020; Tsekouras et al. 2022). The compact nature of MET systems also makes them suitable for retrofitting existing treatment facilities, further enhancing their commercial potential (Ramírez-Vargas et al. 2018b; Li et al. 2023).

Technological Advancements

Ongoing research and development efforts in METs continue to advance the technology's capabilities in wastewater treatment. Novel electrode materials, improved reactor designs, and optimized microbial communities are being explored to enhance treatment efficiency, stability, and scalability. Additionally, advancements in sensor technologies and process control systems contribute to real-time monitoring and optimization of MET-based wastewater treatment processes. These technological advancements create opportunities to commercialize METs in the wastewater treatment industry (Kesari et al. 2021; Das et al. 2021a, b; Mojiri and Bashir 2022).

In conclusion, METs offer significant opportunities for commercialization in wastewater treatment. The potential for enhanced treatment efficiency, resource recovery, energy generation, nutrient removal, decentralized systems, and ongoing technological advancements positions METs as a promising technology for the future of wastewater treatment. As the scalability, cost-effectiveness, and robustness of METs are further demonstrated, their commercial adoption in the wastewater treatment sector is likely to increase, contributing to sustainable and efficient wastewater management practices.

2.1.2 Progress in MET-Based Wastewater Treatment

Microbial electrochemical technologies (METs) have shown significant progress and promise in the field of wastewater treatment. This section highlights the recent advancements and achievements in MET-based wastewater treatment, demonstrating the technology's potential for commercialization and market adoption.

Enhanced Organic Matter Removal

METs have demonstrated remarkable performance in removing organic matter from wastewater. The electroactive microorganisms involved in METs can directly oxidize organic pollutants, promoting efficient degradation and mineralization. This direct electron transfer pathway significantly accelerates the breakdown of complex organic compounds, reducing chemical oxygen demand (COD) and biochemical oxygen demand (BOD). MET-based systems, such as microbial fuel cells (MFCs) and bioelectrochemical systems (BESs), have shown superior organic matter removal capabilities compared to conventional treatment methods (Shahid et al. 2021; Wang et al. 2023a).

Nutrient Removal and Recovery

Effective nutrient removal is a crucial aspect of wastewater treatment. METs offer innovative approaches for nutrient removal, particularly nitrogen and phosphorus.

By integrating appropriate electrode materials and manipulating the operating conditions, METs can facilitate denitrification processes, converting nitrogen into harmless nitrogen gas. Additionally, the electrochemical conditions within METs can promote the selective precipitation and recovery of phosphorus as struvite, a valuable fertilizer. These nutrient removal and recovery advancements contribute to sustainable wastewater treatment practices and resource conservation (Baby and Ahammed 2022).

Microbial Community Engineering

The understanding and engineering of microbial communities are central to the progress of MET-based wastewater treatment. Researchers have focused on optimizing and controlling the composition and activity of electroactive microbial consortia. Strategies such as biofilm development, inoculum selection, and microbial community enrichment have been employed to enhance the performance and stability of MET systems. Advances in metagenomics and microbial ecology have provided insights into the key microbial players and their functions, enabling targeted manipulation and optimization of microbial communities for improved wastewater treatment outcomes (Glaven 2019, 2022b, 2023; Bird et al. 2019, 2021; Lai et al. 2022).

Integration with Conventional Treatment Processes

Integrating METs with existing wastewater treatment processes has shown great potential for enhanced performance and efficiency. METs can be incorporated as a polishing step following conventional treatment methods, allowing for further removal of organic matter, nutrients, and trace contaminants. This integration provides an opportunity to upgrade and optimize the overall treatment process, improving effluent quality and reducing the environmental impact of wastewater discharge. The compatibility of METs with conventional treatment infrastructure makes them attractive for retrofitting existing wastewater treatment plants (Tejedor-Sanz et al. 2016; Ramírez-Vargas et al. 2018a; Das et al. 2021a, b; Zakaria and Dhar 2022).

Technology Demonstration and Pilot-Scale Studies

Successful demonstrations and pilot-scale studies have been conducted to validate the feasibility and effectiveness of MET-based wastewater treatment. These studies have provided crucial data on system performance, stability, and scalability under real-world conditions. Demonstrating the viability and reliability of METs at larger scales is essential for gaining the confidence of investors, regulators, and end-users. The

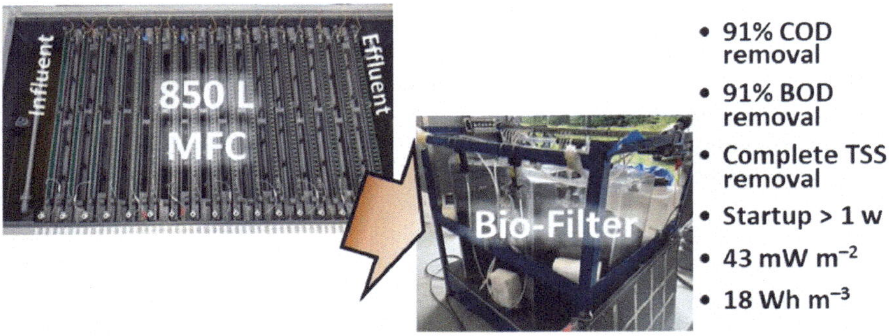

Fig. 3 Pilot-scale 850 L MFC paired with biofilter for wastewater treatment [*Source* Reprinted with permission from Rossi et al. (2022) Copyright {2024} Elsevier]

outcomes of these technology demonstrations have paved the way for the commercialization and market adoption of MET-based wastewater treatment systems (Cusick et al. 2011; Tota-Maharaj and Paul 2015; Muhamad Ng et al. 2021; Bird et al. 2022a; Rossi et al. 2022) (Fig. 3).

Cost Reduction and Process Optimization

Significant efforts have been made to reduce the costs associated with MET-based wastewater treatment and optimize the process parameters. Advances in electrode materials, system design, and operational strategies have contributed to cost savings, energy efficiency, and process optimization (Zhang et al. 2022a). The development of low-cost electrode materials and efficient electrode architectures can potentially reduce the capital and operational expenses of MET systems, making them more economically viable for widespread adoption (Priyadarshini et al. 2022).

In conclusion, the progress made in MET-based wastewater treatment demonstrates the technology's potential for commercialization and market penetration. The advancements in organic matter removal, nutrient removal, and recovery, microbial community engineering, integration with conventional treatment processes, technology demonstrations, cost reduction, and process optimization collectively drive the commercial prospects of METs in the wastewater treatment industry. Continued research, development, and collaboration among researchers, industry stakeholders, and policymakers will further propel the commercialization of MET-based wastewater treatment technologies, leading to sustainable and efficient solutions for global water management challenges.

2.1.3 Market Opportunities for MET-Based Wastewater Treatment

Microbial electrochemical technologies (METs) offer significant market opportunities in wastewater treatment. As the technology advances and demonstrates its effectiveness, several market opportunities are emerging. This section explores the potential market opportunities for MET-based wastewater treatment and highlights their commercialization prospects.

Growing Global Wastewater Treatment Market

The global wastewater treatment market is experiencing significant growth due to increasing population, urbanization, and stringent environmental regulations. The need for sustainable and cost-effective wastewater treatment solutions has created a demand for innovative technologies like METs (Mojiri and Bashir 2022). METs offer unique advantages such as enhanced treatment efficiency, resource recovery, and energy generation, making them attractive options for wastewater treatment plant operators, municipal authorities, and industrial sectors. The growing market size provides ample opportunities for commercializing and widespread adoption of MET-based wastewater treatment technologies (Ramírez-Vargas et al. 2018b).

Industrial Applications

METs have vast potential for wastewater treatment in various industrial sectors, including food and beverage, pharmaceuticals, pulp and paper, and chemical manufacturing. These industries generate high-strength wastewater with complex organic compounds and challenging pollutants. METs can effectively remove organic matter, nutrients, and recalcitrant compounds, providing sustainable and cost-efficient treatment solutions. By addressing different industries' specific wastewater treatment need MET-based technologies can tap into niche markets and offer tailored solutions, driving their commercialization and market expansion (Karchiyappan et al. 2022).

Water Reuse and Resource Recovery

With water scarcity becoming a global concern, the focus on water reuse and resource recovery is increasing. METs play a significant role in facilitating water reuse by achieving high-quality effluent standards. Moreover, METs enable the recovery of valuable resources from wastewater, such as energy, nutrients, and chemicals. These recovered resources can be used for various applications, including energy generation, agricultural fertilizers, and industrial processes. The market demand for sustainable water reuse and resource recovery solutions creates opportunities for MET-based wastewater treatment technologies (Modin and Aulenta 2017; Jain and He 2018; Zhang et al. 2022b).

Decentralized and Off-Grid Applications

In many regions, centralized wastewater treatment infrastructure is limited, making decentralized and off-grid solutions necessary. MET-based wastewater treatment systems can be designed as modular, compact units suitable for decentralized applications. These systems can be installed in small communities, rural areas, or remote locations where conventional centralized treatment plants are not feasible. The ability of METs to operate with minimal external energy requirements and the potential for energy generation make them attractive options for off-grid applications. The market opportunities for decentralized and off-grid wastewater treatment solutions provide a favorable environment for the commercialization of MET-based technologies (Ramírez-Vargas et al. 2018a, b; Chen et al. 2018; Tsekouras et al. 2022; Rahimi and Eicker 2022).

Policy and Regulatory Support

Supportive policies and regulations play a vital role in driving the market adoption of innovative wastewater treatment technologies. Governments and regulatory agencies across the globe are increasingly emphasizing sustainable and efficient wastewater treatment practices. Policies promoting water reuse, nutrient recovery, and environmental protection create a favorable market environment for MET-based wastewater treatment technologies. Access to research funding, grants, and financial incentives further encourages technology development, demonstration, and commercialization. MET-based technologies can capitalize on these policy-driven opportunities to penetrate the market and gain industry acceptance (Thailand 2015; Thebo et al. 2017; Shah and Shah 2020).

Collaborative Partnerships and Technology Transfer

Collaborations between research institutions, industry stakeholders, and technology providers are essential for commercializing MET-based wastewater treatment technologies. Partnerships can foster technology transfer, facilitate demonstration projects, and accelerate market penetration. Collaborative efforts between academia, industry, and government bodies can ensure technology scalability, cost-effectiveness, and regulatory compliance. These partnerships provide avenues for knowledge exchange, funding support, and end-user access, creating market opportunities and accelerating the commercialization of MET-based wastewater treatment solutions (Table 1).

Table 1 Collaborative projects between different countries in the area of microbial electrochemical technology (MET)

Sr. no.	Project/ partnership	Partners	Technology	Application	Benefits	References
1	PAVITRA GANGA	20 partners from Europe and India	BES and other technologies	Wastewater treatment and reuse, resource recovery	Enhance water quality, quantity and security, promote social acceptance and stakeholder involvement	Quesada (2024)
2	BioelectroMET	10 partners from Europe and Latin America	BES	Industrial wastewater treatment and chemical production	Reduce water scarcity, energy consumption and environmental pollution, convert wastewater into valuable products	Dopson (2012)
3	Thermax India and Hindustan Unilever Limited	Thermax India and Hindustan Unilever Limited	BES and MBR	Factory wastewater treatment and reuse, biogas production	Reduce capital expenditure, guarantee performance, generate biogas and clean water	Fukukawa (2014)
4	Electrochaea GmbH and Mahindra and Mahindra Ltd	Electrochaea GmbH and Mahindra and Mahindra Ltd	BEM	Carbon dioxide and hydrogen conversion to methane	Reduce greenhouse gas emissions, provide market opportunity, generate 5renewable methane	Campling (2022)
5	MFC4C	6 partners from India and Germany	MFCs	Domestic and industrial wastewater treatment and reuse, energy and nutrient recovery	Enhance wastewater treatment efficiency, reduce energy consumption and greenhouse gas emissions, recover valuable resources	Kalbar (2021)

(continued)

Table 1 (continued)

Sr. no.	Project/ partnership	Partners	Technology	Application	Benefits	References
6	SWINGS	12 partners from India and Europe	BES and other technologies	Rural and urban wastewater treatment and reuse, resource recovery	Enhance water quality, quantity and security, promote social acceptance and stakeholder involvement	Álvarez et al. (2017)
7	Electrochaea GmbH and Indian Oil Corporation Limited	Electrochaea GmbH and Indian Oil Corporation Limited	BEM	Carbon dioxide and hydrogen conversion to methane	Reduce greenhouse gas emissions, provide market opportunity, generate renewable methane	Klitkou et al. (2017)
8	INNOQUA project	20 consortium partners from 12 countries in four continents (Africa, Asia, Latin America and Europe)	Nature-based processes (earthworms, zooplankton, microalgae)	Wastewater treatment and reuse	Contribute to the UN Sustainable Development Goal 6 by demonstrating and testing the technology in different contexts and climates, engaging with multiple stakeholders	Tompkins et al. (2019)
9	REDE project	Researchers from Germany, Ireland, France, Colombia, Peru, India and South Africa	MFCs and constructed wetlands	Wastewater treatment and reuse	Foster scientific collaboration between the EU and the Global South on sustainable development issues, establish partnerships with local communities, authorities and industries	Clifford et al. (2023)

(continued)

Table 1 (continued)

Sr. no.	Project/ partnership	Partners	Technology	Application	Benefits	References
10	Cambrian Innovation and Lagunitas Brewing Company	Cambrian Innovation and Lagunitas Brewing Company	MFCs	Brewery wastewater treatment and reuse, energy production	Reduce capital expenditure, guarantee performance, generate renewable energy and clean water	Camozzi et al. (2014)
11	Emefcy Group Limited and Ethiopian Ministry of Water, Irrigation and Electricity	Emefcy Group Limited and Ethiopian Ministry of Water, Irrigation and Electricity	MFCs	Industrial park wastewater treatment and reuse, energy production	Reduce upfront investment for the government, provide guaranteed performance for Emefcy Group Limited	Tsehaye et al. (2022)

In conclusion, the market opportunities for MET-based wastewater treatment are substantial and diverse. The growing global wastewater treatment market, industrial applications, water reuse, and resource recovery needs, decentralized/off-grid applications, supportive policies and regulations, and collaborative partnerships collectively contribute to the commercialization prospects of MET-based technologies. By addressing the evolving needs of the wastewater treatment industry and offering sustainable and efficient solutions, METs are well-positioned to capture significant market share and contribute to advancing wastewater treatment practices globally.

2.2 *Biosensors*

2.2.1 Significance of Biosensors in Biotechnology

Biosensors play a crucial role in biotechnology, offering innovative solutions for a wide range of applications. This section highlights biosensors' significance and growing importance in various industries, including healthcare, environmental monitoring, agriculture, and food safety.

Enhanced Sensitivity and Specificity

Biosensors provide enhanced sensitivity and specificity in detecting target analytes. They utilize biological components such as enzymes, antibodies, or microorganisms to recognize and interact with specific molecules of interest. This selective recognition enables biosensors to detect and quantify analytes accurately, even at low concentrations. The ability to precisely measure and monitor analytes is invaluable in fields such as medical diagnostics, environmental monitoring, and quality control in industries (Turner 2013).

Real-Time Monitoring

One of the critical advantages of biosensors is their ability to provide real-time monitoring of analytes. Traditional laboratory-based methods often require time-consuming sample preparation and analysis. In contrast, biosensors offer rapid, on-site detection and continuous monitoring, allowing immediate decision-making and timely intervention. Real-time monitoring is particularly valuable in healthcare applications, enabling early diagnosis, personalized treatment, and disease management (Divyasorubini et al. 2022).

Portability and Point-Of-Care Testing

Biosensors are designed for portability and ease of use, making them suitable for point-of-care testing. These handheld or miniaturized devices provide rapid and accurate results at the patient's bedside, eliminating the need for sample transportation and centralized laboratory analysis. Point-of-care biosensors are instrumental in remote and resource-limited settings, enabling timely diagnosis, monitoring, and treatment without sophisticated laboratory infrastructure (Han et al. 2020; Sharmili Roy et al. 2022).

Multiplexing Capabilities

Biosensors, by using multiplexing, can be designed to detect multiple analytes simultaneously. This capability is valuable in applications where multiple parameters must be monitored simultaneously. For instance, in environmental monitoring, a biosensor can simultaneously detect and quantify various pollutants or contaminants in a single sample. Multiplexing enhances efficiency, reduces costs, and provides a comprehensive understanding of complex systems (Ng et al. 2011; Glatz et al. 2022; Shaimerdenova et al. 2022).

Biocompatibility and Biodegradability

Biosensors utilize biological components that are often biocompatible and biodegradable. This characteristic makes them suitable for applications where compatibility with living organisms or the environment is essential. Biosensors can be integrated into living systems, such as implantable sensors for monitoring physiological parameters or environmental sensors for studying microbial communities. Their biocompatibility and biodegradability minimize adverse effects and promote sustainability (Song et al. 2021; Wang et al. 2021; Ghorbanizamani et al. 2023).

Advances in Nanotechnology

Nanotechnology has revolutionized the field of biosensors, enabling unprecedented sensitivity and miniaturization. Nanomaterials, such as nanoparticles, nanowires, and nanocomposites, are incorporated into biosensor platforms to enhance their performance. The unique properties of nanomaterials, such as large surface area, high reactivity, and quantum effects, significantly improve biosensors' sensitivity, selectivity, and response time. Nanotechnology-based biosensors have found applications in various fields, including medical diagnostics, environmental monitoring, and food safety (Zhang et al. 2009; George Kerry et al. 2021).

Impact on Industry and Society

The utilization of biosensors profoundly affected both industry and society. In healthcare, biosensors facilitate early disease detection, enable personalized medicine, and improve patient outcomes. In environmental monitoring, biosensors contribute to pollution control, water quality management, and ecosystem preservation. Biosensors also play a crucial role in food safety by ensuring the quality and integrity of food products. Their wide-ranging applications across industries contribute to advancements in biotechnology, improving human health, environmental sustainability, and overall quality of life (Turner 2013; Karim 2020; Tetyana et al. 2021).

In conclusion, biosensors have emerged as indispensable tools in biotechnology, offering enhanced sensitivity, real-time monitoring, portability, multiplexing capabilities, and compatibility with living systems. Their significance in various industries, including healthcare, environmental monitoring, agriculture, and food safety, cannot be overstated. Continued advancements in biosensor technology, along with collaborative efforts between researchers, industry stakeholders, and policymakers, will further drive their commercialization and market penetration, contributing to a more sustainable and technologically advanced future.

2.2.2 Progress in MET-Based Biosensors

Microbial electrochemical technologies (METs) have shown tremendous potential in developing biosensors, offering innovative solutions for various applications. This section highlights the progress made in MET-based biosensors, exploring their advantages, recent advancements, and potential for commercialization (Fig. 4).

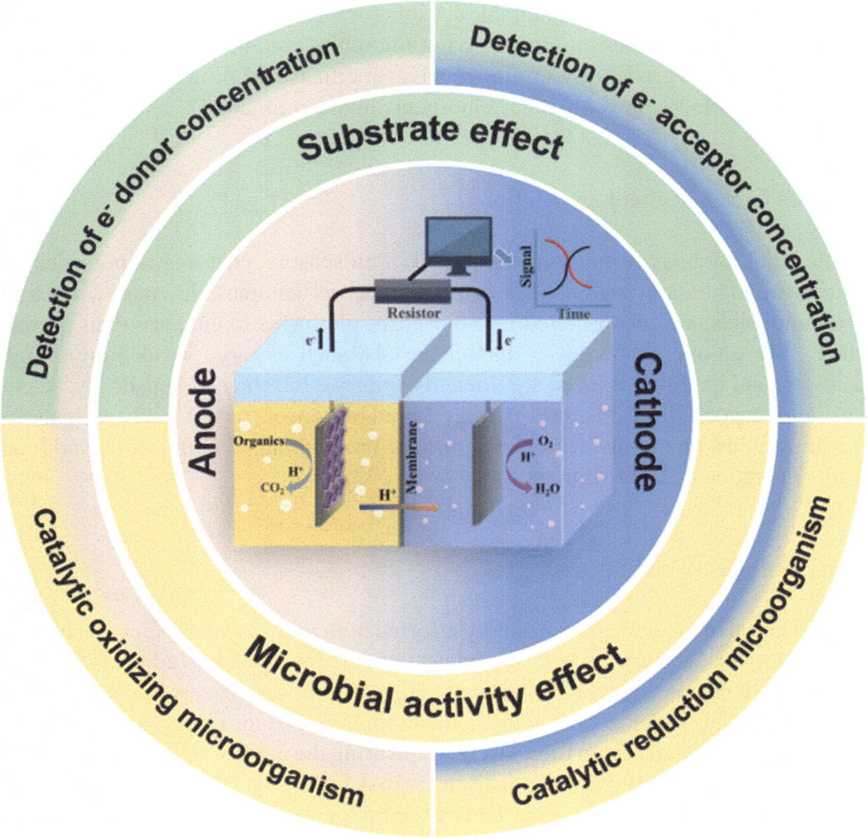

Fig. 4 Schematic for MET-based biosensor [*Source*: Reprinted with permission from Xue et al. (2024) Copyright {2024} Elsevier]

Advantages of MET-Based Biosensors

MET-based biosensors leverage the unique capabilities of microbial communities and electrochemical processes, resulting in several advantages:

- **Enhanced Sensitivity:** MET-based biosensors exhibit enhanced sensitivity due to the metabolic activities of microorganisms. Microbes can recognize and interact with specific analytes, leading to highly sensitive detection and quantification. The electrochemical transduction amplifies microbial responses, enhancing the sensor's sensitivity (Jiang et al. 2017; Ivars-Barceló et al. 2018a; Li et al. 2019).
- **Selectivity and Specificity:** Microorganisms can be engineered or selected to exhibit high selectivity and specificity toward target analytes. This allows for developing biosensors that can detect specific compounds or classes of compounds with minimal interference from other substances. The selectivity and specificity

of MET-based biosensors make them valuable tools for selective detection and monitoring in complex samples (Ivars-Barceló et al. 2018b).

- **Real-time Monitoring:** MET-based biosensors enable real-time monitoring of analytes, providing immediate and continuous data. The electrochemical signals generated by microbial activities can be rapidly measured and translated into quantitative information. Real-time monitoring is particularly beneficial for applications that require timely detection, such as environmental monitoring, food safety, and medical diagnostics (Chouler and Di Lorenzo 2015; Bharatula et al. 2020; Gonzalez et al. 2021; Yao et al. 2023).

Recent Advancements in MET-Based Biosensors

- **Enhanced Performance and Stability:** Recent research has focused on improving the performance and stability of MET-based biosensors. This includes optimizing the selection and engineering of microbial communities to enhance their sensing capabilities. Additionally, electrode materials, reactor designs, and sensor integration advancements have contributed to improved sensor performance and long-term stability (Divyasorubini et al. 2022).
- **Integration of Nanomaterials:** The integration of nanomaterials into MET-based biosensors has shown promising results. Nanomaterials, such as nanoparticles, nanowires, and nanocomposites, can enhance biosensors' sensitivity, selectivity, and response time. These nanomaterials provide a larger surface area for microbial attachment, promote electron transfer, and improve the signal-to-noise ratio, leading to improved sensor performance (Ivars-Barceló et al. 2018a, b).
- **Miniaturization and Portable Designs:** Miniaturization and portable designs are essential for biosensor commercialization and market adoption. Recent advancements in microfabrication techniques and electronics have enabled the development of miniaturized MET-based biosensors. These compact devices offer the advantages of portability, ease of use, and point-of-care testing, making them suitable for on-site and in-field applications (Su et al. 2022).
- **Multiplexing Capabilities:** MET-based biosensors have also made progress in multiplexing capabilities. Multiplexed biosensors can simultaneously detect and quantify multiple analytes in a single sample, providing a comprehensive analysis. This capability is valuable for applications where multiple parameters must be monitored simultaneously, such as environmental monitoring and biomedical diagnostics (Tahernia et al. 2020; Mathur et al. 2023; Zhu et al. 2023).
- **Commercialization Potential**: The progress in MET-based biosensors holds significant commercialization potential. The advantages of MET-based biosensors, including enhanced sensitivity, selectivity, real-time monitoring, and the ability to integrate nanomaterials, make them attractive options for various industries. In environmental monitoring, MET-based biosensors can provide rapid and cost-effective detection of pollutants, facilitating efficient monitoring and remediation. MET-based biosensors can enable early disease detection, personalized

medicine, and point-of-care diagnostics in the healthcare industry. The agricultural and food sectors can benefit from MET-based biosensors for quality control, pathogen detection, and monitoring of agricultural processes. Several challenges need to be addressed to realize the commercial potential of MET-based biosensors. These include optimization of sensor performance, standardization of protocols and procedures, scalability of manufacturing processes, and compliance with regulatory requirements. Collaborations between academia, industry, and regulatory bodies are essential for advancing the field, overcoming challenges, and accelerating the commercialization of MET-based biosensors.

In conclusion, the progress in MET-based biosensors has demonstrated their potential to revolutionize the field of biosensing. The advantages METs offer, along with recent advancements in performance, miniaturization, and multiplexing capabilities, position these biosensors as promising tools for various applications. Continued research, technological innovation, and collaborative efforts will pave the way for the commercialization and widespread adoption of MET-based biosensors, driving advancements in biosensing technologies and benefiting multiple industries (Lin and Wang 2005).

2.2.3 Market Opportunities for MET-Based Biosensors

Microbial electrochemical technologies (METs) offer significant market opportunities in biosensors. The unique capabilities of METs, such as enhanced sensitivity, selectivity, real-time monitoring, and integration of nanomaterials, make them attractive to a wide range of industries. This section discusses the market opportunities and potential applications of MET-based biosensors.

Environmental Monitoring

MET-based biosensors have great potential in environmental monitoring applications. They can be employed for the rapid and cost-effective detection of pollutants in air, water, and soil, including heavy metals, organic compounds, and microbial contaminants. MET-based biosensors can provide real-time data, enabling efficient environmental quality monitoring, pollution control, and early detection of environmental risks. The increasing global focus on sustainability and environmental protection creates a strong market demand for MET-based biosensors in this sector (Maria da Costa Silva et al. 2011; Justino et al. 2017; Tovar-Sánchez et al. 2019; Gavrilaş et al. 2022).

Healthcare and Medical Diagnostics

The healthcare industry represents a significant market opportunity for MET-based biosensors. These biosensors can be utilized for various medical diagnostic applications, including disease detection, monitoring, and personalized medicine. MET-based biosensors offer real-time monitoring of biomarkers, enabling early detection of diseases such as infections, cancer, and metabolic disorders. Their portability and point-of-care testing capabilities make them valuable tools for remote and resource-limited settings, improving healthcare accessibility and patient outcomes (Han et al. 2010).

Food Safety and Quality Control

MET-based biosensors can play a crucial role in ensuring food safety and quality control. These biosensors can detect food products' pathogens, toxins, and contaminants, preventing foodborne illnesses and facilitating rapid intervention. MET-based biosensors enable real-time monitoring throughout the food production and distribution chain, ensuring compliance with regulations and standards. With the increasing demand for safe and high-quality food products, MET-based biosensors have a substantial market opportunity in the food industry (Radeef and Ismail 2019).

Industrial Process Monitoring

MET-based biosensors have the potential to revolutionize industrial process monitoring and optimization. They can be employed to monitor parameters such as pH, temperature, and chemical concentrations in various industrial processes, including bioreactors, fermentation, and wastewater treatment. MET-based biosensors provide real-time feedback, enabling precise control and optimization of process parameters. By improving efficiency, reducing costs, and ensuring product quality, these biosensors can find applications in industries such as pharmaceuticals, biotechnology, and chemical manufacturing (Hao et al. 2020).

Agricultural Applications

MET-based biosensors can be utilized in the agricultural sector for monitoring soil health, detecting pathogens, and optimizing nutrient management. These biosensors can provide real-time information on soil parameters, water quality, and crop health, aiding in sustainable agricultural practices and maximizing crop yield. MET-based biosensors offer opportunities for precision agriculture, where resources can be efficiently allocated based on accurate and timely data, reducing environmental impact and improving agricultural productivity (Suganya et al. 2022; Wang et al. 2023b).

Defense and Security

MET-based biosensors have potential applications in defense and security-related fields. These biosensors can be used to rapidly detect chemical and biological agents, facilitating early warning systems and response measures. MET-based biosensors offer advantages such as portability, sensitivity, and real-time monitoring, making them suitable for applications in defense, homeland security, and emergency response scenarios (Tender 2014).

Several factors need to be considered to fully leverage the market opportunities for MET-based biosensors. These include technology optimization, protocols and performance metrics standardization, cost-effectiveness, regulatory compliance, and establishing strategic partnerships between academic researchers, industry stakeholders, and end-users. Collaboration and knowledge-sharing among different sectors will drive innovation, accelerate commercialization, and create a robust market for MET-based biosensors.

In conclusion, MET-based biosensors present significant market opportunities across various industries, including environmental monitoring, healthcare, food safety, industrial process monitoring, agriculture, and defense. The unique capabilities of METs, coupled with the increasing demand for rapid, accurate, and real-time monitoring solutions, position MET-based biosensors as promising tools for addressing societal challenges and driving technological advancements. Continued research, development, and collaboration among stakeholders are crucial for unlocking the full potential of MET-based biosensors in the market.

2.3 Environmental Remediation

2.3.1 METs for Environmental Remediation

Microbial electrochemical technologies (METs) offer innovative solutions for environmental remediation, addressing the challenges of pollution and contamination in various ecosystems. This section explores the application of METs in environmental remediation and their potential for commercialization.

Soil Remediation

Soil contamination is a widespread environmental issue resulting from industrial activities, improper waste disposal, and agricultural practices. METs have shown promise in soil remediation by facilitating the degradation and removal of pollutants. Electrochemical systems such as microbial fuel cells (MFCs) and microbial electrolysis cells (MECs) can stimulate indigenous microorganisms' activity or introduce exogenous microbial consortia to enhance pollutant degradation. METs can target various soil contaminants, including petroleum hydrocarbons, heavy metals, pesticides, and chlorinated compounds (Wang et al. 2023b).

Groundwater Treatment

Groundwater contamination is a significant concern due to its role in water supply and ecosystem health. METs offer potential solutions for the removal of pollutants from groundwater. Electrochemical processes, such as electrocoagulation and electrooxidation, can be coupled with microbial activities to enhance the removal of organic compounds and heavy metals. MET-based systems can promote the precipitation, immobilization, or transformation of contaminants, improving the quality of groundwater resources (Shah and Shah 2020).

Wastewater Treatment

Wastewater treatment is critical to environmental remediation, as it removes pollutants and protects water bodies. METs can be employed in wastewater treatment processes, such as microbial fuel cells (MFCs), microbial electrolysis cells (MECs), and bioelectrochemical systems (BESs). These systems harness microbial activities and electrochemical reactions to remove organic matter, nutrients, and emerging contaminants from wastewater. MET-based systems offer energy-efficient and sustainable alternatives to conventional wastewater treatment methods (Siddiqui et al. 2021; Rajesh Banu et al. 2021; Muhamad Ng et al. 2021; Zhang et al. 2022b, a; Li et al. 2023).

Marine and Aquatic Remediation

Marine and aquatic ecosystems are prone to pollution from industrial discharges, oil spills, and agricultural runoff. METs have demonstrated potential in the remediation of marine and aquatic environments. Bioelectrochemical systems, such as sediment microbial fuel cells (SMFCs) and constructed wetland systems, can enhance the natural processes of pollutant degradation and nutrient removal. MET-based approaches facilitate the microbial-driven breakdown of organic matter, reduction of oxygen demand, and removal of contaminants, contributing to the restoration and preservation of marine and aquatic ecosystems (Geist and Hawkins 2016).

Air Pollution Control

METs also show promise in air pollution control, specifically for treating volatile organic compounds (VOCs) and odorous gases. MET-based biofilters and biotrickling filters employ microbial communities to oxidize or adsorb gaseous pollutants biologically. The electrochemical stimulation in METs can enhance microbial activities, increasing the efficiency and effectiveness of air pollution control systems (Jung et al. 2008; Prévoteau et al. 2020; Dessì et al. 2021; Zhang et al. 2021).

Commercialization Potential

The application of METs in environmental remediation presents significant commercialization potential. The advantages of METs, such as their sustainability, energy efficiency, and potential for integration with existing treatment systems, make them attractive options for environmental remediation industries.

MET-based technologies offer a cost-effective and environmentally friendly approach to remediation, reducing reliance on chemical-intensive methods. Additionally, integrating METs with renewable energy sources, such as solar or wind power, can enhance their commercial viability and sustainability.

To fully capitalize on the commercial opportunities, key considerations must be addressed. These include optimizing the performance and scalability of MET-based systems, developing standardized protocols and guidelines, conducting thorough economic evaluations, and ensuring regulatory compliance. Collaborations between academic researchers, industry stakeholders, and regulatory bodies are crucial for advancing the commercialization of MET-based environmental remediation technologies.

In conclusion, METs are promising for environmental remediation, providing sustainable and efficient solutions for soil, groundwater, wastewater, marine, and air pollution control. The application of METs in environmental remediation offers significant commercialization opportunities, driven by the need for sustainable remediation technologies and regulatory compliance. Continued research, development, and collaboration will advance the commercialization and market adoption of MET-based environmental remediation technologies, paving the way for a cleaner and healthier environment.

2.3.2 Advancements in MET-Based Remediation Technologies

Microbial electrochemical technologies (METs) have undergone significant advancements in recent years, making them increasingly attractive for environmental remediation. This section explores the latest developments in MET-based remediation technologies and their implications for commercialization.

Enhanced Performance and Efficiency

One of the critical advancements in MET-based remediation technologies is improving performance and efficiency. Researchers have focused on optimizing electrode materials, reactor configurations, and microbial consortia to enhance the remediation capabilities of METs. Using advanced nanomaterials, such as carbon nanotubes, graphene, and conductive polymers, as electrode materials has improved electron transfer rates and enhanced microbial activity. These advancements result in higher removal efficiencies, shorter treatment times, and increased system robustness (Cusick et al. 2011; Tota-Maharaj and Paul 2015; Singh et al. 2019).

Integration of Multiple Processes

Another significant advancement is the integration of METs with other remediation processes to create synergistic effects. METs can be coupled with techniques like electrokinetics, electrocoagulation, and electro-Fenton to enhance the removal of contaminants. For example, the combination of METs with electrokinetics can improve the transport of contaminants to the electrode surface, while the electrochemical processes in METs facilitate their degradation. Integrating multiple processes allows for more comprehensive and efficient remediation of complex contaminant mixtures.

Development of Hybrid Systems

Hybrid systems that combine METs with other treatment technologies have also emerged as a promising advancement. These hybrid systems leverage the strengths of METs and other remediation techniques to achieve superior performance. For instance, coupling METs with bioremediation processes, such as microbial degradation or phytoremediation, can synergistically enhance the removal of organic pollutants. The integration of METs with physical or chemical processes can also improve the overall treatment efficiency and broaden the range of contaminants that can be targeted (Kalbar 2021; Zakaria and Dhar 2022).

Scale-Up and Field Demonstration

Advancements in MET-based remediation technologies have also involved successful scale-up and field demonstration. Researchers and industry stakeholders have focused on scaling up MET-based systems from lab-scale to pilot-scale and even full-scale applications. This progress has provided valuable insights into the operational challenges, system design considerations, and cost implications associated with large-scale implementation. Field demonstrations have showcased the practicality and effectiveness of MET-based remediation technologies in real-world scenarios, validating their commercial potential (Cusick et al. 2011; Tota-Maharaj and Paul 2015; Gonzalez et al. 2021; Muhamad Ng et al. 2021; Bird et al. 2022a; Rossi et al. 2022).

Cost Reduction and Economic Viability

Another essential aspect of advancements in MET-based remediation technologies is the continuous efforts to reduce costs and enhance economic viability. Researchers have been working on optimizing reactor designs, electrode materials, and system configurations to minimize capital and operational expenses. Integrating renewable energy sources, such as solar or wind power, can also contribute to cost reduction and sustainability. Additionally, economic evaluations and life cycle assessments are being conducted to comprehensively understandhe economic benefits and environmental impacts of MET-based remediation technologies (Zhang and Angelidaki 2016; Han et al. 2020).

These advancements in MET-based remediation technologies have significantly strengthened their commercialization potential. The improved performance, integration with other processes, successful scale-up, and cost-reduction efforts have positioned MET-based remediation technologies as competitive alternatives to traditional remediation methods.

However, challenges still exist in terms of standardization, regulatory acceptance, and market adoption. Standardized protocols and guidelines are needed to ensure the reproducibility and comparability of results across different studies and applications. Regulatory frameworks should be established to address the specific requirements and considerations associated with MET-based remediation technologies. Furthermore, market acceptance and adoption will require collaboration between researchers, industry stakeholders, policymakers, and end-users to build confidence, demonstrate effectiveness, and address perceived barriers.

In conclusion, the advancements in MET-based remediation technologies have brought them closer to commercialization. The improved performance, integration with other processes, successful scale-up, and cost-reduction efforts have expanded the potential applications and market opportunities for MET-based remediation technologies. Continued research, collaboration, and regulatory support will be vital to unlocking their full commercial potential and facilitating widespread adoption for effective and sustainable environmental remediation.

2.3.3 Market Prospects for MET-Based Remediation Technologies

Microbial electrochemical technologies (METs) have shown significant promise in environmental remediation, offering sustainable and efficient solutions for pollution mitigation. This section explores the market prospects for MET-based remediation technologies, including current trends and future opportunities.

Growing Environmental Concerns

The increasing awareness and global concerns regarding environmental pollution have driven the demand for innovative and sustainable remediation technologies.

METs offer a unique approach by harnessing the power of microorganisms and electrochemical reactions to remove contaminants from various environmental matrices. The market prospects for MET-based remediation technologies are bolstered by the urgent need for practical solutions to address soil, water, and air pollution challenges.

Regulatory Drivers

Stringent environmental regulations imposed by governmental bodies worldwide act as crucial drivers for adopting MET-based remediation technologies. Compliance with environmental standards and the need to restore contaminated sites to safe and acceptable levels are critical factors shaping the market prospects for METs. The ability of METs to meet regulatory requirements, coupled with their sustainable and cost-effective nature, positions them as attractive options for industries seeking environmentally friendly remediation solutions.

Technological Advancements

The continuous advancements in MET-based remediation technologies have significantly enhanced their market prospects. Improved performance, scalability, and integration with other remediation processes have expanded the applicability and effectiveness of METs. These technological advancements, including developing advanced electrode materials, reactor designs, and hybrid systems, have propelled the commercialization potential of MET-based remediation technologies.

Cost-Effectiveness and Efficiency

The cost-effectiveness and efficiency of MET-based remediation technologies contribute to their strong market prospects. METs offer several advantages, such as lower energy consumption, reduced reliance on chemical reagents, and the potential for resource recovery. The ability of METs to convert contaminants into valuable products, such as electricity or hydrogen, adds economic value to the remediation process—these cost and resource-saving benefits position MET-based technologies favorably in the market.

Diverse Application Areas

MET-based remediation technologies have various application areas, ranging from soil and groundwater remediation to wastewater treatment and air pollution control. This versatility broadens the market prospects for METs, as they can cater to different industries and sectors. The ability of METs to address various types of contaminants, including organic compounds, heavy metals, and nutrients, further expands their market potential.

Collaboration and Partnerships

Collaboration and partnerships among research institutions, industry players, and regulatory bodies are vital in driving the market prospects for MET-based remediation technologies. These collaborations foster knowledge exchange, technology transfer, and joint efforts in research and development. Industry partnerships facilitate the translation of research findings into practical applications, enabling the commercialization and market uptake of MET-based remediation technologies.

Emerging Markets

As the awareness and adoption of MET-based remediation technologies continue to grow, emerging markets present significant opportunities. Developing countries facing significant environmental challenges and industrial growth can benefit from METs as sustainable and cost-effective remediation solutions. These emerging markets provide a fertile ground for the commercialization and market expansion of MET-based remediation technologies.

In conclusion, the market prospects for MET-based remediation technologies are promising and influenced by various factors, including growing environmental concerns, regulatory drivers, technological advancements, cost-effectiveness, diverse applications, collaborations, and emerging markets. With continued research, innovation, and market development efforts, MET-based remediation technologies are poised to play a significant role in addressing environmental pollution challenges and shaping the future of sustainable remediation practices.

2.4 Bioelectrosynthesis

2.4.1 Significance of Electrosynthesis

Electrosynthesis, also known as microbial electrochemical synthesis, is a transformative technology that harnesses the power of microorganisms to convert electrical energy into valuable chemical compounds. This process utilizes microbial electrochemical cells (MECs) or bioelectrochemical systems (BES) to drive the conversion reactions, ranging from simple redox reactions to more complex organic transformations. The significance of electrosynthesis lies in its potential to revolutionize various industries, including energy, environment, and chemical production (Al-Mamun et al. 2023) (Fig. 5).

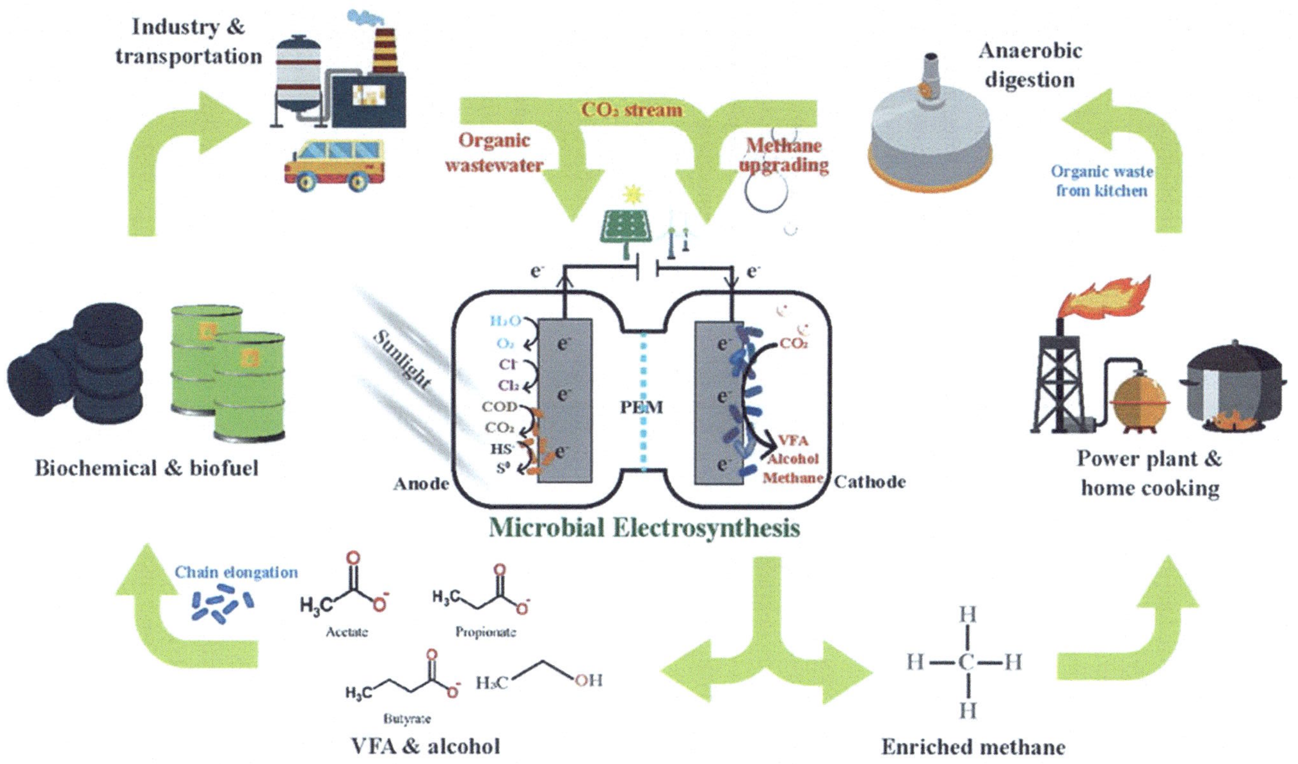

Fig. 5 Schematic for Bioelectrosynthesis of CO_2 to useful chemicals [*Source*: Reprinted with permission from Bian et al. (2020) Copyright {2024} Elsevier]

One of the key advantages of electrosynthesis is its ability to enable sustainable and environmentally friendly production processes. Traditional chemical synthesis methods often rely on non-renewable resources and generate harmful byproducts, contributing to pollution and climate change. Conversely, electrosynthesis offers a greener alternative by utilizing renewable electricity and microorganisms as catalysts. Electrosynthesis can reduce the carbon footprint associated with chemical production by harnessing renewable energy sources such as solar or wind power, making it a valuable tool for achieving sustainability goals (Dessì et al. 2021).

Moreover, electrosynthesis offers a new paradigm for producing high-value chemicals and fuels. Using MECs or BES, a wide range of compounds can be synthesized, including hydrogen, methane, acetate, ethanol, and various organic acids. These chemicals have numerous applications, ranging from energy storage and transportation to producing pharmaceuticals, polymers, and fine chemicals. Integrating electrosynthesis into existing industrial processes caneduce dependence on fossil fuels and transition towards a more diversified and sustainable chemical industry (Nath et al. 2021).

In addition to its potential for chemical production, electrosynthesis also holds promise in remediation and wastewater treatment. Microbial electrochemical systems can be employed to degrade or transform various pollutants, including organic contaminants, heavy metals, and even emerging contaminants such as pharmaceuticals and personal care products. The unique ability of microorganisms to couple their metabolic activities with the transfer of electrons to or from solid electrodes enables efficient removal or transformation of pollutants. This opens up new possibilities for developing cost-effective and energy-efficient remediation technologies, which can contribute to protecting water resources and preserving ecosystems (Gadkari et al. 2021).

Furthermore, electrosynthesis has the potential to enhance the performance and expand the functionality of biosensors. By integrating electroactive microorganisms or biomolecules with sensing elements, microbial electrochemical sensors can be developed to detect and monitor a wide range of analytes. The direct electron transfer between the biological components and the electrode enables sensitive and selective detection, while the biocatalytic activity of microorganisms offers the advantage of signal amplification. This combination of electrochemical sensing and microbial catalysis can lead to the development of highly efficient and versatile biosensors for applications in environmental monitoring, medical diagnostics, and food safety (Hassan et al. 2021).

Overall, the significance of electrosynthesis lies in its ability to address pressing challenges in energy, environment, and chemical production. By harnessing the power of microorganisms and electrical energy, this technology offers a sustainable and efficient approach to chemical synthesis, pollutant remediation, and biosensing. As research and development in microbial electrochemical technologies continue to advance, the commercialization and widespread adoption of electrosynthesis hold the potential to transform industries and contribute to a more sustainable and resilient future (Moscoviz et al. 2018).

2.4.2 Progress in MET-Based Bioelectrosynthesis

Microbial electrochemical technologies (METs) have garnered significant attention in recent years due to their potential for bioelectrosynthesis, which involves the production of valuable chemicals and fuels using microorganisms as catalysts and electrical energy as the driving force (Nevin et al. 2010). This section focuses on the progress made in MET-based bioelectrosynthesis and highlights the key advancements and challenges.

One of the notable achievements in MET-based bioelectrosynthesis is the diversification of the range of chemicals that can be synthesized. Initially, METs were primarily explored for the production of hydrogen gas through microbial electrolysis cells (MECs). However, extensive research efforts have expanded the scope of bioelectrosynthesis, enabling the production of a wide array of valuable compounds. This includes organic acids, alcohols, hydrocarbons, and even complex molecules such as pharmaceutical intermediates and biopolymers. The ability to produce such diversehemicals using renewable energy sources and sustainable microbial processes presents exciting opportunities for reducing dependence on fossil fuels and transitioning towards a bio-based economy (Tian et al. 2019).

The optimization of microbial communities and electrode materials has been a key focus of research in MET-based bioelectrosynthesis. Efforts have been made to understand the metabolic capabilities of different microorganisms and their interactions within the biofilm or planktonic communities. This knowledge has led to identifying and engineering specific microbial strains or consortia with enhanced electroactivity and desired metabolic pathways for target compound production. Additionally, developing advanced electrode materials with improved conductivity, stability, and selectivity has played a crucial role in enhancing the performance and efficiency of METs. Novel electrode materials, such as carbon-based nanomaterials, conductive polymers, and metal catalysts, have been investigated to optimize electron transfer kinetics and increase product yields (Mahmoud et al. 2021).

Another significant aspect of progress in MET-based bioelectrosynthesis is the integration of synthetic biology approaches. By employing genetic engineering and synthetic biology tools, researchers have manipulated microbial pathways and optimized enzyme activities to enhance the production of target compounds. Metabolic engineering strategies have increased precursor availability, redirected metabolic fluxes, and eliminated competing pathways, resulting in improved yields and titers. Synthetic biology also offers the potential for developing novel pathways and introducing non-natural functionalities into microbial hosts, expanding the range of chemicals that can be produced using METs.

Scale-up and process optimization are critical steps toward the commercialization of MET-based bioelectrosynthesis. Efforts have been made to upscale the production of target compounds from lab-scale to pilot-scale and beyond. Scaling up METs involves addressing challenges such as mass transport limitations, electrode fouling, and system stability. Various reactor designs, including membrane-based systems, flow cells, and stacked configurations, have been explored to improve performance, increase productivity, and enhance process control. Additionally, advanced process monitoring and control strategies, such as real-time monitoring of key parameters and feedback control algorithms, are being developed to optimize bioelectrosynthesis and ensure reproducibility and scalability (Rabaey and Rozendal 2010).

Despite the significant progress made in MET-based bioelectrosynthesis, several challenges and barriers must be addressed for successful commercialization. The high cost of electrode materials, limited microbial diversity, and the need for further understanding of the complex microbial-electrode interactions are some of the challenges that need to be overcome. Additionally, optimizing the energy efficiency and overall process economics and demonstrating the scalability and reliability of MET-based bioelectrosynthesisare essential for attracting industrial investment and market adoption (Hernandez and Osma 2020; Omidi et al. 2021).

In conclusion, the progress in MET-based bioelectrosynthesis has expanded the possibilities for sustainable chemical production using renewable energy and microbial catalysis. Advances in microbial community engineering, electrode materials, synthetic biology, and process optimization have propelled the development of METs as a promising technology for bioelectrosynthesis. However, further research and development, along with the demonstration of scalability, economic viability, and environmental sustainability, are necessary to facilitate the commercialization and widespread adoption of MET-based bioelectrosynthesis in various industries.

2.4.3 Market Opportunities for MET-Based Bioelectrosynthesis

Bioelectrosynthesis, the process of using microorganisms and electrical energy to produce valuable compounds, holds tremendous market potential across various industries. This section explores the market opportunities and potential applications for bioelectrosynthesis, highlighting its significance in driving economic growth, sustainability, and innovation.

Sustainable Chemical Production

Bioelectrosynthesis offers a sustainable and environmentally friendly approach to chemical production. With the increasing demand for greener alternatives, there is a significant market opportunity for synthesizing high-value chemicals using microbial electrochemical technologies (METs). Bioelectrosynthesis enables the production of organic acids, alcohols, specialty chemicals, and even complex molecules such as pharmaceutical intermediates and biopolymers. These compounds find applications in diverse industries, including pharmaceuticals, bioplastics, agriculture, and personal care products. Bioelectrosynthesis provides a unique advantage by utilizing renewable electricity and microbial catalysts, reducing reliance on fossil fuels and contributing to a circular economy (Nevin et al. 2010; Moscoviz et al. 2018; Gadkari et al. 2021; Dessì et al. 2021).

Renewable Energy Conversion and Storage

Bioelectrosynthesis is vital in renewable energy conversion and storage technologies. The ability to produce hydrogen, methane, and other energy-rich compounds using METs presents opportunities for sustainable energy storage and renewable energy conversion systems. Hydrogen, in particular, holds significant market potential as a clean fuel for transportation, power generation, and industrial processes. Bioelectrosynthesis can contribute to the advancement of technologies such as fuel cells, electrolyzers, and microbial fuel cells, offering efficient and renewable methods for energy storage and conversion (Tian et al. 2019; Nath et al. 2021; Omidi et al. 2021).

Environmental Remediation and Wastewater Treatment

Applying bioelectrosynthesis in environmental remediation and wastewater treatment provides market opportunities for addressing pollution and resource recovery challenges. METs have demonstrated potential in degrading or transforming various pollutants, including organic contaminants, heavy metals, and emerging contaminants like pharmaceuticals and personal care products. By integrating bioelectrochemical systems into existing treatment processes, pollutant removal efficiency can be enhanced, energy consumption reduced, and valuable resources recovered. The demand for cost-effective and sustainable solutions for pollution control and water treatment drives the market for bioelectrosynthesis in environmental remediation (Dange et al. 2022).

Bioremediation and Bioaugmentation

Bioelectrosynthesis offers possibilities for bioremediation and bioaugmentation strategies. METs can facilitate the degradation or transformation of pollutants in contaminated environments by leveraging the metabolic activities of microorganisms coupled with electron transfer to or from solid electrodes (Wang et al. 2020). This provides market opportunities for the remediation of contaminated soil and water, including brownfield sites, oil spills, and industrial wastewater. Bioelectrosynthesis-based bioremediation approaches can offer cost-effective and environmentally friendly solutions for cleaning up polluted sites (Omokhagbor Adams et al. 2020; Muter 2023).

In conclusion, bioelectrosynthesis presents substantial market opportunities across various industries. The demand for sustainable chemical production, renewable energy conversion and storage, environmental remediation, biosensors, and bioremediation drives the commercialization prospects for bioelectrochemical technologies (Logan and Rabaey 2012). As research and development continue to advance, and as the scalability, cost-effectiveness, and reliability of bioelectrosynthesis are further demonstrated, the market opportunities in this field are poised to expand, fostering economic growth, environmental sustainability, and technological innovation.

3 Challenges and Opportunities for the Commercialization of METs

While microbial electrochemical technologies (METs) are promising in various applications, their commercialization faces challenges and opportunities. This section explores the key challenges that need to be addressed and the potential opportunities for the successful commercialization of METs.

3.1 Challenges for the Commercialization of METs

3.1.1 Regulation and Standardization

METs must comply with the environmental and safety regulations and standards in different countries and regions. They also need to establish standard protocols and methods for testing and evaluating of their performance and quality.

3.1.2 Technology Scalability

One of the primary challenges for commercializing METs is scaling up the technology from laboratory-scale to industrial-scale operations. The transition from small-scale experimental setups to large-scale production systems involves addressing challenges related to mass transport limitations, reactor design, electrode scaling, and system stability. Further research and engineering efforts are needed to optimize and scale up MET-based processes for efficient and cost-effective production (Jadhav et al. 2022).

3.1.3 Microbial Diversity and Stability

METs heavily rely on microbial communities and their electroactivity. However, maintaining stable and productive microbial communities over extended periods can be challenging. Factors such as competition, contamination, and environmental conditions can impact the performance and stability of METs. Efforts are needed to understand and manipulate microbial diversity, enhance electron transfer kinetics, and develop strategies to maintain stable and efficient microbial communities (Bird et al. 2019, 2021, 2022b).

3.1.4 Economic Viability

The economic viability of METs is crucial for their commercialization. The cost of electrode materials, energy consumption, and maintenance can significantly impact the competitiveness of MET-based processes. Developing cost-effective electrode materials, optimizing system design, and reducingnergy requirements are necessary to make METs economically viable compared to conventional technologies (Zhang and Angelidaki 2016; Savla et al. 2021).

3.1.5 Market Development and Awareness

METs need to identify and target their products and services' potential customers and end-users. They also need to raise the awareness and acceptance of their technology among the public and stakeholders (Kehrein et al. 2020; Savla et al. 2021).

3.1.6 Technological Complexity and Integration

METs involve the integration of diverse disciplines, including microbiology, electrochemistry, materials science, and process engineering. This interdisciplinary nature adds complexity to technology development, operation, and optimization. Bridging the knowledge gap and fostering collaboration between experts from different fields is essential to overcome technical challenges and facilitate the commercialization of METs (Savla et al. 2021).

3.2 Opportunities

3.2.1 Sustainable and Green Market Demand

A growing global demand for sustainable and environmentally friendly technologies exists. METs align well with this demand by offering greener chemical production, energy storage, and environmental remediation alternatives. The emphasis on reducing greenhouse gas emissions, transitioning to renewable energy sources, and adopting circular economy principles creates significant market opportunities for METs as sustainable and economically viable solutions (Das et al. 2021a; b).

3.2.2 Policy and Regulatory Support

Supportive policies and regulations that promote sustainable technologies and incentivize their adoption can significantly accelerate the commercialization of METs. Governments and regulatory bodies are crucial in driving market adoption by providing financial incentives, research funding, and supportive frameworks for technology demonstration and deployment. Collaborative efforts between policymakers, researchers, and industry stakeholders are essential to creating an enabling environment for the commercialization of METs [(ESMAP) 2022].

3.2.3 Technological Advancements

Continued research and development efforts in METs have the potential to address existing challenges and drive technological advancements. Developing novel electrode materials, improved microbial strains, and advanced reactor designs can enhance MET-based processes' performance, stability, and scalability. Additionally, integrating emerging technologies such as synthetic biology, artificial intelligence, and advanced sensors can further optimize and expand the capabilities of METs (Lai et al. 2022).

3.2.4 Collaborative Partnerships and Market Penetration

Collaborations between research institutions, industry players, and potential end-users can facilitate the commercialization of METs. Collaborative research projects, technology transfer initiatives, and partnerships can accelerate MET-based technologies' development, optimization, and market penetration. Engaging with stakeholders early on and demonstrating the economic and environmental benefits of METs can help build market demand and pave the way for successful commercialization (Fukukawa 2014; Quesada 2024; Camozzi et al. 2014; Tompkins et al. 2019; Campling 2022; Clifford et al. 2023).

4 Conclusions

This chapter has extensively explored the commercialization and current market status of microbial electrochemical technologies (METs) in biosensing, remediation, and wastewater treatment. The findings demonstrate the immense potential of METs as innovative and sustainable solutions for various industries and sectors. METs have evolved from promising laboratory-scale concepts to advanced pilot-scale demonstrations, signaling their readiness for commercialization. The market opportunities in biosensing, remediation, and wastewater treatment are significant and driven by global needs, environmental regulations, and the demand for cost-effective and sustainable technologies. MET-based systems offer enhanced performance, resource recovery, energy generation, and compatibility with existing infrastructure, making them attractive options for end-users. Technological advancements have played a crucial role in advancing METs toward commercialization. Ongoing research efforts in electrode materials, reactor designs, microbial community engineering, and sensor technologies continually improve MET-based systems' performance, stability, and scalability.

Collaboration between researchers, industry stakeholders, and policymakers is essential to drive these technological advancements and ensure successful commercialization. While METs offer tremendous market opportunities, several challenges need to be addressed. Cost-effectiveness, scalability, long-term stability, and standardization are critical factors that impact market acceptance. Additionally, managing public acceptance, regulatory compliance, and technology integration with existing infrastructure are essential considerations for successful commercialization. Continued research and development, collaboration, standardization, and supportive policies will be pivotal in driving the commercialization and market adoption of MET-based technologies.

Further understanding fundamental processes, interdisciplinary collaborations, and real-world demonstrations will provide crucial data and validation for market acceptance. Standardized procedures, test methods, and certification frameworks will ensure comparability, reliability, and quality assurance. Economic incentives and policy frameworks that support sustainable technologies and resource recovery will create an enabling environment for the widespread adoption of METs. In conclusion, the commercialization of METs represents a significant opportunity to address global challenges in biosensing, remediation, and wastewater treatment. With their unique capabilities and advantages, METs offer innovative and sustainable solutions that can drive industry transformation and contribute to a more sustainable future. By capitalizing on the current market opportunities and addressing the challenges ahead, MET-based technologies can play a vital role in shaping a greener and more efficient world.

Acknowledgements Authors express appreciation to the Technology Mission Division (Energy, Water and others), Department of Science and Technology (DST), Government of India for providing the funding for the project entitled "Selection of exoelectrogens using novel polymer-supported nanocomposite based functionalized flexible electrodes in a microbial fuel cell for energy positive treatment of textile wastewater" under Water Technologies Cell [Sanction Order No. WTC/OWUIS-2021/RS-02 (C)].

References

A Review of Marine Oil Spill Recovery and Remediation Techniques | Geoengineer.org. https://www.geoengineer.org/education/web-class-projects/ce-176-environmental-geotechnics/assignments/review-of-marine-oil-spill-recovery-remediation-techniques. Accessed 7 July 2023

Al-Mamun A, Ahmed W, Jafary T et al (2023) Recent advances in microbial electrosynthesis system: metabolic investigation and process optimization. Biochem Eng J 196:108928. https://doi.org/10.1016/j.bej.2023.108928

Álvarez JA, Ávila C, Otter P et al (2017) Constructed wetlands and solar-driven disinfection technologies for sustainable wastewater treatment and reclamation in rural India: SWINGS project. Water Sci Technol 76:1474–1489. https://doi.org/10.2166/WST.2017.329

Anwer AH, Khan MD, Khan MZ, Joshi R (2020) Microbial electrochemical cell: an emerging technology for waste water treatment and carbon sequestration. In: Modern age waste water problems. Springer, Cham, pp 339–360

Baby MG, Ahammed MM (2022) Nutrient removal and recovery from wastewater by microbial fuel cell-based systems—a review. Water Sci Technol 86:29–55. https://doi.org/10.2166/WST.2022.196

Bharatula LD, Marsili E, Kwan JJ (2020) Impedimetric detection of *Pseudomonas aeruginosa* attachment on flexible ITO-coated polyethylene terephthalate substrates. Electrochim Acta 332:135390. https://doi.org/10.1016/J.ELECTACTA.2019.135390

Bian B, Bajracharya S, Xu J et al (2020) Microbial electrosynthesis from CO_2: challenges, opportunities and perspectives in the context of circular bioeconomy. Bioresour Technol 302:122863. https://doi.org/10.1016/j.biortech.2020.122863

Bioelectrochemical Systems for Metal Recovery | BIOELECTROMET | Project | News & Multimedia | FP7 | CORDIS | European Commission. https://cordis.europa.eu/project/id/282970/reporting. Accessed 6 July 2023

Bird LJ, Onderko EL, Phillips DA et al (2019) Engineered living conductive biofilms as functional materials. MRS Commun 9:505–517. https://doi.org/10.1557/MRC.2019.27

Bird LJ, Kundu BB, Tschirhart T et al (2021) Engineering wired life: synthetic biology for electroactive bacteria. ACS Synth Biol 10:2808–2823. https://doi.org/10.1021/ACSSYNBIO.1C00335/ASSET/IMAGES/LARGE/SB1C00335_0005.JPEG

Bird H, Heidrich ES, Leicester DD, Theodosiou P (2022a) Pilot-scale microbial fuel cells (MFCs): a meta-analysis study to inform full-scale design principles for optimum wastewater treatment. J Clean Prod 346:131227. https://doi.org/10.1016/j.jclepro.2022.131227

Bird LJ, Otero FJ, Yates MD et al (2022b) Engineered living conductive biofilms. Eng Liv Mater. https://doi.org/10.1007/978-3-030-92949-7_4/COVER

Bird LJ, Leary DH, Hervey J et al (2023) Marine biofilm engineered to produce current in response to small molecules. ACS Synth Biol. https://doi.org/10.1021/ACSSYNBIO.2C00417/ASSET/IMAGES/LARGE/SB2C00417_0008.JPEG

Camozzi M, Guttchen K, Robertson A (2014) Lagunitas brewing company. Case studies in agribusiness

Campling P (2022) Wastewater treatment and resource recovery for a sustainable water future in India. Open Access Gov 36:396–397. https://doi.org/10.56367/OAG-036-10514

Chen S, Brown RK, Patil SA et al (2018) Aerobic microbial electrochemical technology based on the coexistence and interactions of aerobes and exoelectrogens for synergistic pollutant removal from wastewater. Environ Sci (Camb) 5:60–69. https://doi.org/10.1039/C8EW00530C

Chouler J, Di Lorenzo M (2015) Water quality monitoring in developing countries: Can microbial fuel cells be the answer? Biosensors 5:450–470. https://doi.org/10.3390/BIOS5030450

Chung TH, Dhar BR (2021) A mini-review on applications of 3D printing for microbial electrochemical technologies. Front Energy Res. https://doi.org/10.3389/fenrg.2021.679061

Clifford E, Dussaussois JB, Schellenberg T, Sodemann C (2023) Research networks and novel partnerships for sustainable development—how an EU research project on innovative wastewater technology developed to an international network for nature-based solutions. Res Develop. https://doi.org/10.1007/978-981-19-6802-0_22/COVER

Cusick RD, Bryan B, Parker DS et al (2011) Performance of a pilot-scale continuous flow microbial electrolysis cell fed winery wastewater. Appl Microbiol Biotechnol 89:2053–2063. https://doi.org/10.1007/s00253-011-3130-9

Dange P, Pandit S, Pant D (2022) Microbial electrosynthesis (MES)-based wastewater treatment and resource recovery. Novel Approach towards Wastew Treat Resource Rec Technol. https://doi.org/10.1016/B978-0-323-90627-2.00006-X

Das S, Diels L, Pant D et al (2020) Review—microbial electrosynthesis: a way towards the production of electro-commodities through carbon sequestration with microbes as biocatalysts. J Electrochem Soc 167:155510. https://doi.org/10.1149/1945-7111/abb836

Das S, Raj R, Das S, Ghangrekar MM (2021a) A sustainable approach for the production of green energy with the holistic treatment of wastewater through microbial electrochemical technologies: a review. Front Sustain 2:792028. https://doi.org/10.3389/FRSUS.2021.792028/BIBTEX
Das S, Raj R, Das S, Ghangrekar MM (2021b) A sustainable approach for the production of green energy with the holistic treatment of wastewater through microbial electrochemical technologies: a review. Front Sustain. https://doi.org/10.3389/frsus.2021.792028
Decentralised Wastewater Treatment (DWWT). https://www.cseindia.org/decentralised-wastewater-treatment-dwwt--1714. Accessed 5 July 2023
Deng S, Wang C, Ngo HH et al (2023) Comparative review on microbial electrochemical technologies for resource recovery from wastewater towards circular economy and carbon neutrality. Bioresour Technol 376:128906. https://doi.org/10.1016/j.biortech.2023.128906
Dessì P, Rovira-Alsina L, Sánchez C et al (2021) Microbial electrosynthesis: towards sustainable biorefineries for production of green chemicals from CO_2 emissions. Biotechnol Adv 46:107675. https://doi.org/10.1016/j.biotechadv.2020.107675
Divyasorubini S, Kandage SM, Liyanage S et al (2022) Microbial biosensors for real-time monitoring of the bioremediation processes. Bioremed Environ Pollut. https://doi.org/10.1007/978-3-030-86169-8_5/COVER
Dopson M (2012) Bioelectrochemical systems for metal recovery
(ESMAP) ESMAP (2022) Regulatory indicators for sustainable energy (RISE) 2022: building resilience
Fudge T, Bulmer I, Bowman K et al (2021) Microbial electrolysis cells for decentralised wastewater treatment: the next steps. Water 13:445. https://doi.org/10.3390/W13040445
Fukukawa K (2014) Corporate social responsibility and local community in Asia. Routledge, London
Gadkari S, Mirza Beigi BH, Aryal N, Sadhukhan J (2021) Microbial electrosynthesis: Is it sustainable for bioproduction of acetic acid? RSC Adv 11:9921–9932. https://doi.org/10.1039/D1RA00920F
Ganiyu SO, Martínez-Huitle CA (2020) The use of renewable energies driving electrochemical technologies for environmental applications. Curr Opin Electrochem 22:211–220. https://doi.org/10.1016/j.coelec.2020.07.007
Gavrilaş S, Ursachi C, Perţa-Crişan S, Munteanu FD (2022) Recent trends in biosensors for environmental quality monitoring. Sensors 22:1513. https://doi.org/10.3390/S22041513
Geist J, Hawkins SJ (2016) Habitat recovery and restoration in aquatic ecosystems: current progress and future challenges. Aquat Conserv 26:942–962. https://doi.org/10.1002/AQC.2702
George Kerry R, Ukhurebor KE, Kumari S et al (2021) A comprehensive review on the applications of nano-biosensor-based approaches for non-communicable and communicable disease detection. Biomater Sci 9:3576–3602. https://doi.org/10.1039/D0BM02164D
Ghangrekar MM, Duteanu NM, Surampalli RY, Zhang TC (2023) Microbial electrochemical technologies. Wiley, London
Ghorbanizamani F, Moulahoum H, Guler Celik E, Timur S (2023) Material design in implantable biosensors toward future personalized diagnostics and treatments. Appl Sci 13:4630. https://doi.org/10.3390/APP13074630
Glatz RT, Ates HC, Mohsenin H et al (2022) Designing electrochemical microfluidic multiplexed biosensors for on-site applications. Anal Bioanal Chem 414:6531–6540. https://doi.org/10.1007/S00216-022-04210-4/FIGURES/5
Glaven SM (2019) Bioelectrochemical systems and synthetic biology: more power, more products. Microb Biotechnol 12:819. https://doi.org/10.1111/1751-7915.13456
Gonzalez L, Ab O, Di Lorenzo M (2021) Microbial fuel cells for in-field water quality monitoring. RSC Adv 11:16307–16317. https://doi.org/10.1039/d1ra01138c
Guo Y, Wang J, Shinde S et al (2020) Simultaneous wastewater treatment and energy harvesting in microbial fuel cells: an update on the biocatalysts. RSC Adv 10:25874–25887. https://doi.org/10.1039/D0RA05234E

Gupta PL, Choi HJ, Lee SM (2016) Enhanced nutrient removal from municipal wastewater assisted by mixotrophic microalgal cultivation using glycerol. Environ Sci Pollut Res 23:10114–10123. https://doi.org/10.1007/S11356-016-6224-1/TABLES/3

Han Y, Yu C, Liu H (2010) A microbial fuel cell as power supply for implantable medical devices. Biosens Bioelectron 25:2156–2160. https://doi.org/10.1016/J.BIOS.2010.02.014

Han YD, Chun HJ, Yoon HC (2020) Low-cost point-of-care biosensors using common electronic components as transducers. Biochip J 14:32–47. https://doi.org/10.1007/S13206-020-4104-8/FIGURES/6

Hao S, Sun X, Zhang H et al (2020) Recent development of biofuel cell based self-powered biosensors. J Mater Chem B 8:3393–3407. https://doi.org/10.1039/C9TB02428J

Hassan RYA, Febbraio F, Andreescu S (2021) Microbial electrochemical systems: principles, construction and biosensing applications. Sensors 21:1279. https://doi.org/10.3390/S21041279

Hernandez CA, Osma JF (2020) Microbial electrochemical systems: deriving future trends from historical perspectives and characterization strategies. Front Environ Sci 8:522049. https://doi.org/10.3389/FENVS.2020.00044/BIBTEX

INNOQUA—Innovative Wastewater Treatment. https://innoqua-project.eu/. Accessed 6 July 2023

Ivars-Barceló F, Zuliani A, Fallah M et al (2018a) Novel applications of microbial fuel cells in sensors and biosensors. Appl Sci 8:1184. https://doi.org/10.3390/app8071184

Ivars-Barceló F, Zuliani A, Fallah M et al (2018b) Novel applications of microbial fuel cells in sensors and biosensors. Appl Sci 8:1184. https://doi.org/10.3390/APP8071184

Jadhav DA, Das I, Ghangrekar MM, Pant D (2020) Moving towards practical applications of microbial fuel cells for sanitation and resource recovery. J Water Process Eng 38:101566. https://doi.org/10.1016/J.JWPE.2020.101566

Jadhav DA, Park S-G, Pandit S et al (2022) Scalability of microbial electrochemical technologies: applications and challenges. Bioresour Technol 345:126498. https://doi.org/10.1016/j.biortech.2021.126498

Jain A, He Z (2018) "NEW" resource recovery from wastewater using bioelectrochemical systems: moving forward with functions. Front Environ Sci Eng 12:1–13. https://doi.org/10.1007/S11783-018-1052-9/METRICS

Jiang Y, Liang P, Liu P et al (2017) Enhancement of the sensitivity of a microbial fuel cell sensor by transient-state operation. Environ Sci Water Res Technol 3:472–479. https://doi.org/10.1039/C6EW00346J

Jung RK, Dec J, Bruns MA, Logan BE (2008) Removal of odors from swine wastewater by using microbial fuel cells. Appl Environ Microbiol 74:2540–2543. https://doi.org/10.1128/AEM.02268-07/ASSET/CE477D4E-2F7D-4627-BEF7-5C15531012C5/ASSETS/GRAPHIC/ZAM0080887470003.JPEG

Justino CIL, Duarte AC, Rocha-Santos TAP (2017) Recent progress in biosensors for environmental monitoring: a review. Sensors 17:2918. https://doi.org/10.3390/S17122918

Kalbar PP (2021) Hybrid treatment systems: a paradigm shift to achieve sustainable wastewater treatment and recycling in India. Clean Technol Environ Policy 23:1365–1373. https://doi.org/10.1007/S10098-021-02034-X/TABLES/2

Kandpal R, Nara S, Shahadat M et al (2021) Impedance spectroscopic study of biofilm formation on pencil lead graphite anode in microbial fuel cell. J Taiwan Inst Chem Eng 128:114–123. https://doi.org/10.1016/j.jtice.2021.09.002

Kandpal R, Shahadat M, Adnan R et al (2022a) Polyaniline-based flexible nanocomposite materials, pp 367–395

Kandpal R, Shahadat M, Ali SW, Ahammad SZ (2022b) Recent technologies for modern and future industry. In: Green nanomaterials for industrial applications. Elsevier, London, pp 1–20

Kandpal R, Shahadat M, Ali SW, Ahammad SZ (2022c) Alternating current based electrochemically polymerized polyaniline modified flexible conducting paper as an exoelectrogenic biofilm platform for a microbial fuel cell. J Water Process Eng 46:102459. https://doi.org/10.1016/j.jwpe.2021.102459

Kandpal R, Shahadat M, Ali SW et al (2023) Material specific enrichment of electroactive microbes on polyaniline-supported anodes in a single chamber multi-anode assembly microbial fuel cell. Mater Res Bull 157:111983. https://doi.org/10.1016/j.materresbull.2022.111983

Karchiyappan T, Karri RR, Dehghani MH (eds) (2022) Industrial wastewater treatment, p 106. https://doi.org/10.1007/978-3-030-98202-7

Karim ME (2020) Biosensors: ethical, regulatory, and legal issues. Handb Cell Biosens. https://doi.org/10.1007/978-3-319-47405-2_23-1

Kehrein P, Van Loosdrecht M, Osseweijer P et al (2020) A critical review of resource recovery from municipal wastewater treatment plants—market supply potentials, technologies and bottlenecks. Environ Sci (Camb) 6:877–910. https://doi.org/10.1039/C9EW00905A

Kesari KK, Soni R, Jamal QMS et al (2021) Wastewater treatment and reuse: a review of its applications and health implications. Water Air Soil Pollut 232:1–28. https://doi.org/10.1007/S11270-021-05154-8/FIGURES/5

Klitkou A, Borup M, Fevolden A, Nikoleris A (2017) General rights Scandinavian demonstration projects for sustainable energy and transport their contribution to the formation of broad and aligned networks. Downloaded from orbit.dtu.dk on

Lai B, Glaven S, Song H (2022) Editorial: electrobiotechnology towards sustainable bioeconomy: fundamental, optimization and applications. Front Bioeng Biotechnol 10:901072. https://doi.org/10.3389/FBIOE.2022.901072/BIBTEX

Li J, Hu J, Yang C et al (2019) Enhanced detection of toxicity in wastewater using a 2D smooth anode based microbial fuel cell toxicity sensor. RSC Adv 9:8700–8706. https://doi.org/10.1039/C8RA10337B

Li Z, Zheng L, Koottatep T, Vinnerås B (2023) Editorial: Decentralized wastewater treatment technologies. Front Environ Sci 11:1199552. https://doi.org/10.3389/FENVS.2023.1199552/BIBTEX

Lin CT, Wang SM (2005) Biosensor commercialization strategy—a theoretical approach. Front Biosci 10:99–106. https://doi.org/10.2741/1512/PDF

Logan BE, Rabaey K (2012a) Conversion of wastes into bioelectricity and chemicals by using microbial electrochemical technologies. Science (1979) 337:686–690. https://doi.org/10.1126/science.1217412

Logan BE, Hamelers B, Rozendal R et al (2006) Microbial fuel cells: methodology and technology. Environ Sci Technol 40:5181–5192. https://doi.org/10.1021/es0605016

Mahmoud RH, Samhan FA, Ibrahim MK et al (2021) Formation of electroactive biofilms derived by nanostructured anodes surfaces. Bioprocess Biosyst Eng 44:759–768. https://doi.org/10.1007/S00449-020-02485-4

Mariada Costa Silva L, Farias Melo A, Medeiros Salgado A (2011) Biosensors for environmental applications. Environ Biosens 2011:3–16

Mathur S, Singh D, Ranjan R (2023) Genetic circuits in microbial biosensors for heavy metal detection in soil and water. Biochem Biophys Res Commun 652:131–137. https://doi.org/10.1016/J.BBRC.2023.02.031

Microbial Electrochemical Technologies—Google Books. https://books.google.co.in/books?hl=en&lr=&id=rmjIDwAAQBAJ&oi=fnd&pg=PA2016-IA21&dq=Microbial+Electrolysis+Cell+(MEC):+A+Versatile+Technology+for+Hydrogen,+Value-added+Chemicals+Production+and+Wastewater+Treatment&ots=DJBfz2sVHL&sig=A3hoxBy8mTUOpuWSE_d5NfS7ueg&redir_esc=y#v=onepage&q=Microbial%20Electrolysis%20Cell%20(MEC)%3A%20A%20Versatile%20Technology%20for%20Hydrogen%2C%20Value-added%20Chemicals%20Production%20and%20Wastewater%20Treatment&f=false. Accessed 5 July 2023

Modin O, Aulenta F (2017) Three promising applications of microbial electrochemistry for the water sector. Environ Sci (Camb) 3:391–402. https://doi.org/10.1039/C6EW00325G

Mojiri A, Bashir MJK (2022) Wastewater treatment: current and future techniques. Water 14:448. https://doi.org/10.3390/W14030448

Mojiri A, Bashir MJK (2022) Wastewater treatment: current and future techniques. Water (Basel) 14:448. https://doi.org/10.3390/w14030448

Moscoviz R, Trably E, Bernet N, Carrère H (2018) The environmental biorefinery: state-of-the-art on the production of hydrogen and value-added biomolecules in mixed-culture fermentation. Green Chem 20:3159–3179. https://doi.org/10.1039/C8GC00572A

Muhamad Ng SN, Idrus S, Ahsan A et al (2021) Treatment of wastewater from a food and beverage industry using conventional wastewater treatment integrated with membrane bioreactor system: a pilot-scale case study. Membranes (Basel) 11:456. https://doi.org/10.3390/membranes11060456

Muter O (2023) Current trends in bioaugmentation tools for bioremediation: a critical review of advances and knowledge gaps. Microorganisms 11:710. https://doi.org/10.3390/microorganisms11030710

Nara S, Kandpal R, Jaiswal V et al (2020) Exploring *Providencia rettgeri* for application to eco-friendly paper based microbial fuel cell. Biosens Bioelectron 165:112323. https://doi.org/10.1016/j.bios.2020.112323

Nath D, Chakraborty I, Ghangrekar MM (2021) Integrating microbial electrochemical technologies for methane-to-bioelectricity and water-splitting to impart self-sustainability to wastewater treatment plants. Bioresour Technol Rep 13:100644. https://doi.org/10.1016/j.biteb.2021.100644

Nevin KP, Woodard TL, Franks AE et al (2010) Microbial electrosynthesis: feeding microbes electricity to convert carbon dioxide and water to multicarbon extracellular organic compounds. Mbio. https://doi.org/10.1128/mBio.00103-10

Ng JK, Chong S, Ng JK, Chong S (2011) Multiplexing capabilities of biosensors for clinical diagnostics. Biosens Health Environ Biosecur. https://doi.org/10.5772/17187

Nguyen LN, Aditya L, Vu HP et al (2022) Nutrient removal by algae-based wastewater treatment. Curr Pollut Rep 8:369–383. https://doi.org/10.1007/S40726-022-00230-X/FIGURES/3

Omidi M, Mashkour M, Biswas JK et al (2021) From electricity to products: recent updates on microbial electrosynthesis (MES). Top Catal 1:1–18. https://doi.org/10.1007/S11244-021-01503-3

Omokhagbor Adams G, Tawari Fufeyin P, Eruke Okoro S, Ehinomen I (2020) Bioremediation, biostimulation and bioaugmention: a review. Int J Environ Bioremediat Biodegrad 3:28–39. https://doi.org/10.12691/ijebb-3-1-5

Prévoteau A, Carvajal-Arroyo JM, Ganigué R, Rabaey K (2020) Microbial electrosynthesis from CO_2: Forever a promise? Curr Opin Biotechnol 62:48–57. https://doi.org/10.1016/j.copbio.2019.08.014

Priyadarshini M, Ahmad A, Das S, Ghangrekar MM (2022) Application of microbial electrochemical technologies for the treatment of petrochemical wastewater with concomitant valuable recovery: a review. Environ Sci Pollut Res 29:61783–61802. https://doi.org/10.1007/s11356-021-14944-w

Public–Private Partnerships for Wastewater Treatment Solutions | Water Technology. https://www.watertechonline.com/wastewater/article/15550527/publicprivate-partnerships-for-wastewater-treatment-solutions. Accessed 6 July 2023

Quesada (Ihe-Delft), Breitenmoser L Anshuman (TERI), Nathaniel B. Dkhar (TERI), and Paul Campling (VITO)

Rabaey K, Rozendal RA (2010) Microbial electrosynthesis—revisiting the electrical route for microbial production. Nat Rev Microbiol 8:706–716. https://doi.org/10.1038/nrmicro2422

Radeef AY, Ismail ZZ (2019) New application of microbial fuel cell-based biosensor for monitoring the quality of actual potato chips' processing wastewater. Waste Dispos Sustain Energy 1:227–235. https://doi.org/10.1007/S42768-019-00017-9/METRICS

Rahimi N, Eicker U (2022) Renewable electricity and hydrogen production via decentralized wastewater treatment systems. Energies 15:8787. https://doi.org/10.3390/EN15238787

Rajesh Banu J, Merrylin J, Kavitha S et al (2021) Trends in biological nutrient removal for the treatment of low strength organic wastewaters. Curr Pollut Rep 7:1–30. https://doi.org/10.1007/s40726-020-00169-x

Ramírez-Vargas C, Prado A, Arias C et al (2018a) Microbial electrochemical technologies for wastewater treatment: principles and evolution from microbial fuel cells to bioelectrochemical-based constructed wetlands. Water (Basel) 10:1128. https://doi.org/10.3390/w10091128
Ramírez-Vargas CA, Prado A, Arias CA et al (2018b) Microbial electrochemical technologies for wastewater treatment: principles and evolution from microbial fuel cells to bioelectrochemical-based constructed wetlands. Water 10:1128. https://doi.org/10.3390/W10091128
Rossi R, Hur AY, Page MA et al (2022) Pilot scale microbial fuel cells using air cathodes for producing electricity while treating wastewater. Water Res 215:118208. https://doi.org/10.1016/j.watres.2022.118208
Roy S, Arshad F, Eissa S et al (2022) Recent developments towards portable point-of-care diagnostic devices for pathogen detection. Sens Diagnost 1:87–105. https://doi.org/10.1039/D1SD00017A
Safeguarding Water Resources in India with Green and Sustainable Technologies | SWINGS | Project | Fact sheet | FP7 | CORDIS | European Commission. https://cordis.europa.eu/project/id/308502. Accessed 6 July 2023
Savla N, Suman PS et al (2021) Techno-economical evaluation and life cycle assessment of microbial electrochemical systems: a review. Curr Res Green Sustain Chem 4:100111. https://doi.org/10.1016/j.crgsc.2021.100111
Shah A, Shah M (2020) Characterisation and bioremediation of wastewater: a review exploring bioremediation as a sustainable technique for pharmaceutical wastewater. Groundw Sustain Dev. https://doi.org/10.1016/j.gsd.2020.100383
Shahid A, Khan AZ, Malik S et al (2021) Advances in green technologies for the removal of effluent organic matter from the urban wastewater. Curr Pollut Rep 7:463–475. https://doi.org/10.1007/S40726-021-00203-6/FIGURES/3
Shaimerdenova M, Ayupova T, Bekmurzayeva A et al (2022) Spatial-division multiplexing approach for simultaneous detection of fiber-optic ball resonator sensors: applications for refractometers and biosensors. Biosensors 12:1007. https://doi.org/10.3390/BIOS12111007
Siddiqui S, Bhatnagar P, Dhingra S et al (2021) Wastewater treatment and energy production by microbial fuel cells. Biomass Convers Biorefinery 13:3569–3592. https://doi.org/10.1007/S13399-021-01411-2
Simoska O, Gaffney EM, Minteer SD et al (2021) Recent trends and advances in microbial electrochemical sensing technologies: an overview. Curr Opin Electrochem 30:100762. https://doi.org/10.1016/j.coelec.2021.100762
Singh A, Sawant M, Kamble SJ et al (2019) Performance evaluation of a decentralized wastewater treatment system in India. Environ Sci Pollut Res 26:21172–21188. https://doi.org/10.1007/S11356-019-05444-Z/TABLES/6
Solon K, Volcke EIP, Spérandio M, Van Loosdrecht MCM (2019) Resource recovery and wastewater treatment modelling. Environ Sci (Camb) 5:631–642. https://doi.org/10.1039/C8EW00765A
Song M, Lin X, Peng Z et al (2021) Materials and methods of biosensor interfaces with stability. Front Mater 7:583739. https://doi.org/10.3389/FMATS.2020.583739/BIBTEX
Su Y, Lu L, Zhou M (2022) Wearable microbial fuel cells for sustainable self-powered electronic skins. ACS Appl Mater Interfaces 14:8664–8668. https://doi.org/10.1021/ACSAMI.2C00313/SUPPL_FILE/AM2C00313_SI_001.PDF
Suganya P, Divya NJ, Lavanya A, Mrinal R (2022) Design and development of microbial fuel cells based low power energy harvesting mechanism for ecological monitoring and farming of agricultural applications. https://doi.org/10.1142/S0218126623501128
Tahernia M, Mohammadifar M, Gao Y et al (2020) A 96-well high-throughput, rapid-screening platform of extracellular electron transfer in microbial fuel cells. Biosens Bioelectron 162:112259. https://doi.org/10.1016/J.BIOS.2020.112259
Tejedor-Sanz S, Bacchetti de Gregoris T, Salas JJ et al (2016) Integrating a microbial electrochemical system into a classical wastewater treatment configuration for removing nitrogen from low COD effluents. Environ Sci (Camb) 2:884–893. https://doi.org/10.1039/C6EW00100A

Tender LM (2014) Microbial fuel cells for powering navy devices. https://apps.dtic.mil/sti/citations/ADA594746; https://www.sciencetheearth.com/uploads/2/4/6/5/24658156/2014_tender_microbial_fuel_cells_for_powering_navy_devices.pdf
Tetyana P, Shumbula PM, Njengele-Tetyana Z et al (2021) Biosensors: design, development and applications. Nanopores. https://doi.org/10.5772/INTECHOPEN.97576
Thailand B II I Policy Guidance Manual on Wastewater Management Bibliographic Reference United Nations Economic and Social Commission for Asia and the Pacific (ESCAP), United Nations Human Settlements Programme (UN-Habitat) and Asian Institute of Technology (AIT) (2015) Policy guidance manual on wastewater management with a special emphasis on decentralized wastewater treatment systems
Thebo AL, Drechsel P, Lambin EF, Nelson KL (2017) A global, spatially-explicit assessment of irrigated croplands influenced by urban wastewater flows. Environ Res Lett. https://doi.org/10.1088/1748-9326/AA75D1
Tian J, Lacroix R, Desmond-Le Quéméner E et al (2019) Upscaling of microbial electrolysis cell integrating microbial electrosynthesis: insights, challenges and perspectives. Biorxiv. https://doi.org/10.1101/609909
Tiquia-Arashiro SM, Pant D (2019) Microbial electrochemical technologies. CRC Press, London, p 519
Tompkins D, Bumbac C, Clifford E et al (2019) EU horizon 2020 research for a sustainable future: INNOQUA—a nature-based sanitation solution. Water 11:2461. https://doi.org/10.3390/W11122461
Tota-Maharaj K, Paul P (2015) Performance of pilot-scale microbial fuel cells treating wastewater with associated bioenergy production in the Caribbean context. Int J Energy Environ Eng 6:213–220. https://doi.org/10.1007/s40095-015-0169-x
Tovar-Sánchez E, Suarez-Rodríguez R, Ramírez-Trujillo A et al (2019) The use of biosensors for biomonitoring environmental metal pollution. Biosens Environ Monit. https://doi.org/10.5772/INTECHOPEN.84309
Tsehaye MT, Assayie AA, Besha AT et al (2022) Membrane technology for water and wastewater treatment in Ethiopia: current status and future prospects article info. J Membr Sci Res 8:540001
Tsekouras GJ, Deligianni PM, Kanellos FD et al (2022) Microbial fuel cell for wastewater treatment as power plant in smart grids: Utopia or reality? Front Energy Res 10:843768. https://doi.org/10.3389/FENRG.2022.843768/BIBTEX
Turner APF (2013) Biosensors: sense and sensibility. Chem Soc Rev 42:3184–3196. https://doi.org/10.1039/C3CS35528D
Wang X, Aulenta F, Puig S et al (2020) Microbial electrochemistry for bioremediation. Environ Sci Ecotechnol 1:100013. https://doi.org/10.1016/j.ese.2020.100013
Wang Y, Huang Y, Bai H et al (2021) Biocompatible and biodegradable polymer optical fiber for biomedical application: a review. Biosensors 11:472. https://doi.org/10.3390/BIOS11120472
Wang B, Bonné R, Zhang Y et al (2022) Renewable energy driving microbial electrochemistry toward carbon neutral. Sustain Horizons 4:100031. https://doi.org/10.1016/j.horiz.2022.100031
Wang M, Ding M, Yuan Y (2023a) Bioengineering for the microbial degradation of petroleum hydrocarbon contaminants. Bioengineering 10:347. https://doi.org/10.3390/bioengineering10030347
Wang SH, Wang JW, Zhao LT et al (2023b) Soil microbial fuel cell based self-powered cathodic biosensor for sensitive detection of heavy metals. Biosensors (Basel) 13:145. https://doi.org/10.3390/BIOS13010145/S1
Xue J, Wang Y, Jing Y et al (2024) Recent advances in microbial fuel cell–based self-powered biosensors: a comprehensive exploration of sensing strategies in both anode and cathode modes. Anal Bioanal Chem. https://doi.org/10.1007/s00216-024-05230-y
Yao H, Xiao J, Tang X (2023) Microbial fuel cell-based organic matter sensors: principles. Struct Appl Bioeng 10:886. https://doi.org/10.3390/bioengineering10080886
Zakaria B, Dhar B (2022) A review of stand-alone and hybrid microbial electrochemical systems for antibiotics removal from wastewater. Processes 10:714. https://doi.org/10.3390/pr10040714

Zhang Y, Angelidaki I (2016) Microbial electrochemical systems and technologies: it is time to report the capital costs. Environ Sci Technol 50:5432–5433. https://doi.org/10.1021/acs.est.6b01601

Zhang X, Guo Q, Cui D (2009) Recent advances in nanotechnology applied to biosensors. Sensors 9:1033–1053. https://doi.org/10.3390/S90201033

Zhang Z, Sarkar D, Li L, Datta R (2022a) Contaminant removal and resource recovery in bioelectrochemical wastewater treatment. Curr Pollut Rep 8:159–176. https://doi.org/10.1007/s40726-022-00218-7

Zhang Z, Sarkar D, Li L, Datta R (2022b) Contaminant removal and resource recovery in bioelectrochemical wastewater treatment. Curr Pollut Rep 8:159–176. https://doi.org/10.1007/S40726-022-00218-7

Zhang S, Jiang J, Wang H et al (2021) A review of microbial electrosynthesis applied to carbon dioxide capture and conversion: the basic principles, electrode materials, and bioproducts. J CO_2 Util 51:101640. https://doi.org/10.1016/j.jcou.2021.101640

Zhu J, Wang B, Zhang Y et al (2023) Living electrochemical biosensing: engineered electroactive bacteria for biosensor development and the emerging trends. Biosens Bioelectron 237:115480. https://doi.org/10.1016/J.BIOS.2023.115480

SPRINGER NATURE

GPSR Compliance

The European Union's (EU) General Product Safety Regulation (GPSR) is a set of rules that requires consumer products to be safe and our obligations to ensure this.

If you have any concerns about our products, you can contact us on ProductSafety@springernature.com

In case Publisher is established outside the EU, the EU authorized representative is:

Springer Nature Customer Service Center GmbH
Europaplatz 3
69115 Heidelberg, Germany

The manufacturer's authorised representative in the EU is Springer Nature Customer Service Centre GmbH, Europaplatz 3, 69115 Heidelberg, Germany. If you have any concerns regarding our products, please contact ProductSafety@springernature.com

Printed and bound by CPI Group (UK) Ltd, Croydon, CR0 4YY
08/10/2026
02229867-0002